Modern Energy Storage

Modern Energy Storage

The Key to Energy System Transformation

Edited by

Huaiguang Jiang
School of Future Technology
South China University of Technology
Guangzhou, China

Xiangmin Xu
Foshan University
Foshan, China

Jun Zhang
School of Electrical Engineering and Automation
Wuhan University
Wuhan, Hubei Province, China

Wenzhong Gao
Department of Electrical & Computer Engineering
University of Denver
Denver, CO, United States

Fei-Yue Wang
Institute of Automation
Chinese Academy of Sciences
Beijing, China
Macau Institute of Intelligent Science and Interdisciplinary Engineering
Faculty of Innovative Engineering
Macau University of Science and Technology
Taipa, Macau, China

ELSEVIER

ACADEMIC PRESS
An imprint of Elsevier

Academic Press is an imprint of Elsevier
125 London Wall, London EC2Y 5AS, United Kingdom
50 Hampshire Street, 5th Floor, Cambridge, MA 02139, United States

ISBN: 978-0-323-95419-8

For information on all Academic Press publications
visit our website at https://www.elsevier.com/books-and-journals

Publisher: Priscilla Braglia
Acquisitions Editor: Edward Payne
Editorial Project Manager: Avni Gupta
Production Project Manager: Sujatha Thirugnana Sambandam
Cover Designer: Mark Rogers

Typeset by VTeX

Contents

7. Aqueous ion batteries: fundamentals, innovations, and applications

Jiayan Zhu, Haobo Dong, and Huaiguang Jiang

Contributors

Gui Chen, School of Materials Science and Engineering, South China University of Technology, Guangdong Provincial Key Laboratory of Advanced Energy Storage Materials, Guangzhou, China

Haobo Dong, School of Future Technology, South China University of Technology, Guangzhou, China

Yi Guo, School of Automation, Beijing Institute of Technology, Beijing, China

Renzong Hu, School of Materials Science and Engineering, South China University of Technology, Guangdong Provincial Key Laboratory of Advanced Energy Storage Materials, Guangzhou, China

Jiesheng Huang, School of Future Technology, South China University of Technology, Guangzhou, China

Huaiguang Jiang, School of Future Technology, South China University of Technology, Guangzhou, China

Yupeng Jiang, School of Future Technology, South China University of Technology, Guangzhou, China

Xin Lei, School of Future Technology, South China University of Technology, Guangzhou, China

Yingjun Lei, China Academy of Space Technology, Beijing, China

Haoqin Li, School of Future Technology, South China University of Technology, Guangzhou, China

Shijie Li, School of Future Technology, South China University of Technology, Guangzhou, China

Ce Liang, School of Future Technology, South China University of Technology, Guangzhou, China

Haotian Liang, School of Materials Science and Engineering, South China University of Technology, Guangdong Provincial Key Laboratory of Advanced Energy Storage Materials, Guangzhou, China

Fei Lin, The Department of Engineering Science, Faculty of Innovation Engineering, Macau University of Science and Technology, Taipa, Macau, China

Huai-Jun Lin, Institute of Advanced Wear & Corrosion Resistance and Functional Materials, Jinan University, Guangzhou, China

Ziyun Liu, School of Future Technology, South China University of Technology, Guangzhou, China

Zhen Mei, School of Future Technology, South China University of Technology, Guangzhou, China

Lin Shu, School of Future Technology, South China University of Technology, Guangzhou, China

Fei-Yue Wang, The State Key Laboratory for Management and Control of Complex Systems, Institute of Automation, Chinese Academy of Sciences, Beijing, China; Macau Institute of Intelligent Science and Interdisciplinary Engineering, Faculty of Innovative Engineering, Macau University of Science and Technology, Taipa, Macau, China

Hui Wang, School of Materials Science and Engineering, South China University of Technology, Guangdong Provincial Key Laboratory of Advanced Energy Storage Materials, Guangzhou, China

Yufeng Wang, School of Future Technology, South China University of Technology, Guangzhou, China

Lin Wu, School of Future Technology, South China University of Technology, Guangzhou, China

Yuhao Xia, School of Materials Science and Engineering, South China University of Technology, Guangdong Provincial Key Laboratory of Advanced Energy Storage Materials, Guangzhou, China

Wei Xiong, School of Biology and Biological Engineering, South China University of Technology, Guangzhou, China

Zihan Xiong, School of Materials Science and Engineering, South China University of Technology, Guangdong Provincial Key Laboratory of Advanced Energy Storage Materials, Guangzhou, China

Ying Xue, School of Electric Power Engineering, South China University of Technology, Guangzhou, China

Hongfei Yang, School of Future Technology, South China University of Technology, Guangzhou, China

Jing Yang, The State Key Laboratory for Management and Control of Complex Systems, Institute of Automation, Chinese Academy of Sciences, Beijing, China

Yiwen Zhang, School of Future Technology, South China University of Technology, Guangzhou, China

Xinhu Zheng, The Hong Kong University of Science and Technology (Guangzhou), Guangzhou, China; Thrust of Intelligent Transportation, Systems Hub, Guangzhou, China; Thrust of Internet of Things, Information Hub, Guangzhou, China

Jiayan Zhu, College of Physics, Jilin University, Changchun, China

Acknowledgments

The editors would like to express their sincere gratitude to all individuals and institutions whose support and contributions made this book possible. This volume has benefited significantly from a robust academic environment and sustained research efforts in energy storage technologies, power and energy systems, and intelligent energy management. In particular, the editors acknowledge the steadfast institutional support from South China University of Technology as the coordinating institution. We also extend our appreciation for the valuable support received from the Guangdong Artificial Intelligence and Digital Economy Laboratory (Pazhou Lab), Wuhan University, and the University of Denver, which provided essential academic resources, research infrastructure, and interdisciplinary platforms.

The editors also gratefully acknowledge the collaborative contributions of all participating research institutions, including Jinan University, Jilin University, Hong Kong University of Science and Technology (Guangzhou), ETH Zurich, China Academy of Space Technology, and Macau University of Science and Technology. We are deeply indebted to all contributing authors, whose expertise and dedication ensured the technical depth and interdisciplinary scope of this volume. Furthermore, the editors appreciate the professional guidance and assistance of the editorial and production teams at Elsevier Academic Press throughout the publication process. Any remaining errors or omissions are solely the responsibility of the editors.

Chapter 1

Fundamentals of energy storage systems

Haoqin Li[a], Yupeng Jiang[a], Shijie Li[a], Yufeng Wang[a], Huaiguang Jiang[a], and Wei Xiong[b]

[a]*School of Future Technology, South China University of Technology, Guangzhou, China,* [b]*School of Biology and Biological Engineering, South China University of Technology, Guangzhou, China*

Contents

1.1 Introduction

The global demand for energy is steadily increasing, presenting the energy sector with two major challenges: the scarcity of reliable and cost-effective energy options and the environmental degradation resulting from fossil fuel usage. Against this backdrop, the pace of transitioning the energy landscape towards sustainability and clean energy has significantly quickened. This transformation encompasses the rapid expansion of intermittent Variable Renewable Energy (VRE), such as wind and solar power. According to projections by BloombergNEF (BNEF), by 2030, renewable energy is expected to account for 51% of the electricity supply, rising to 63% by 2040 and 70% by 2050 [1]. While this shift bolsters the sustainability of the energy sector, the inherent variability and unpredictability of renewable energy sources present a significant challenge to the stability of power systems.

Modern Energy Storage. https://doi.org/10.1016/B978-0-32-395419-8.00007-6

As a fundamental component of modern energy systems, Energy Storage Systems (ESSs) are increasingly recognized as vital in addressing the imbalance between energy supply and demand, enhancing the flexibility and reliability of energy systems, and driving significant transformations in energy structures. ESSs not only effectively mitigate the volatility of renewable energy generation and provide stability and reliability to the grid but also optimize the coordinated interaction among generation, the grid, load, and storage. This optimization further facilitates the efficient operation of diverse sectors, including power systems, transportation, industrial production, and residential energy consumption. Consequently, ESSs enhance energy utilization efficiency and economic benefits while supporting broader energy structural transitions and advancing the goals of sustainable development.

1.1.1 Description of energy storage system

ESS refers to a device that uses specific technologies to transform electrical energy into a form suitable for storage, retain it in various media, and later convert it back into electrical energy when needed [2]. As illustrated in Fig. 1.1, ESS typically operates in four distinct states: idle, energy input, storage, and energy output [3].

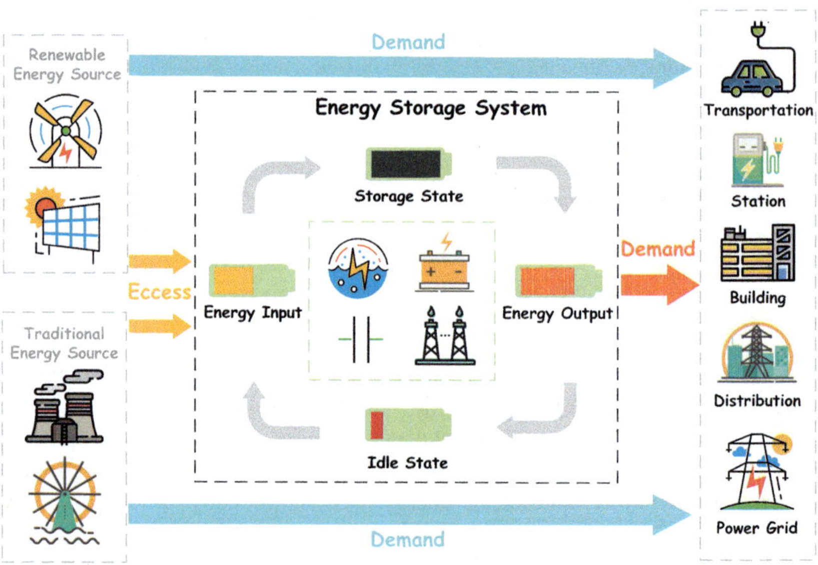

FIGURE 1.1 Schematic of energy storage system.

In the idle state, ESS does not perform energy input or output operations but remains in a low-power standby mode. This state typically occurs during periods of system load balance or when the ESS is not in use. In the energy input state, the system absorbs energy from external sources and stores it. The energy sources can encompass renewable generation methods, such as wind or solar power, surplus electricity during grid off-peak periods, or other flexibly dispatched energy forms, ensuring a reliable supply for future use. Once energy input is complete, the system transitions to the storage state, where the energy is stored in a specific form (e.g., chemical, thermal, or mechanical) within the storage medium. The key objective in this state is to minimize energy loss and ensure the long-term stability of the stored energy. When energy demand increases, the system enters the energy output state, converting the stored energy back into electrical energy and supplying it to the grid or end-use devices to meet real-time electricity needs. The energy output state is widely used to balance grid loads, address peak electricity demand, or provide backup power during emergencies. These four operational states—idle, energy input, storage, and energy output—complement each other and collectively constitute the core functionality of ESS, offering crucial support for the efficient operation and flexible deployment of energy systems.

Beyond the four operational states described above, ESS not only plays a role at the level of individual units for storing and releasing energy but also plays a pivotal role across various tiers of the energy system. From centralized storage within transmission systems to distributed storage within regional grids, and even to residential and building-level storage on the user side, ESS supports modern energy systems by balancing supply and demand, facilitating energy dispatch, and optimizing efficiency. Table 1.1 further summarizes the key functions and roles of energy storage across various application contexts [4].

TABLE 1.1 The role of energy storage systems in energy supply systems.

Role	Centralized Storage	Regional/ Distribution Systems	End Users (Buildings/ Residential)
Balancing Supply	Handles long- and short-term fluctuations, supports intermittent renewables	Manages daily and hourly variations, reduces peak demand	Optimizes daily energy use at the consumer level
Power Transmission	Ensures grid stability, supports electricity trading	Enhances distribution efficiency, maintains voltage/frequency	Reduces distribution stress, lowers energy losses
Improving Efficiency	Boosts overall energy utilization with time-shifting	Improves distribution performance and load management	Promotes efficient energy use and behavioral changes

1.1.2 Advantages of energy storage system

As a crucial component of contemporary energy systems, ESS offers substantial advantages due to its diverse functionalities and extensive applicability. By balancing energy supply and demand, optimizing system efficiency, and supporting the proliferation of renewable energy, ESS provides essential technical support for constructing a clean, low-carbon energy infrastructure. The primary benefits of energy storage technologies across various applications are as follows:

(a) Enhancing the Efficient Utilization of Renewable Energy: Given that renewable energy sources, such as wind and solar power, are inherently intermittent and unstable, they cause fluctuations in electricity supply. ESS can store excess energy when generation is high and discharge it during periods of scarcity, thereby stabilizing output and ensuring the stable functioning of power networks [5]. Furthermore, the implementation of ESS increases the proportion of renewable energy in the global energy mix, thereby playing a vital role in achieving energy transition objectives.

(b) Optimizing Grid Operations through Peak Shaving and Valley Filling: In power systems, variations in electricity demand often result in high load pressure during peak periods and energy waste during off-peak times [6]. ESS addresses this challenge by capturing electricity when demand is low and releasing it when demand is high. This approach not only balances power supply and demand but also substantially enhances grid efficiency and stability while diminishing the necessity for grid infrastructure expansion.

(c) Improving Energy Utilization Efficiency and Reducing Waste: Energy waste from unused heat is a common issue in industrial production and building heating applications. Thermal ESS can effectively recover and store excess heat generated during production for later use [7]. This not only significantly enhances energy utilization efficiency but also minimizes energy waste, ultimately reducing energy costs for both businesses and consumers.

(d) Strengthening the Emergency Response Capability of Power Systems: Natural disasters, grid failures, and other emergencies can cause power outages that severely affect critical infrastructure, such as hospitals and data centers. ESS serves as a reliable backup power source, ensuring a continuous electricity supply during emergencies. This supports the uninterrupted operation of essential facilities and enhances the safety and resilience of power systems.

(e) Supporting Distributed Energy Systems and Smart Grids: The effective implementation of distributed energy systems heavily depends on advanced ESSs. Through intelligent control technologies, energy storage enables the localized storage and release of energy, optimizing regional energy supply. Furthermore, the integration of energy storage with smart grids facilitates more efficient and precise dynamic scheduling of power, significantly improving overall energy system performance [8].

(f) Driving Low-Carbon Energy Development: ESS plays a pivotal role in reducing reliance on fossil fuels and lowering carbon emissions by enabling renewable energy adoption and enhancing energy system efficiency. These advancements are essential for addressing global climate change and achieving carbon neutrality goals. Energy storage technologies not only underpin the transition to a low-carbon energy future but also drive the optimization of future energy structures.

1.2 Historical development and trends of energy storage system

1.2.1 Early beginnings of energy storage system

The history of energy storage can be traced back to the formative stages of human civilization. In ancient times, individuals achieved basic energy storage through the direct utilization of natural conditions. For instance, they stored firewood during colder seasons for heating and preserved food by storing ice and snow in the summer. This direct use of natural resources represents humanity's earliest practices in energy storage.

During the agricultural era, the concepts of energy storage and management underwent significant development. Ancient Chinese hydraulic projects, such as the Dujiangyan Irrigation System, exemplified the effective utilization and storage of water resources through the regulation of water flow, serving as an early prototype of mechanical energy storage. Concurrently, ancient Greek water wheels and Roman milling equipment converted the kinetic energy of water into mechanical power used for grain processing and irrigation. Although these practices relied on relatively simple technology, they signified an early recognition of the importance of energy storage and regulation within human societies.

1.2.2 Milestones in energy storage system

(1) The Discovery of Electricity and the Emergence of Battery Technology

As shown in Fig. 1.2, the discovery of electricity and the invention of chemical batteries represent significant milestones in the history of energy storage systems. In 1800 Alessandro Volta, a renowned Italian scientist, developed the Voltaic Pile, the first device capable of delivering a stable output of electrical energy, thus laying the foundation for modern electrochemical energy storage technologies.

In 1839 Sir William Robert Grove developed an early version of a fuel cell that generated electricity and water through by combining hydrogen and oxygen in the presence of an electrolyte [10]. Although early fuel cells did not see widespread application, they established a theoretical foundation for subsequent advancements in fuel cell technology, particularly in the late 20th century, when

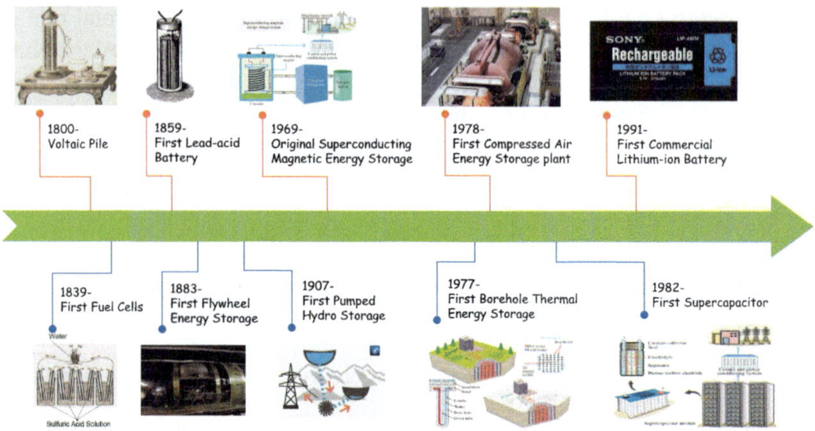

FIGURE 1.2 Historical milestones in energy storage system.

fuel cells began to be utilized extensively in power generation and automotive power systems.

In 1859 Gaston Planté, a French physicist, developed the lead-acid battery, which became the first commercially rechargeable battery, representing a major step forward in electrochemical energy storage, marking a significant advancement in electrochemical energy storage [11]. Due to its low cost and mature technology, lead-acid batteries continue to play a crucial role today, particularly in automotive starting systems and backup power applications.

(2) Commercialization of Mechanical Energy Storage

At the beginning of the 20th century, the rapid growth of the electricity sector led to an increasing need for large-scale energy storage. Pumped hydro storage (PHES) became the first widely adopted mechanical energy storage solution. In 1907 Switzerland launched the Engeweihr storage plant, marking the initial practical implementation of PHES. This system, which stores water in an elevated reservoir when electricity demand is low and releases it during high-demand periods to produce power, facilitated "time-shifting" and efficient energy dispatch [12].

Later, in the 1970s, compressed air energy storage (CAES) was introduced, leading to the establishment of the first-ever CAES power station in Germany in 1978 [13]. By compressing air and storing it in subterranean caverns, which could then be released to drive turbines, CAES provided an innovative method for mechanical energy storage.

Concurrently, flywheel energy storage developed gradually during the late 19th and 20th centuries. As early as 1883, American inventor John A. Howell created the first flywheel energy storage system for military applications [14]. However, it was not until the late 20th century that flywheel energy storage began to demonstrate its advantages in modern energy systems. Despite its limited early applications, the technology later found practical use in areas such as grid frequency regulation and energy recovery in rail transit, ultimately becoming a critical complement to mechanical energy storage.

(3) Rapid Development of Electrochemical Energy Storage

By the mid-20th century, electrochemical energy storage technology had entered a phase of rapid development. The invention of cadmium and nickel-metal hydride batteries provided reliable energy storage solutions for portable electronic devices.

In the 1970s, significant advancements were achieved in foundational research concerning lithium-ion batteries. By 1991, Sony Corporation successfully had commercialized lithium-ion batteries [15]. Due to their high energy density, extended lifespan, and lightweight structure, these batteries rapidly became fundamental to the development of modern energy storage technologies.

(4) The Rise and Engineering Applications of Thermal Energy Storage

In the late 20th century, thermal energy storage systems made significant advancements in industrial waste heat recovery and solar thermal utilization. Sensible heat storage, which involves heating solid or liquid substances (such as water or rocks) to retain thermal energy, represents one of the earliest methods of thermal energy storage. On the other hand, latent heat storage uses Phase Change Materials (PCMs) to absorb and release significant quantities of thermal energy during phase transitions between solid and liquid states, offering higher energy density and efficiency.

In 1977 the first-ever Borehole Thermal Energy Storage (BTES) system was established in Sigtuna, Sweden. This system consisted of 42 boreholes designed for storing and recovering thermal energy, presenting an innovative solution for building heating and energy savings [16]. This development marked the transition of thermal energy storage from laboratory research to practical applications, particularly in regions rich in geothermal resources, where it demonstrated significant potential.

(5) The Emergence of Electrical Energy Storage

In the mid-to-late 20th century, electrical energy storage technologies emerged as a significant branch of the energy storage sector. These technologies were distinguished by their ability to directly store electrical energy without converting it to other forms, enabling efficient and rapid charging and discharging. In 1969 Ferrier introduced the concept of Superconducting Magnetic Energy Storage (SMES) [17]. This technology retains energy in the magnetic field generated by a superconducting coil, which operates at low temperatures. Despite its high efficiency and rapid response capabilities, the substantial cooling costs and technical complexity at that time limited its large-scale application. Nonetheless, SMES provided valuable insights for high-power, short-duration energy storage scenarios.

In 1982 the Pinnacle Research Institute in the United States developed the first generation of supercapacitors, which were initially employed for instantaneous power compensation in military applications [18]. This development marked a breakthrough in supercapacitor energy storage technology. With their ultra-long cycle life and exceptionally high power density, supercapacitors have become a critical technology for modern short-term energy storage and high-frequency charge-discharge applications, finding widespread use in grid frequency regulation, rail transit energy recovery, and consumer electronics.

1.2.3 Development trend and future directions

As the global transition to clean energy accelerates, ESSs, crucial to maintaining the flexibility and stability of energy infrastructure, are becoming increasingly pressing. These technologies not only facilitate the large-scale integration of renewable energy sources into the grid but also serve as key elements in the transformation of energy systems towards low-carbon, smart, and sustainable models. In the near future, the development of ESSs is expected to show the following prominent trends.

(1) Scaling Up

As renewable energy sources are progressively incorporated into power grids, demand for ESSs is expected to grow. To accommodate the increasing need for grid storage, it will be essential to deploy devices with greater capacities. Large-capacity energy storage solutions are expected to become the primary approach in the future, facilitating the efficient storage and transfer of substantial electricity volumes. This trend toward expanded capacity will not only help reduce costs through economies of scale but also improve the overall economic feasibility of energy storage, thus strongly supporting the broader adoption of these technologies.

(2) Diversification

The diversification of energy storage technologies and their application scenarios represents a critical future trend. Currently, energy storage technologies coexist in a multifaceted landscape, encompassing electrochemical storage (e.g., lithium-ion and sodium-ion batteries), mechanical storage (e.g., pumped hydro and compressed air energy storage), and emerging innovations such as liquid-air energy storage and thermal energy storage. Moving forward, these technologies will continue to undergo performance enhancements and collaborative development, converging into a complementary, flexible, and efficient integrated energy storage system. This diversification will better address the challenges of fluctuating power outputs and meet the multilayered, cross-sectoral demands for energy storage, thereby advancing grid flexibility, stability, and decarbonization objectives.

(3) Intelligent Development

The intelligent, automated, and collaborative evolution of ESS will serve as a central focus for future development. By incorporating advanced control systems, algorithms, and big data analytics, these devices will enable real-time monitoring and intelligent regulation of grid operations [9], effectively optimizing energy utilization and enhancing response times. Intelligent energy storage systems can seamlessly integrate with power systems and interact efficiently, thereby improving the reliability and flexibility of grid operations. For instance, smart energy storage systems can dynamically adjust their charging and discharging strategies in response to fluctuations in electricity prices, load demand, and weather conditions, maximizing both economic and environmental benefits. The intelligent development of these systems represents one of the most critical and transformative trends in the field of energy storage.

(4) Green Development

Environmental protection and sustainability will be fundamental principles guiding the future evolution of energy storage technologies. To reduce their environmental impact, energy storage systems must comply with rigorous ecological standards throughout their lifecycle, including production, operation, and disposal. At the same time, combining ESS with renewable energy sources will be crucial to creating a sustainable energy supply chain. In this regard, hydrogen storage technologies, which are seen as key solutions for long-duration energy storage, are gaining increasing focus. Their high energy density, zero carbon emissions, and ability to store surplus renewable energy make them particularly promising. Hydrogen storage, facilitated by water electrolysis, converts excess electricity from photovoltaic or wind power during off-peak hours into hydrogen, which can later be used to generate electricity via fuel cells or other technologies during peak demand. This method not only supports effective energy usage across different time periods but also boosts the flexibility and stability of the grid. Additionally, innovations in solid-state, liquid, and organic liquid hydrogen storage systems will further enhance the safety, efficiency, and economic feasibility of hydrogen storage and transport, providing ideal solutions for the environmentally sustainable growth of energy storage systems.

In summary, scaling up, diversification, intelligent development, and green development represent four key trends that will drive the future of energy storage technologies. These trends are interrelated, forming a critical foundation for innovation in energy storage. Collectively, these advancements will contribute to the development of a highly efficient, clean, and sustainable energy system, offering strong technological support to enhance energy security and achieve carbon-neutrality targets.

1.3 Classification of energy storage systems

1.3.1 Classification based on storage mechanisms

Energy storage systems can be categorized into five types based on their physical or chemical storage mechanisms: mechanical energy storage, electrochemical energy storage, electrical energy storage, thermal energy storage, and chemical energy storage, as illustrated in Fig. 1.3. Each type of energy storage system has its specific working principles, technical characteristics, and application

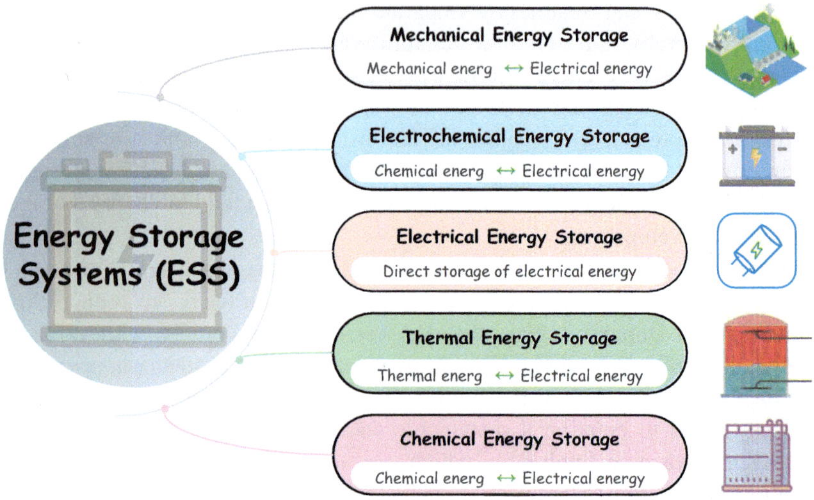

FIGURE 1.3 Classification of ESSs based on storage mechanisms.

scenarios. Table 1.2 summarizes the key performance indicators for each type of energy storage system.

(1) Mechanical Energy Storage System

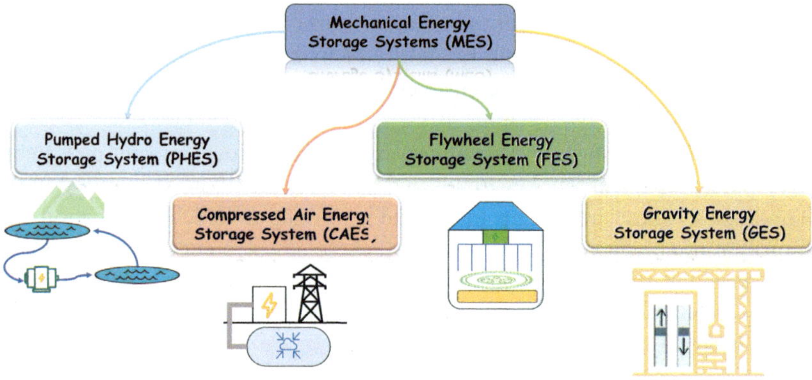

FIGURE 1.4 Schematic of mechanical energy storage classification.

Mechanical energy storage systems function by transforming electrical energy into mechanical energy. The primary technologies in this category include pumped hydro storage, compressed air energy storage, flywheel energy storage, and gravity energy storage. With their well-established engineering practices and reliable operational characteristics, mechanical energy storage systems play an important role in large-scale energy storage applications, as shown in Fig. 1.4.

TABLE 1.2 Comparison of energy storage types and their characteristics.

Energy Storage Type	Energy Density (Wh/kg)	Power Density (W/kg)	Response Time	Cycle Life (cycles)	System Efficiency
		Mechanical Energy Storage			
Pumped Hydro Storage	0.5–1.5	1–2	Minutes	>50,000	70–85%
Compressed Air Energy Storage	2–6	50–200	Minutes	>10,000	40–70%
Flywheel Energy Storage	5–130	1000–10,000	Milliseconds	>20,000	90–95%
Gravity Energy Storage	0.2–2	1–2	Minutes	>10,000	75–90%
		Electrochemical Energy Storage			
Lead-Acid Batteries	30–50	75–300	Seconds	500–2000	70–90%
Lithium-Ion Batteries	100–300	250–340	Milliseconds	500–10,000	85–95%
Sodium-Ion Batteries	75–200	1000	Milliseconds	Hundreds to thousands	Up to 92%
Flow Batteries (e.g., Vanadium Redox)	10–50	100–2000	Milliseconds	>10,000	60–85%
		Electrical Energy Storage			
Supercapacitors	1–15	10,000–100,000	Milliseconds	>1,000,000	90–98%
Superconducting Magnetic Energy Storage	0.1–10	1000–10,000	Instantaneous	>100,000	90–95%
		Thermal Energy Storage			
Molten Salt Energy Storage	80–150	N/A	Minutes	>10,000	70–90%
Phase Change Material Storage	50–150	N/A	Minutes	>10,000	75–90%
Sensible Heat Storage	10–50	N/A	Minutes	>10,000	50–90%
		Chemical Energy Storage			
Hydrogen Energy Storage	33–39 (based on LHV)	N/A	Seconds	N/A	25–45%
Synthetic Fuels	13 (Methanol)	N/A	Seconds	N/A	30–50%

Pumped hydro storage is the dominant form of mechanical energy storage. Its basic principle involves using electricity to drive water pumps that transport water to a higher-elevation reservoir, converting electrical energy into the potential energy of water. During periods of electricity demand, water from the upper reservoir is released to drive turbines connected to generators, thereby converting potential energy back into electrical energy. With high efficiency and large-scale capabilities, pumped hydro storage plays a key role in grid regulation. As of 2024, the global installed capacity of pumped hydro storage has exceeded 180 GW [19].

Compressed air energy storage uses electricity to drive compressors, which store compressed air in underground caverns or specially designed storage tanks. When electricity demand arises, the compressed air is released and expanded to power turbines, generating electricity. Studies indicate that advanced adiabatic CAES systems can achieve efficiencies exceeding 70% [20]. This technology demonstrates promising potential for large-scale energy storage applications.

Flywheel energy storage operates by storing kinetic energy in a rapidly spinning rotor. An electric motor accelerates the flywheel, transforming electrical energy into kinetic energy, which can later be reconverted into electricity by a generator during the discharge phase. With its rapid response capability and high power density, this technology is particularly suitable for short-duration, high-power energy storage scenarios. Thanks to breakthroughs in carbon fiber composite materials, modern flywheel systems can achieve rotational speeds exceeding 100,000 revolutions per minute [21].

Gravity energy storage is an emerging mechanical energy storage solution. It works by using electricity to lift heavy objects to higher positions, thereby storing potential energy. When energy is needed, the descent of these objects drives a generation system to produce electricity. This technology offers advantages such as relatively low construction costs and flexible site selection. Common implementations include using abandoned mines, buildings, or specially constructed towers as track systems for lifting and lowering heavy objects. Gravity energy storage systems can achieve a round-trip efficiency of 62%–80% [22] and have a long service life. Compared to traditional pumped hydro storage, gravity energy storage is less dependent on specific geographic conditions and can be deployed in urban and industrial areas, offering a new pathway for distributed energy storage.

(2) Electrochemical Energy Storage System

Electrochemical energy storage is a technology system that stores and releases energy through electrochemical reactions, primarily encompassing various types of secondary batteries and flow batteries. This technology finds widespread application in portable devices, electric transportation, and grid energy storage.

Traditional lead-acid batteries are one of the oldest electrochemical energy storage devices, operating based on the reversible redox reactions of lead and its compounds during charge and discharge cycles. With advantages such as high technological maturity and low manufacturing costs, lead-acid batteries hold a significant position in backup power systems. To overcome the limitation of short cycle life in traditional lead-acid batteries, advanced lead-carbon batteries incorporate carbon materials, significantly improving charge-discharge performance and extending service life.

Lithium-ion batteries store and release energy through the intercalation and deintercalation of lithium ions between anode and cathode materials. Known for their high energy density and long cycle life, they have become the leading choice for portable electronics and electric vehicles. By 2024, commercial lithium-ion batteries have reached an energy density of 350 Wh/kg. Meanwhile, third-generation solid-state battery technology, employing solid electrolytes, has achieved a breakthrough energy density of 500 Wh/kg, alongside significant improvements in safety performance [23].

Flow batteries use liquid electrolytes as energy carriers, with their core principle relying on redox reactions of active materials in the flowing electrolytes to store and release energy. Among flow battery technologies, the all-vanadium flow battery is the most mature. Its distinguishing feature is the independent design of energy capacity and power output, making it particularly suitable for large-scale energy storage applications. This technology has already been commercialized at scale [24].

(3) Electrical Energy Storage System

Electrical energy storage technologies store electrical energy directly through electric or magnetic fields. The two primary forms of this technology are supercapacitors and superconducting magnetic energy storage. With their rapid response characteristics, these technologies play a crucial role in specific application scenarios.

Supercapacitors store energy through the electric double-layer phenomenon occurring at the interface between the electrode and the electrolyte. The working principle involves the accumulation of charges at the electrode/electrolyte interface, creating a potential difference. This technology is characterized by high power density, fast charge-discharge rates, and long cycle life [36]. Although its energy density is relatively low, significant performance improvements are being achieved through the development and application of novel materials, such as graphene-based materials. Supercapacitors have demonstrated clear economic and technical advantages in applications like regenerative braking energy recovery in rail transit systems [30].

Superconducting magnetic energy storage technology stores electrical energy through superconductors operating in a zero-resistance state. Its fundamental principle involves generating a strong magnetic field in a superconducting coil to store energy. This technology is renowned for its extremely fast response speed and high efficiency. However, due to the need for ultra-low-temperature environments to maintain superconducting states, the current application costs are high. Advances in high-temperature superconducting materials offer new pathways to reduce the operating costs of SMES systems [25].

(4) Thermal Energy Storage System

Thermal energy storage is a technological system that stores thermal energy through temperature changes or phase transition processes. Energy storage and release can be achieved using sensible heat, latent heat, or thermochemical reactions. This technology offers unique advantages in the integrated utilization of energy systems.

Molten salt energy storage is a representative thermal energy storage technology, widely used in Concentrated Solar Power (CSP) generation. Its basic principle involves heating molten salt with solar energy and storing the thermal energy in molten salt tanks. The stored heat can then be released at night or during periods without sunlight to drive steam turbines for electricity generation. In recent years, advanced molten salt materials have raised system operating temperatures to 565 °C, achieving an energy storage density of 450 kWh/m^3. A 50 MW CSP plant equipped with a 19-hour molten salt energy storage system can achieve stable, round-the-clock power generation, with an annual utilization of over 3138 hours [26].

PCM energy storage relies on the latent heat absorbed or released during phase transitions to store energy. This technology is characterized by a high energy storage density per unit volume. It has promising applications in building energy efficiency and heating/cooling systems. Research into organic phase change materials and composite phase change materials has opened new avenues for enhancing PCM energy storage performance.

Sensible heat storage, such as heating rocks or water, and ice thermal storage technology, are also widely used in industrial and residential applications. These technologies provide effective and reliable solutions for energy storage in diverse settings.

(5) Chemical Energy Storage System

Chemical energy storage is a technological pathway that stores energy by converting electrical energy into chemical energy, typically based on the reversibility of chemical reactions. This technology offers unique advantages for long-term, large-scale energy storage applications.

Hydrogen energy storage is a representative form of chemical energy storage. It involves producing hydrogen through water electrolysis, converting electrical energy into the chemical energy stored in hydrogen. The stored hydrogen can then be reconverted into electrical energy using fuel cells, enabling energy release. The efficiency of hydrogen production is a critical factor for this technology. Currently, three primary methods are used for electrolytic hydrogen production: Solid Oxide Electrolyzers (SOEs), Proton Exchange Membrane (PEM) electrolyzers, and Alkaline Water Electrolyzers (AWEs) [35]. Recent advancements in high-efficiency electrolyzers include PEM systems with reported efficiencies of 70%–80% [27].

Chemical energy storage also includes technologies that store energy by synthesizing methanol or natural gas using electricity. These processes typically involve reacting hydrogen with carbon dioxide to generate chemical fuels, thereby not only achieving energy storage but also contributing to greenhouse gas reduction. Although chemical energy storage systems boast high energy density and long storage durations, they often suffer from relatively low conversion efficiencies. Breakthroughs in advanced catalytic technologies are playing a crucial role in improving the overall efficiency of these systems.

1.3.2 Classification based on application scale

Energy storage systems can be categorized into three levels according to their application scale and usage scenarios: grid-scale, distributed, and mobile. This classification method reflects the differences in technical characteristics, economic viability, and operational modes of ESSs in different application scenarios. Table 1.3 summarizes the key technical requirements and application scenarios for energy storage of different scales, while Fig. 1.5 illustrates this classification schematically.

TABLE 1.3 Classification of ESSs based on application scale.

Application Type	Typical Scale	Key Technical Requirements	Main Application Scenarios
Grid-scale Energy Storage	>10 MW	Large capacity, high reliability, long lifespan, high efficiency, fast response, intelligent management	Peak shaving, renewable energy integration, backup power, black start, frequency regulation
Distributed Energy Storage	100 kW–10 MW	Flexibility, cost-effectiveness, modular design, high safety, intelligent control	User-side management, demand response, microgrid support, industrial and commercial storage
Mobile Energy Storage	<100 kW	High energy density, high safety, fast charging, lightweight design, environmental adaptability	Electric transportation, portable devices, emergency power, outdoor activities

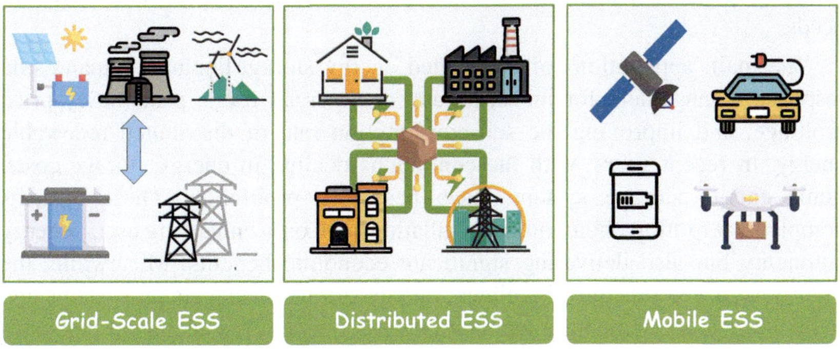

FIGURE 1.5 Classification of ESSs based on application scale.

(1) Grid-Scale Energy Storage

Grid-scale energy storage refers to large-scale energy storage facilities directly serving power systems, typically with installed capacities in the megawatt range or higher. These systems are critical for maintaining the safe and stable operation of power grids while enhancing the capacity to integrate renewable energy sources, characterized by their large capacity, fast response, and high reliability.

The core functions of grid-scale energy storage include peak shaving, energy arbitrage, and auxiliary services such as frequency and voltage regulation for power grids. In peak shaving applications, ESSs store electricity during off-peak periods and release it during peak periods, effectively flattening the load curve. Research data show that properly configured grid-scale energy storage, particularly Long-Duration Energy Storage (LDES), can significantly impact grid operations. For example, implementing a 20-terawatt-hour (TWh) LDES mandate by 2050 could lead to a 92% reduction in curtailment, a 10% decrease in total installed power capacity, a 75% reduction in transmission infrastructure requirements, and a 70% drop in electricity prices during peak periods [28]. Large-scale pumped hydro storage power plants are the most mature form of grid-scale energy storage, with individual plants capable of reaching installed capacities of several gigawatts. With technological advancements and continuously declining costs, electrochemical energy storage has also gradually developed to the megawatt scale, playing an increasingly prominent role in grid regulation. In particular, the development of coordinated control and intelligent dispatch technologies for large-scale energy storage power station clusters has significantly improved the overall operational efficiency and economic benefits of these systems.

(2) Distributed Energy Storage System

Distributed energy storage systems refer to small and medium-sized energy storage facilities deployed on the user side, typically with installed capacities ranging from kilowatts to megawatts. These systems directly serve end users, offer high flexibility and ease of deployment, and can meet personalized user needs.

The main applications of distributed energy storage include demand-side response management for industrial and commercial users, peak-valley price arbitrage, and improving the self-consumption rate of distributed renewable energy. In recent years, with the continuous decline in energy storage costs, home energy storage systems have begun to proliferate. These systems complement rooftop photovoltaic installations, not only enhancing users' energy autonomy but also delivering significant economic benefits. Meanwhile, the innovative development of intelligent energy storage management systems and multienergy complementary technologies has opened new technical pathways for the efficient utilization of distributed energy storage [29].

(3) Mobile Energy Storage

Mobile energy storage refers to ESSs applied in transportation vehicles and portable devices. These energy storage facilities must simultaneously meet stringent requirements for high energy density, high safety, and lightweight design, and their technological development directly influences the innovation progress of the electric transportation and mobile device industries.

Electric Vehicle (EV) battery systems are the most representative application of mobile energy storage. In addition to meeting the energy demands of vehicle operation, EV batteries must also adapt to complex and variable working environments. Continuous innovation in new battery materials and advanced battery management systems is driving comprehensive improvements in the energy density, cycle life, and safety of EV batteries. Additionally, battery systems in devices like smartphones and laptops are also classified as mobile energy storage. These energy storage devices require higher portability and user-friendliness, with key areas of technological development including fast charging and lightweight design. Breakthroughs in next-generation fast-charging technologies and lightweight solutions have significantly improved the practical application experience of portable energy storage devices.

1.3.3 Classification based on energy output timeframe

Energy storage systems can be categorized into three types based on the duration of energy output: long-term, medium-term, and short-term. This classification method reflects the essential differences in energy-release characteristics, application scenarios, and technical-economic aspects across different energy storage systems. These technologies complement each other in the time dimension, collectively building a complete energy storage application system.

(1) Long-Term Energy Storage System

Long-term energy storage systems can achieve energy storage over days, months, or even seasons, making them a key supporting technology for building high-proportion renewable energy power systems. These systems are characterized by their large storage capacity and low unit storage cost.

Pumped hydro storage and compressed air energy storage are traditional long-term energy storage solutions. These technologies leverage natural geographical conditions to achieve large-scale, low-cost energy storage. In-depth research shows that long-term energy storage plays an irreplaceable role in balancing seasonal fluctuations in renewable energy, significantly improving the utilization efficiency of renewable energy [33]. Chemical energy storage technologies such as hydrogen storage are considered important pathways for achieving large-scale, long-term storage of renewable energy due to their ultra-high energy density (up to 500–3000 Wh/kg) and extremely long storage duration. These technologies store energy through reversible chemical reactions, offering advantages such as convenient transportation and stable storage. Breakthroughs in new energy storage materials and system integration technologies are opening new directions for the development of long-term energy storage systems.

(2) Medium-Term Energy Storage System

Medium-term energy storage systems typically have discharge time scales of hours and are primarily used for intra-day load regulation and for short-term integration of renewable energy into power systems. These systems require a balance between energy density, power characteristics, and economic viability.

Lithium-ion batteries are the leading medium-term energy storage systems, known for their high energy density (typically 150–300 Wh/kg), fast response times, and long cycle life (typically 3000–8000 cycles). Practical data show that properly configured medium-term energy storage systems can increase the integration rate of renewable energy by more than 10%–20%, significantly improving the economic viability of renewable energy generation. Flow batteries, with their unique advantage of independently designing energy capacity and power, are becoming a highly promising technology choice in the medium-term energy storage field. This technology stores energy through the circulation of electrolytes, theoretically enabling longer lifespans and more flexible system designs.

(3) Short-Term Energy Storage System

Short-term energy storage systems are facilities with discharge times ranging from seconds to minutes, characterized by ultra-high power density and extremely fast response times. These systems are primarily used for instantaneous power balancing and power quality improvement in power systems, effectively suppressing transient disturbances in the grid.

Supercapacitors and flywheel energy storage are the most representative short-term energy storage technologies [31,32]. These technologies offer millisecond-level response times and exceptional cycle life (often exceeding hundreds of thousands of cycles), but they are characterized by relatively low energy density and high cost per unit of storage. In practical applications, short-term energy storage systems can achieve millisecond-level responses in scenarios such as grid frequency regulation and uninterruptible power supplies, providing strong support for the stable operation of power systems. In recent years, the development of new energy storage materials and the application of advanced control strategies have driven continuous breakthroughs in the power density, response speed, and lifespan of short-term energy storage systems. In particular, innovations in carbon-based materials and new composite electrode materials have significantly improved the performance indicators of supercapacitors.

1.4 Safety considerations in energy storage system

The safety of ESS is crucial to maintaining the secure and dependable operation of energy networks, representing a major challenge for their large-scale implementation and adoption. From the perspective of overall energy system security [34], ESS must achieve efficiency, reliability, and environmental friendliness during operation, while also demonstrating resilience against extreme operating conditions and external threats.

The physical safety of ESS is essential for its reliable operation. Many energy storage systems function under conditions of high temperature, high pressure, or other complex environments, necessitating that the materials and engineering designs of key components can withstand extreme loads. For instance, in energy storage systems utilizing high-temperature molten media, tanks, pipelines, and sealing materials must demonstrate exceptional heat resistance and long service life. Mechanical failure under high-stress conditions could result in medium leakage or structural damage, posing substantial risks to the surrounding environment, personnel safety, and energy supply. Furthermore, storage systems installed in specialized locations, such as mountainous or coastal areas, may encounter physical risks from geological instability, natural disasters, or extreme weather conditions, all of which require heightened attention.

The chemical safety of ESS is a critical factor influencing its operational reliability. Many energy storage systems depend on high-energy-density chemical materials to store and release energy, making them vulnerable to risks such as thermal runaway, accelerated chemical reactions, and even explosions under extreme conditions. For instance, during events such as overload, short circuits, or external impacts, the storage medium may undergo chain reactions, potentially causing device malfunction or catastrophic failure. These chemical risks are particularly significant in storage systems designed for long-term operation or high-frequency charge-discharge cycles. Additionally, the environmental impact of the chemical materials used in ESS—throughout their production, transportation, use, and disposal—poses serious concerns. In systems involving rare metal resources, the production and decommissioning processes can exert additional pressure on ecosystems.

The cybersecurity of energy storage systems has become increasingly critical due to the deep integration of digitalization and intelligent technologies. Modern ESS often rely heavily on intelligent control systems, such as Energy Management Systems (EMS) and internal monitoring algorithms, to achieve dynamic regulation and condition monitoring. While these systems optimize operations through real-time data collection and analysis, they are also vulnerable to potential cyberattack threats. Malicious actors may tamper with critical parameters or disrupt control signals, resulting in abnormal device operations that can lead to regional energy outages or safety incidents. As energy storage systems grow in scale and integrate more deeply into the energy internet, cybersecurity will become a key focus for ensuring their safety.

The environmental safety of ESS is of paramount importance. Some energy storage systems may directly or indirectly affect their surrounding environment. For instance, certain systems can threaten local ecosystems due to thermal leakage or the release of storage media. Similarly, technologies such as pumped hydro storage may have long-term implications for water resource allocation and ecological balance. As the adoption of energy storage technologies intensifies, potential environmental risks will expand, necessitating stricter environmental and sustainability requirements for these systems.

The long-term operational safety of ESS necessitates significant attention. High-intensity operating conditions can lead to performance degradation or aging of critical components, which creates potential safety hazards. For instance, energy storage materials may lose stability at high temperatures or under high pressure, or during repeated cycles, while mechanical components may fail due to prolonged fatigue. To ensure stable long-term operation, it is essential to comprehensively consider material properties, engineering strength, and operational environments during the design phase, thereby extending the device lifespan and mitigating the risk of failure.

In conclusion, ESS safety is a complex and multifaceted issue that involves physical, chemical, cybersecurity, environmental, and long-term operational considerations. To enable the widespread use of ESS, it is essential to address these safety challenges comprehensively. Ongoing improvements in the design, operation, and management of storage systems are crucial to maintaining the stability and long-term viability of energy networks, ultimately contributing to the broader goals of energy security and achieving carbon neutrality.

References

[1] BNEF, New energy outlook 2024 [Online], https://about.bnef.com/new-energy-outlook/, 2024. (Accessed 21 May 2024).

[2] X. Luo, J. Wang, M. Dooner, J. Clarke, Overview of current development in electrical energy storage technologies and the application potential in power system operation, Applied Energy 137 (2015) 511–536.

[3] M. Qi, Y. Liu, R.S. Landon, Y. Liu, I. Moon, Assessing and mitigating potential hazards of emerging grid-scale electrical energy storage systems, Process Safety and Environmental Protection 149 (2021) 994–1016.

[4] A.A. Kebede, T. Kalogiannis, J. Van Mierlo, M. Berecibar, A comprehensive review of stationary energy storage devices for large scale renewable energy sources grid integration, Renewable and Sustainable Energy Reviews 159 (2022) 112213.

[5] P. Ralon, M. Taylor, A. Ilas, H. Diaz-Bone, K. Kairies, Electricity Storage and Renewables: Costs and Markets to 2030, vol. 164, International Renewable Energy Agency, Abu Dhabi, United Arab Emirates, 2017.

[6] S. Li, W. Li, L. Chen, H. Jiang, J. Zhang, D.W. Gao, Real-time robust state estimation for large-scale low-observability power-transportation system based on meta physics-informed graph timesnet, IEEE Transactions on Smart Grid 15 (6) (2024) 5500–5513.

[7] W. Chen, Z. Huang, K.J. Chua, Sustainable energy recovery from thermal processes: a review, Energy, Sustainability and Society 12 (1) (2022) 46.

[8] M. Aneke, M. Wang, Energy storage technologies and real life applications – a state of the art review, Applied Energy 179 (2016) 350–377.

[9] S. Li, W. Li, L. Chen, H. Li, Y. Zhang, H. Jiang, A short-term load forecasting method for large-scale power distribution systems based on a novel spatio-temporal neural network, in: 2023 China Automation Congress (CAC), 2023, pp. 9235–9241.

[10] J. Mitali, S. Dhinakaran, A. Mohamad, Energy storage systems: a review, Energy Storage and Saving 1 (3) (2022) 166–216.

[11] P. Kurzweil, Gaston Planté and his invention of the lead–acid battery—the genesis of the first practical rechargeable battery, Journal of Power Sources 195 (14) (2010) 4424–4434.
[12] Wikipedia contributors, Pumped-storage hydroelectricity-Wikipedia, the free encyclopedia [Online]. Available: https://en.wikipedia.org/w/index.php?title=Pumped-storagehydroelectricity&oldid=126503, 2024. (Accessed 9 January 2025).
[13] H. Chen, T.N. Cong, W. Yang, C. Tan, Y. Li, Y. Ding, Progress in electrical energy storage system: a critical review, Progress in Natural Science 19 (3) (2009) 291–312.
[14] E. Hossain, H.M.R. Faruque, M.S.H. Sunny, N. Mohammad, N. Nawar, A comprehensive review on energy storage systems: types, comparison, current scenario, applications, barriers, and potential solutions, policies, and future prospects, Energies 13 (14) (2020).
[15] M.V. Reddy, A. Mauger, C.M. Julien, A. Paolella, K. Zaghib, Brief history of early lithium-battery development, Materials 13 (8) (2020).
[16] B. Zalba, J.M. Marın, L.F. Cabeza, H. Mehling, Review on thermal energy storage with phase change: materials, heat transfer analysis and applications, Applied Thermal Engineering 23 (3) (2003) 251–283.
[17] Z. Abdin, K.R. Khalilpour, Chapter 4 – single and polystorage technologies for renewable-based hybrid energy systems, in: K.R. Khalilpour (Ed.), Polygeneration with Polystorage for Chemical and Energy Hubs, Academic Press, 2019, pp. 77–131.
[18] Wikipedia contributors, Supercapacitor-Wikipedia, the free encyclopedia [Online]. Available: https://en.wikipedia.org/w/index.php?title=Supercapacitor&oldid=1256882093, 2024. (Accessed 9 January 2025).
[19] International Hydropower Association, 2024 World Hydropower Outlook, International Hydropower Association, 2024 [Online]. Available: https://www.hydropower.org/publications/2024-world-hydropower-outlook.
[20] D. Wu, W. Wei, J. Bai, S. Mei, Energy and exergy efficiency analysis of advanced adiabatic compressed air energy storage based trigeneration energy hub, CSEE Journal of Power and Energy Systems 9 (6) (2023) 2409–2422.
[21] F. Xu, X. Dai, Y. Wang, D. Hu, H. Zhang, H. Chen, Research progress on permanent magnet machines for flywheel energy storage, Energy Storage Science and Technology 13 (10) (2024) 3423.
[22] X. Zhang, K. Zhang, Research status and prospect analysis of gravity energy storage, in: International Conference on Advances in Energy Resources and Environment Engineering, Springer, 2022, pp. 153–160.
[23] M. Burton, S. Narayanan, B. Jagger, L.F. Olbrich, S. Dhir, M. Shibata, M.J. Lain, R. Astbury, N. Butcher, M. Copley, et al., Techno-economic assessment of thin lithium metal anodes for solid-state batteries, Nature Energy 10 (1) (2025) 135–147.
[24] C. Zhang, Z. Yuan, X. Li, Designing better flow batteries: an overview on fifty years' research, ACS Energy Letters 9 (7) (2024) 3456–3473.
[25] I. Patel, A. Shah, B. Shen, H. Wei, L. Hao, J. Hu, Q. Wang, T. Coombs, Stochastic optimisation and economic analysis of combined high temperature superconducting magnet and hydrogen energy storage system for smart grid applications, Applied Energy 341 (2023) 121070.
[26] SolarPACES, Gonghe CSP in Qinghai province runs a record 19 of 24 hours [Online]. Available: https://www.solarpaces.org/gonghe-csp-in-qinghai-province-runs-a-record-19-of-24-hours/, 2024.
[27] Y. Song, H. Chen, X. Wang, C. Weng, K. Zou, C. Wang, Y. Yuan, Y. Ma, X. Yang, W. Lin, Engineering Ir-based catalysts for high current density applications in proton exchange membrane water electrolyzers, Energy & Environmental Science 18 (1) (2025) 130–154.
[28] M. Staadecker, J. Szinai, P.A. Sánchez-Pérez, S. Kurtz, P. Hidalgo-Gonzalez, The value of long-duration energy storage under various grid conditions in a zero-emissions future, Nature Communications 15 (1) (2024) 9501.
[29] B. Zakeri, G.C. Gissey, P.E. Dodds, D. Subkhankulova, Centralized vs. distributed energy storage–benefits for residential users, Energy 236 (2021) 121443.

[30] Z. Zhang, T. Ding, Q. Zhou, Y. Sun, M. Qu, Z. Zeng, Y. Ju, L. Li, K. Wang, F. Chi, A review of technologies and applications on versatile energy storage systems, Renewable and Sustainable Energy Reviews 148 (2021) 111263.
[31] W. Ji, F. Hong, Y. Zhao, L. Liang, H. Du, J. Hao, F. Fang, J. Liu, Applications of flywheel energy storage system on load frequency regulation combined with various power generations: a review, Renewable Energy 223 (2024) 119975.
[32] J.M. Lim, Y.S. Jang, H.V.T. Nguyen, J.S. Kim, Y. Yoon, B.J. Park, D.H. Seo, K.-K. Lee, Z. Han, K.K. Ostrikov, et al., Advances in high-voltage supercapacitors for energy storage systems: materials and electrolyte tailoring to implementation, Nanoscale Advances 5 (3) (2023) 615–626.
[33] H. Blanco, A. Faaij, A review at the role of storage in energy systems with a focus on power to gas and long-term storage, Renewable and Sustainable Energy Reviews 81 (2018) 1049–1086.
[34] A. Azzuni, C. Breyer, Energy security and energy storage technologies, in: International Renewable Energy Storage Conference, IRES 2018, 13–15 March 2018, Dusseldorf, Germany, in: Energy Procedia, vol. 155, 2018, pp. 237–258.
[35] Y. Xia, H. Cheng, H. He, W. Wei, Efficiency and consistency enhancement for alkaline electrolyzers driven by renewable energy sources, Communications Engineering 2 (1) (2023) 22.
[36] C. Choi, T.G. Yun, B. Hwang, Dispersion stability of carbon nanotubes and their impact on energy storage devices, Inorganics 11 (10) (2023) 383.

Chapter 2

Mechanical and thermal energy storage technologies

Ying Xue
School of Electric Power Engineering, South China University of Technology, Guangzhou, China

Contents

2.1 Introduction

This chapter focuses on two key energy storage technologies: mechanical energy storage and thermal (both heating and cooling) energy storage. Mechanical energy storage achieves the storage of other forms of energy through the interaction of materials, deformations arising from mechanical motion, and inertia. In contrast, thermal energy storage technologies utilize energy storage materials as thermal carriers, taking advantage of their ability to absorb or release significant amounts of energy, thereby storing energy in specific media. These two technologies play crucial roles in demand-side management, eliminating diurnal peak-to-valley differences and smoothing out load fluctuations. Subsequently, this chapter will elaborate on the overall

Modern Energy Storage. https://doi.org/10.1016/B978-0-32-395419-8.00008-8

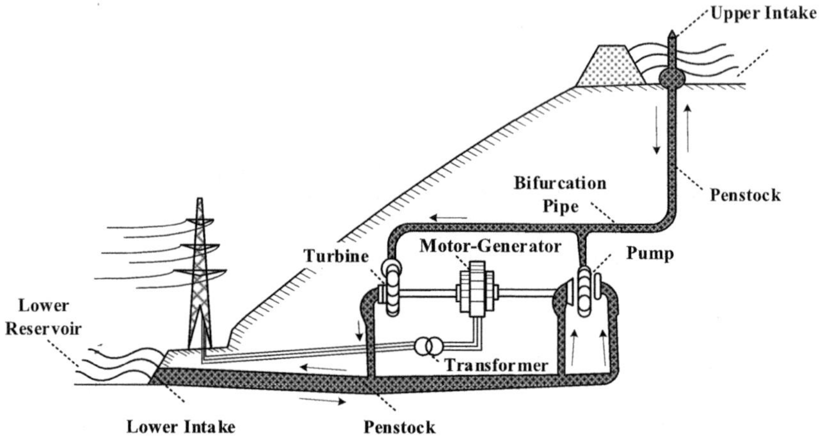

FIGURE 2.1 Structure of a pumped hydro storage power station.

research status of mechanical and thermal energy storage, respectively, thereby clarifying their research foundations. It will delve into key technologies and ultimately present a comprehensive picture of the performance of different energy storage methods through integrated application demonstrations both domestically and internationally.

2.2 Pumped storage hydro power

Currently, pumped hydro storage technology is a mainstream technology in large-scale energy storage. Its working principle is illustrated in Fig. 2.1. It uses surplus electrical energy to pump water from a downstream reservoir to an upstream reservoir. During peak grid load, water from the upstream reservoir is released back to the downstream reservoir to drive turbines for electricity generation. As the electrical energy storage technology with the largest storage capacity currently available, pumped hydro storage boasts a comprehensive operating efficiency exceeding 70%. The time from unit startup to full-load operation is less than 3 minutes, and from no-load to full-load is less than 1 minute. It offers advantages such as large storage capacity, high system efficiency, and a mature technical system [1]. Therefore pumped hydro storage technology is widely used for peak shaving, frequency regulation, phase regulation, voltage regulation, and other aspects of power systems. It also has great potential to facilitate the smooth and large-scale output of renewable energy [2].

As of the end of 2023, the global installed capacity of pumped hydro storage reached 179,130 MW, with China's installed capacity accounting for approximately 29% of the global total at 50,940 MW, ranking first worldwide. According to the medium- and long-term development plan for pumped storage

(2021–2035) issued by the National Energy Administration, China aims to achieve a total installed capacity of pumped hydro storage exceeding 62,000 MW by 2025 and double that figure by 2030, reaching a total installed capacity of 120,000 MW.

2.2.1 Types and performance of units

Pumped Storage Units (PSUs) can be classified into four-machine, three-machine, and two-machine types based on their configurations. Early PSUs adopted a four-machine separate design, in which the turbine and generator formed a turbine-generator set, and the electric motor and pump formed a pump set, totaling four machines. Since the generator set and pump set were independent of each other, both sets could be designed and operated near their theoretically optimal working conditions, resulting in high efficiency. However, this design required a large footprint, had a complex structure, and was costly to invest in, so it is no longer commonly used in China [3]. Subsequently, three-machine tandem units emerged, in which the pump, turbine, and motor operated coaxially, as shown in Fig. 2.2. Three-machine PSUs can be further classified into horizontal and vertical configurations. The pump and turbine within the unit rotate in the same direction, enabling seamless switching between pumping and power generation modes, greatly improving the flexibility of unit regulation. In practical engineering, they are mostly used in high-head pumped storage projects, such as pumped storage power stations designed for heads ranging from 600 to 800 meters or higher.

With the advancement of technology, three-machine PSUs are gradually being replaced by two-machine units. Two-machine units are also known as reversible units, where the pump and turbine are combined into a reversible pump-turbine, and the generator and electric motor are integrated into a reversible motor-generator. The motor-generator and pump-turbine shafts are connected through a coupling, as shown in Fig. 2.3. When the pump-turbine rotates in one direction, it operates in turbine mode; when it rotates in the opposite direction, it operates in pump mode. Two-machine units have a simple structure and lower cost, making them the mainstream type of PSUs currently.

With the continuous construction and development of pumped storage power stations, the global trend in pump-turbine development exhibits characteristics of high head, large capacity, and high speed, closely linked to the construction of pumped storage power stations. Power stations strive for high head to increase the rotational speed of the units, thereby reducing the size of the pump-turbine and motor. At a fixed installed capacity, high head can decrease the water intake flow rate, lower the demand for reservoir capacity, and simultaneously suppress water level fluctuations, ensuring efficient and stable operation of the pump-turbine, see Fig. 2.4.

Due to constraints related to hydropower engineering and specialized technology, the main unit types for pumped storage power stations with ultra-

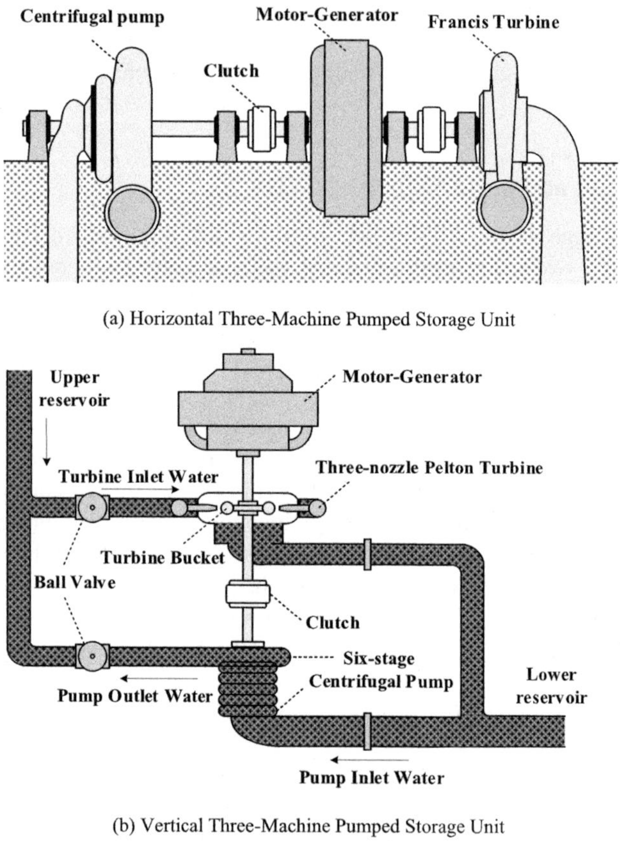

FIGURE 2.2 Structure of three-machine pumped storage unit.

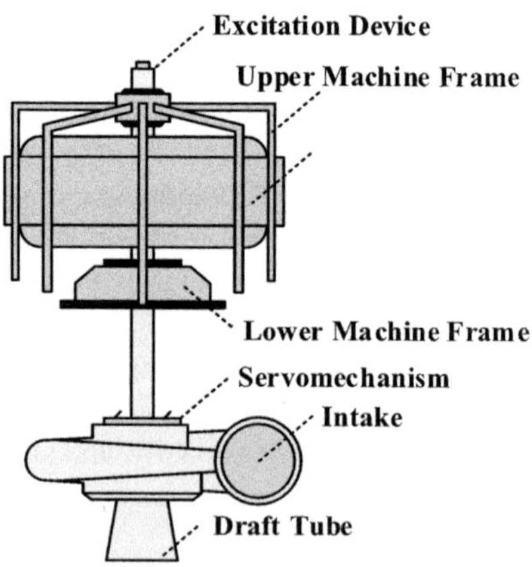

FIGURE 2.3 The structure of a two-machine pumped storage.

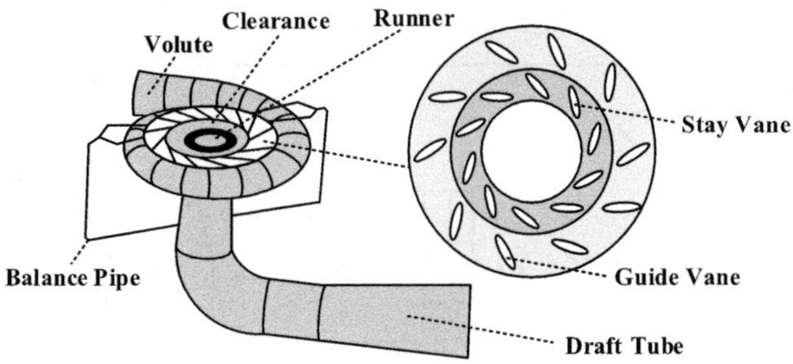

FIGURE 2.4 Flow field diagram of pump-turbine.

high heads above 800 meters include the four-machine separate design, the three-machine tandem design, and the multistage reversible design. The four-machine separate design, which requires duplicating piping system equipment, is rarely adopted due to its high cost. The three-machine tandem design pairs an impulse turbine with a multistage pump, with the turbine and pump independently designed to allow both machines to operate within their optimal efficiency ranges. Currently, the Bieudron pumped storage hydropower station in Switzerland holds the record for the highest single-unit capacity and head, with a rated capacity of 423.13 MW and a rated head of 1883 meters for its impulse turbines. In the future, similar three-machine PSUs are expected to increase operating heads to 2000 meters.

Modern PSUs integrate multiple functions, including peak shaving, valley filling, frequency regulation, phase adjustment, and black-start capabilities. As a critical component, the turbine undergoes more than ten startups and shutdowns per day. The startup and shutdown processes are susceptible to multiple factors, which can easily lead to issues such as severe oscillations and pressure pulsations. Among them, the flow characteristics of the pump-turbine directly affect the working performance of the pumped storage unit. Flow characteristics mainly include turbulence, vortices, cavitation, and other aspects, which exhibit different behaviors under pump and turbine operating conditions. The velocity and pressure of the water body are tightly coupled and are the fundamental influencing factors of flow characteristics. Under pump operating conditions, the water body is accelerated by the impeller, increasing its velocity, and the pressure gradually increases from the inlet to the outlet. Conversely, under turbine operating conditions, the velocity of the water body decreases as kinetic energy is converted to mechanical energy, and the pressure gradually decreases from the inlet to the outlet. The uniformity of the water velocity and pressure distribution directly impacts the efficiency of energy conversion. Uneven velocity distribution can lead to increased energy losses and reduced efficiency, while uneven pressure distribution can cause phenomena such as fluid cavitation and vortices, further exacerbating energy losses and reducing equipment stability.

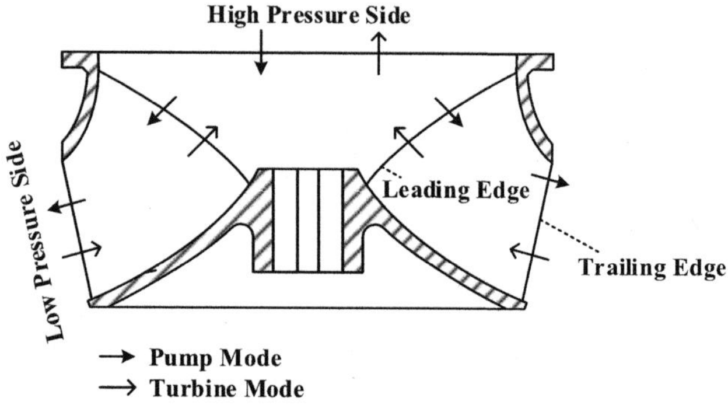

FIGURE 2.5 Axial cross-sectional diagram of pump-turbine.

Many researchers have focused on exploring the flow characteristics of pump-turbines. Utilizing CFD (Computational Fluid Dynamics) numerical simulation technology, they have simulated the dynamics of PSUs during turbine mode startup, revealing the complex sequence of movements of the movable guide vanes during the startup phase, including the initial slow opening, subsequent rapid closing, and maintenance of the no-load opening. During this period, the hydraulic losses at the movable guide vanes show a trend of first decreasing and then increasing. After the unit is connected to the grid, these losses gradually decrease and eventually reach a stable state. Related research highlights that there is a close intrinsic link between the flow characteristics of PSUs and the real-time adjustment of movable guide vanes. Fig. 2.5 intuitively presents the flow dynamics of the pump-turbine. To further analyze the root causes of instability during startup and operation, the following discussion will examine flow instability in pump-turbines. The emergence of the saddle-shaped characteristic region is closely related to the double-peak feature on the flow rate-head curve. This region is often accompanied by significant increases in backflow, rotating stall, and pressure pulsations. Within this region, a reduction in flow rate promotes flow separation near the suction surface of the blades, triggering rotating stall, which disrupts the stability of the entire flow field in a periodically rotating manner. At the same time, backflow and recirculation phenomena may occur at the impeller inlet and around the guide vanes, further disrupting the uniformity and stability of the fluid. In unstable flow scenarios, pressure pulsations are particularly prominent, with the superposition of low-frequency components and blade-passing frequencies, which can easily induce mechanical vibration and noise issues.

During the slow opening and rapid closing stages of the movable guide vanes, the fluid flow field undergoes drastic changes, which may lead to transient occurrences of flow separation and rotating stall. This unsteady flow can induce pressure pulsations within the guide vane and impeller passages, increasing hydraulic losses. When the guide vanes reach the no-load opening, the fluid

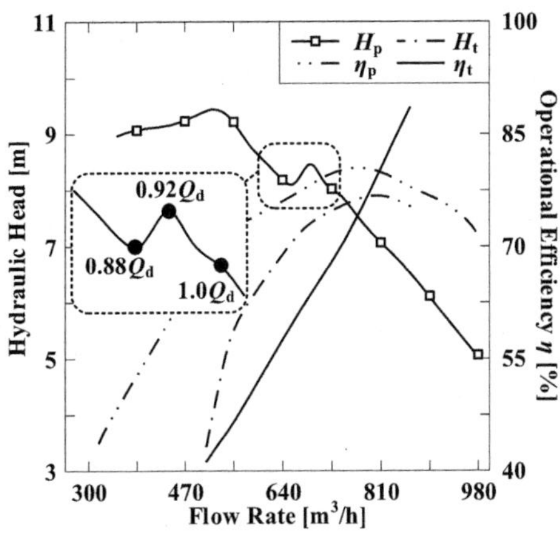

FIGURE 2.6 External characteristics of pump-turbine.

conditions gradually stabilize, reducing the amplitude of pressure pulsations and starting to decrease the hydraulic losses.

The dynamic adjustment of the movable guide vanes is also closely related to backflow and recirculation phenomena. During the initial stage of guide vane opening, fluid flow is restricted, potentially forming localized backflow regions upstream and downstream of the guide vanes. This backflow can interfere with the mainstream flow at the impeller inlet, further affecting hydraulic performance. During the rapid closing stage, due to sudden changes in flow rate, recirculation vortices near the guide vanes may develop and extend into the blade passages, causing short-term pressure fluctuations and abnormal fluctuations in head. However, as the guide vanes gradually reach the operating opening and enter a stable operating condition, backflow and recirculation phenomena gradually disappear, and the flow in the guide vane and impeller passages becomes stable, thereby reducing hydraulic losses.

By studying the saddle-shaped characteristic region, as shown in Fig. 2.6, it can be observed that the dynamic adjustment process of the guide vanes may span multiple operating points, some of which happen to be located within the saddle-shaped region of the flow rate-head curve. This implies that during the startup process, the unit may experience a positive slope interval of head, leading to instability in system operation. Such instability may further increase the intensity of rotating stall and pressure pulsations, posing a threat to the safety and reliability of the equipment.

Therefore during the startup and grid connection phases of PSUs, optimizing the opening strategy of movable guide vanes and controlling the dynamic adjustment process of the guide vanes can effectively mitigate flow instability and reduce hydraulic losses. Combining CFD numerical simulations with experimental research can deeply reveal the relationship between the dynamic adjustment of guide vanes and flow characteristics, thereby providing a scientific basis and technical support for the efficient operation and reliability improvement of PSUs.

2.2.2 Performance evaluation and fault diagnosis of the unit

PSUs face severe challenges due to their high rotational speeds and frequent operational mode switches. As units continue to serve, components of pump-turbines are prone to issues such as cavitation, wear, and aging, leading to gradual performance decline and potentially serious faults, which pose hidden risks to the safe and stable operation of the units and the power quality of the power grid. Therefore accurately evaluating unit performance is crucial for ensuring stable operation and enhancing power quality.

To achieve this goal, the first step is to establish a scientific and reasonable comprehensive performance evaluation index system and select key indicators that can effectively reflect performance degradation or fault conditions. Next, through in-depth analysis of typical faults in pump-turbines, the objective laws behind these faults are explored, and key state indicators are extracted to form the basis of the performance evaluation index system.

Based on the characteristics of pump-turbine faults, they can be divided into gradual faults and sudden faults. Gradual faults are usually caused by the accumulation of wear and fatigue in components, such as efficiency reduction due to component wear or cavitation. Their development follows a certain pattern and can be reflected through operational parameters such as vibration and temperature. Sudden faults, on the other hand, occur randomly, such as sudden component fracture or severe vibration fluctuations, causing relevant state parameters to exceed normal ranges. The core purpose of conducting comprehensive condition assessments of pump-turbines is to determine the current operating state of the units, thereby facilitating the implementation of condition-based maintenance. Through routine maintenance, equipment can be maintained in optimal operating conditions, delaying the progression of gradual faults and preventing them from escalating into sudden faults under unexpected circumstances.

Fig. 2.7 illustrates a multiindex evaluation system that aligns with practical engineering for pumped-storage power stations. The construction of this evaluation system needs to consider factors that affect the operating state of pump-turbines. These factors can influence the operating state of pump-turbines at different levels, aspects, and degrees. Below is an analysis of these factors in conjunction with classic faults of pump-turbines: (a) Abnormal oscillations in pump-turbines are generally caused by significant wear and aging of components after long-term use, manifesting as collisions and friction between fixed parts and rotating parts, excessive bearing clearance, loose connecting screws, deviation of the axial line, uneven mass distribution of the rotor, and other characteristics. (b) Excessively high bearing temperatures often

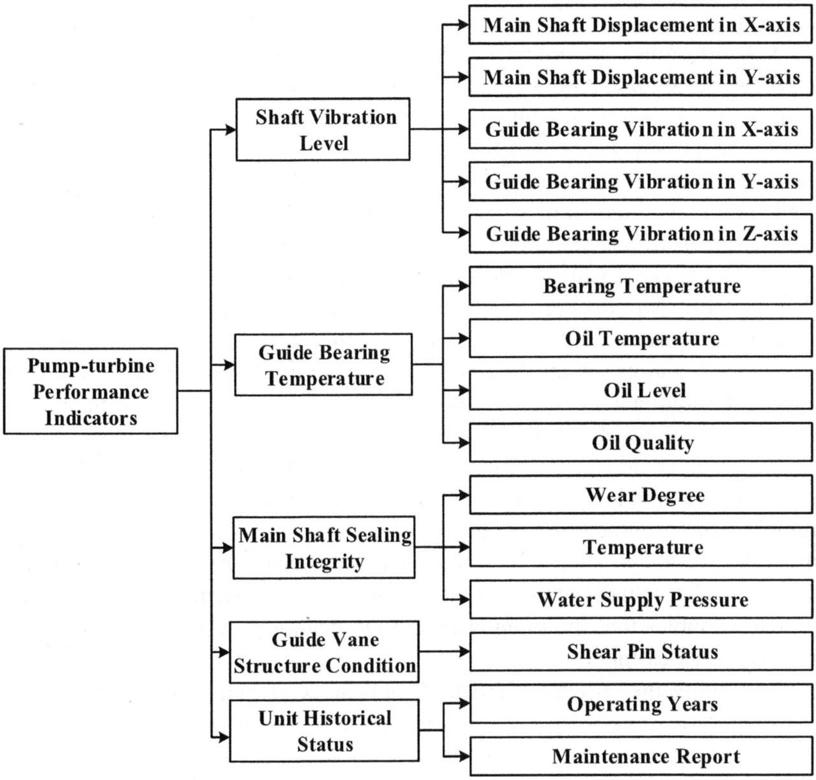

FIGURE 2.7 Multiple index system for pump-turbine performance evaluation.

occur when bearing insulation ages or oil quality in the oil tank deteriorates, leading to increased leakage current that disrupts the oil film. This results in the heat generated during bearing operation exceeding the heat dissipation capacity of the oil cooling system, causing excessively high bearing temperatures. (c) Simultaneous malfunction of the head cover drainage system and failure of the main shaft seal allow water to invade the bearings, leading to severe consequences. (d) Failure of pump-turbines to shut down properly and inability to reduce to braking speed within a reasonable time are primarily caused by poor sealing between guide vanes or shearing of shear pins, preventing effective cut-off of incoming water flow. (e) Under nearly constant water flow conditions, a significant decrease in unit output often occurs due to solid debris blocking guide vanes, extensive clogging of trash racks, or damage to tailpipes. This results in a significant reduction in the effective flow through the guide vanes, subsequently reducing runner speed and unit output. (f) Shear pins can be sheared when blocked by a large amount of debris or when guide vane opening and closing operations are too fast, causing excessive shear forces.

The normal operation of the established comprehensive performance evaluation index system relies on monitoring devices. The monitoring system serves two main functions: First, it monitors the various state indicators of the pump-turbine in real time to determine whether they are within normal fluctuation ranges. Second, it provides data support for comprehensive condition assessments. Based on collected monitoring data, combined with typical fault characteristics of the equipment and state assessment algorithm models, the system analyzes the current operating condition, helping maintenance personnel to promptly address faults and ensure stable system operation, thereby maximizing power plant benefits. Monitoring information primarily originates from the existing monitoring system of the unit, which is the most data-rich database for hydropower stations, containing numerous operating-state variables and fault alarm information for PSUs. If the existing monitoring system cannot meet data acquisition requirements in terms of progress or speed, additional sensors can be installed to supplement data samples. The measured data from the monitoring system is crucial for comprehensive condition assessment of pump-turbines. The accuracy and acquisition frequency of monitoring data directly affect the accuracy of the state assessment. The selection of indicators should align with the actual capabilities of the monitoring system, avoiding the use of state variables that the monitoring system cannot provide or are unclear in meaning as assessment indicators.

Combining the structural and fault indicators of pump-turbines, below is an example of condition-monitoring indicators from an actual project — data sampled and monitored by the monitoring device of Unit 1 at Bailianhe Pumped Storage Power Station. It includes measurement point numbers, measurement point status descriptions, measurement point names, and measurement point data thresholds. The sampled data from the monitoring device support unit performance evaluation and fault diagnosis, see Table 2.1.

As a complex nonlinear system closely coupled with hydraulics, mechanics, and electromagnetism, pump-turbines today face increasingly heavy tasks such as frequency regulation, peak shaving, and spinning reserve. Relying solely on expert experience to make reasonable assessments of unit performance has become increasingly difficult. Traditional scheduled maintenance approaches struggle to scientifically determine maintenance intervals, often resulting in insufficient or excessive maintenance, and failing to achieve rapid and precise fault location. With the improvement of monitoring devices, physically modeled and data-driven prediction models are playing a crucial role in assessing the degradation state and predicting degradation trends of pump-turbines.

By integrating a composite prediction model that incorporates the maximum information coefficient, light gradient boosting machine, variational mode decomposition, and GRUs (Gated Recurrent Units), the health status of pump-turbines is evaluated, and degradation trends are predicted based on operational parameters measured by monitoring devices, such as active power, reactive power, excitation current, excitation voltage, operating head, and guide vane

TABLE 2.1 Monitoring indicators table for Bailianhe pumped storage power station.

Number	Description of Measurement Point Status	Measurement point name	Unit of measurement point	Upper limit	Lower limit
float289	Generator. Lower Bracket/Frame	Y-axis displacement	mm/s	20	0
float290	Generator. Lower Bracket/Frame	Z-axis displacement	mm/s	20	0
float291	Turbine. Measurement	Operating head	m	400	0
float292	Turbine. Measurement	Upper reservoir water level	m	315	285
float293	Turbine. Measurement	Lower reservoir water level	m	115	85
float294	Turbine. Measurement	Turbine operating flow rate	m^3/s	239.8	0
float295	Turbine. Measurement	Pump operating flow rate	m^3/s	192	0
float296	Turbine. Measurement	Sealing ring pressure 2	bar	40	0
float297	Turbine. Measurement	Guide vane pressure	bar	40	0
float298	Turbine. Measurement	Sealing ring pressure 1	bar	40	0

opening. This approach aims to obtain the most accurate health model and the best-performing prediction model. Below is the main framework of the degradation trend prediction model. The GRU submodel, composed of GRUs, serves as the core component of the degradation trend prediction model. Each decomposed mode of the performance degradation index sequence is fed into the GRU submodel, which effectively extracts the long-term dependencies within each mode. The output of the last GRU layer is then fed to a fully connected layer to obtain the prediction for that mode. Finally, the prediction values from all modes are summed to obtain the future Performance Degradation Index (PDI), thereby enabling the prediction of degradation trends for PSUs [4], see Fig. 2.8.

Currently, PSUs are widely equipped with online monitoring devices to monitor their operational status. A commonly adopted method is based on

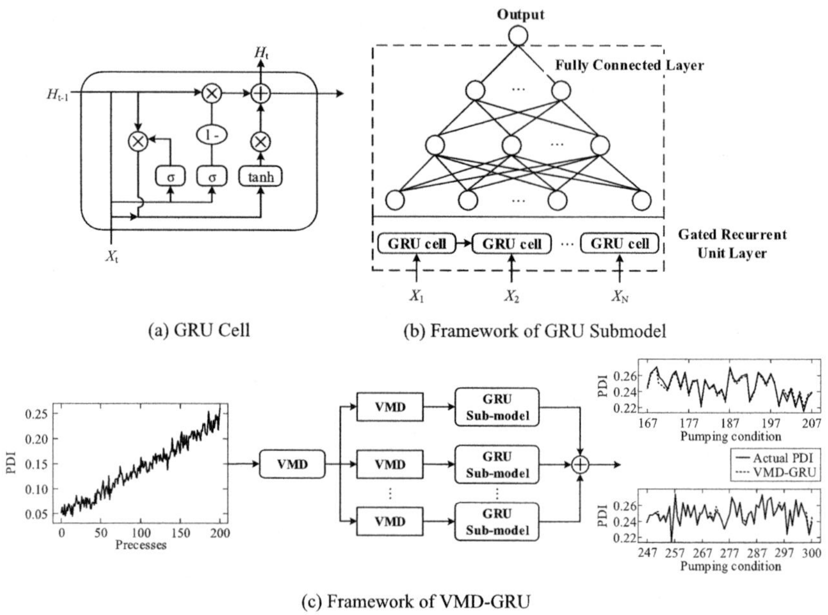

FIGURE 2.8 Degradation prediction model for pump-turbines.

vibration and swing threshold alarming, which involves monitoring the swing of the hydro-generator and issuing an alarm signal when an abnormality is detected. However, this method has obvious limitations. First, vibration and swing threshold alarming relies on preset thresholds, which may not cover all potential fault modes, resulting in some faults not being identified in a timely manner. Second, pump-turbine units are subject to complex coupling effects from hydraulic, mechanical, and electrical factors during operation. Single-point monitoring of vibration and swing cannot comprehensively reflect the operational status of the unit, and can easily lead to false alarms or missed alarms.

Noise signals contain abundant fault characteristics, providing more comprehensive information for unit condition assessment than single vibration and swing monitoring. They can capture subtle changes in unit operation, which are often signs of early faults and contribute to early warning and preventive maintenance. For example, under typical operating conditions of a certain unit, by comparing the baseline spectrogram and real-time spectrogram in Fig. 2.9, it was found that the frequency components in the 50–60 Hz band increased by about 25%, the 70–80 Hz band increased by about 21%, and the 100–110 Hz band increased by about 22%. Therefore it can be concluded that the fault-sound source frequencies causing abnormalities in the unit are mainly concentrated in these three frequency bands. By combining typical vibration frequency com-

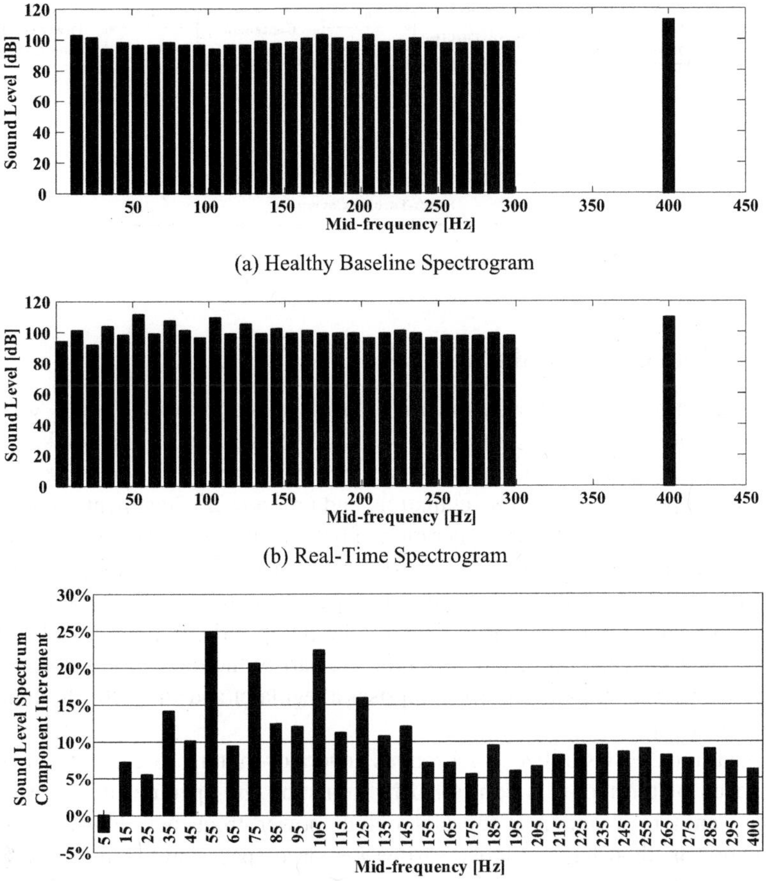

(c) Relationship Between Spectrogram Band Increment and Frequency Range Center

FIGURE 2.9 Noise spectrogram of PSUs.

ponents generated by hydraulics, mechanics, and electromagnetism, the fault source can be accurately located.

Performance degradation prediction is a proactive maintenance measure that provides early warning for fault diagnosis. Its purpose is to identify and address potential issues in advance, so that the unit can take corresponding measures before faults occur, thereby reducing the frequency and complexity of fault diagnosis. Fault diagnosis, on the other hand, is a postmaintenance measure aimed at resolving existing faults. Its diagnostic results can be fed back into the performance degradation prediction model to improve prediction accuracy. The combination of these two measures can effectively ensure the safe and stable operation of PSUs.

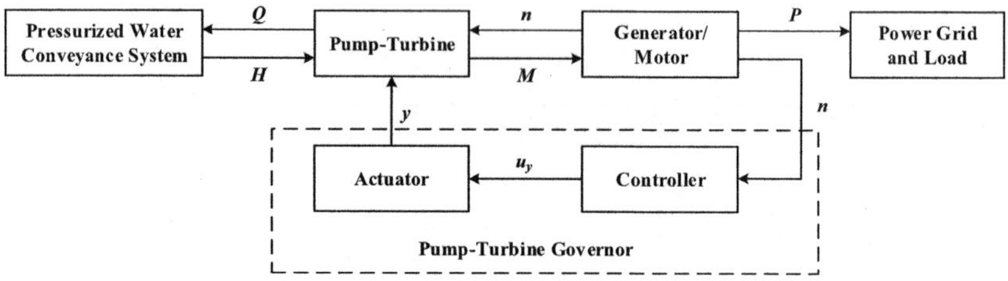

FIGURE 2.10 Regulation system for PSUs.

2.2.3 Unit control and scheduling

The regulation system for PSUs serves as a critical control system within the automation center of pumped-storage power stations. It undertakes the important tasks of controlling unit startups and shutdowns, operating mode transitions, primary frequency regulation, and unit power adjustment, playing a vital role in ensuring the stable operation of both the units and the power system. The regulation system for PSUs consists of two major parts: the controlled objects and the regulation system. The specific components are shown in Fig. 2.10. The controlled objects include the pump-turbine, pressurized water conveyance system, and generator/motor. As part of the regulation system, the pump-turbine governor primarily comprises a regulator and an electro-hydraulic servo system [5].

Traditional PSUs utilize synchronous generator-motors. In generator mode, the speed of the motor is controlled by adjusting the opening of the turbine guide vanes to maintain a constant output power frequency. However, this regulation method has a relatively slow response speed, and in motor mode, the motor has a limited ability to regulate input power. In contrast, variable-speed pumped storage power stations employ AC-excited generator-motors and provide controllable AC excitation to the rotor windings through inverters. This configuration allows for independent adjustment of both the frequency and amplitude of the rotor current, enabling the motor to operate at different speeds while maintaining a constant output frequency. This significantly enhances the system's efficiency and response capabilities across various operating conditions, offering clear advantages compared to traditional synchronous generator systems.

The variable-speed pumped storage unit, as illustrated, see Fig. 2.11, comprises components such as control devices, frequency converters, generator-motors, and pump-turbines with guide vanes. The control devices serve dual purposes: on the one hand, they adjust the output current of the frequency converter to control the active and reactive power output on the stator side of the generator-motor; on the other hand, they regulate the opening of the guide vanes to control the rotational speed of the pump-turbine, which further controls the rotational speed of the generator-motor rotating coaxially with the pump-

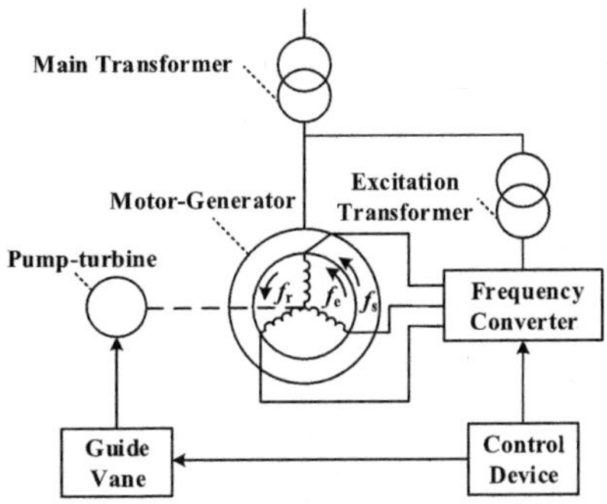

FIGURE 2.11 Structure of an AC-excited machine.

turbine. In the AC-excited variable-speed pumped storage unit, the generator-motor employs a Doubly Fed Induction Generator (DFIG), with three-phase windings on both the stator and rotor. Three lead terminals are connected to the power grid from the stator side, while the rotor is connected to three additional lead terminals via slip rings, which serve to connect to the AC excitation frequency converter. The frequency converter provides excitation current to the rotor, enabling real-time adjustment of the amplitude, frequency, and phase of the rotor current, thereby achieving control of active and reactive power output on the basis of Variable Speed Constant Frequency (VSCF).

During grid dispatch and operation, PSUs frequently switch between generating and pumping conditions, with frequent start-stop operations. During startup, it is essential to ensure that the electromagnetic traction force between the stator and rotor is sufficiently large to overcome the static inertia of the unit and the mechanical resistance of the bearing pads on the rotating shaft. The transition from a steady stopped state to an operating state is the most vulnerable part of the entire condition conversion process, with a higher failure rate. Therefore the widespread application of large synchronous motors urgently requires effective startup control strategies. In practical engineering applications, the most common startup methods currently include asynchronous startup, back-to-back synchronous startup, semisynchronous startup, coaxial small motor startup, and Static Frequency Converter (SFC) startup, see Fig. 2.12.

Large pumped storage power stations, both domestically and internationally, generally employ SFC startup as the primary method, with back-to-back synchronous startup serving as a backup. The SFC startup method uses a specific frequency converter to convert line-frequency AC power into stator current that meets the required specifications through AC/DC/AC current conversion, thereby initiating the motor. A typical connection diagram is shown in Fig. 2.13. In the event of unexpected failures in the power grid or frequency converter, the back-to-back synchronous startup method utilizes existing equipment to

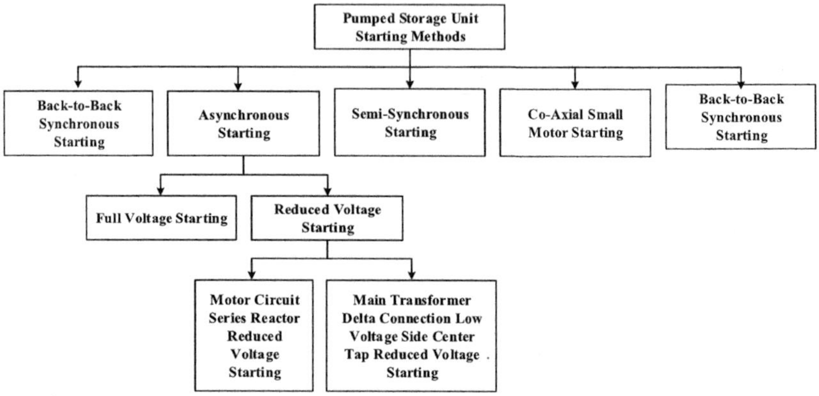

FIGURE 2.12 Startup methods for PSUs.

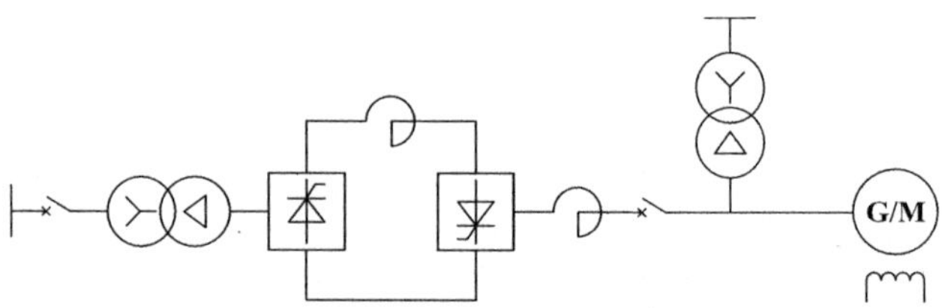

FIGURE 2.13 Typical wiring configuration for initiating with the SFC.

safely and reliably connect the unit to the grid for pumping or pumping-phase modulation.

The architecture of the SFC startup control system can be divided into two core units: the rectifier speed control unit and the self-synchronous inversion control unit. In the rectifier speed control unit, the operation of the rectifier bridge is regulated by a dual-loop control unit for current and speed. This control unit precisely controls the motor speed by adjusting the triggering angle of the rectifier bridge. The self-synchronous inversion control unit employs a fixed commutation advance angle triggering mechanism to ensure stable operation of the inversion bridge. To avoid commutation failure, the inversion bridge adopts two methods—pulse commutation and load commutation—based on different motor frequencies, to adapt to various operating conditions [6], see Fig. 2.14.

The effective sequential triggering of the thyristors in the inversion bridge is crucial for the implementation of load commutation. Specifically, the thyristors in the inversion bridge are triggered by the rotor position. Considering the need to maintain a certain margin for the commutation angle in practical engineering, and the fact that an excessively large advance triggering angle γ_0 can result in a decrease in the average electromagnetic torque and an increase in pulsating harmonic components, this section adopts $\gamma_0 = 60°$, while ignoring the influence of the commutation overlap angle. The control system detects

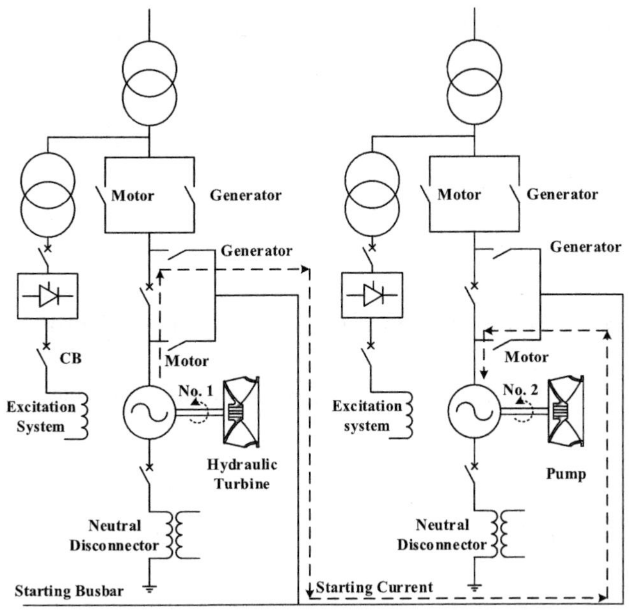

FIGURE 2.14 Typical wiring for back-to-back synchronous startup.

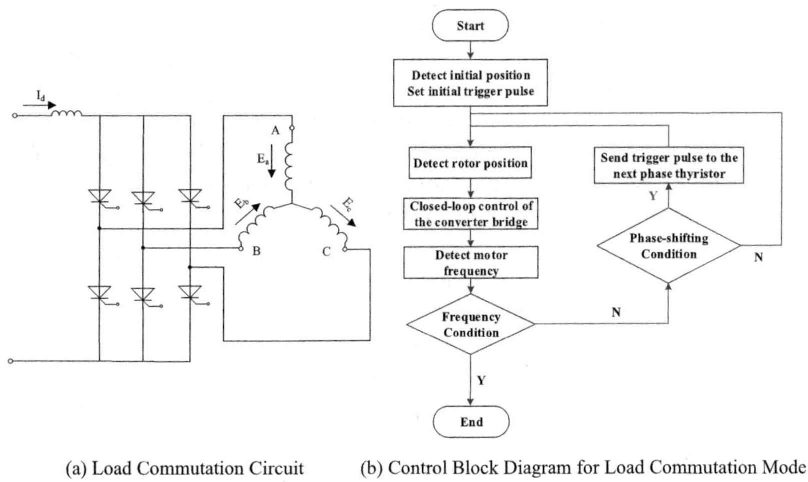

FIGURE 2.15 Load commutation strategy for SFCs.

the rotor position to trigger the thyristors according to Table 2.2. The control strategy for the inversion bridge under load commutation mode is illustrated in Fig. 2.15.

TABLE 2.2 The conduction sequence of thyristors in the inversion bridge ($\gamma_0 = 60°$).

Rotor position	0°–60°	60°–120°	120°–180°	180°–240°	240°–300°	300°–360°
Conducting phase	CB	AB	AC	BC	BA	CA
Conducting thyristor	V5V6	V1V6	V1V2	V2V3	V3V4	V4V5

TABLE 2.3 The sequence of thyristor conduction in the inversion bridge ($\gamma_0 = 0°$).

Rotor position	0°–60°	60°–120°	120°–180°	180°–240°	240°–300°	300°–360°
Conducting phase	CA	CB	AB	AC	BC	BA
Conducting thyristor	V4V5	V5V6	V1V6	V1V2	V2V3	V3V4

Pulse commutation belongs to the category of forced commutation and does not suffer from the commutation failure issues associated with load commutation. In practical engineering, to increase the electromagnetic torque and thereby enhance the rotational speed during the initial startup phase, γ_0 is generally set to 0°. During pulse commutation, the triggering angle α is first set to 120° to make the six-pulse rectifier bridge operate in an inverting state. When the DC, I_d, decreases to 0 kA, the triggering angle α is adjusted to switch the rectifier bridge to a rectifying state. Subsequently, the next set of thyristors in the inversion bridge is triggered to complete one round of commutation. The triggering sequence of the thyristors in the inversion bridge during this process is shown in Table 2.3.

During a back-to-back synchronous startup, the unit to be started is first connected to the already running unit through an electrical shaft to the starting busbar, and appropriate excitation is applied to both units. At this point, the already running unit acts as a generator, producing a low-frequency voltage at its stator winding terminals, which is transmitted to the unit to be started through the starting busbar. This causes the unit to start generating electromagnetic torque, driving the rotor to rotate until it operates synchronously with the already running unit. When the speed increases to nearly 80% of the rated speed, the excitation regulators of each unit are engaged to continue synchronous acceleration to the rated speed. Subsequently, the synchronization and grid connection operations are completed, see Fig. 2.16.

After exploring the micro-control level of PSUs, let us turn our attention to their macro-scheduling situation in power grid operation. By the end of September 2024, China's cumulative wind and photovoltaic power generation

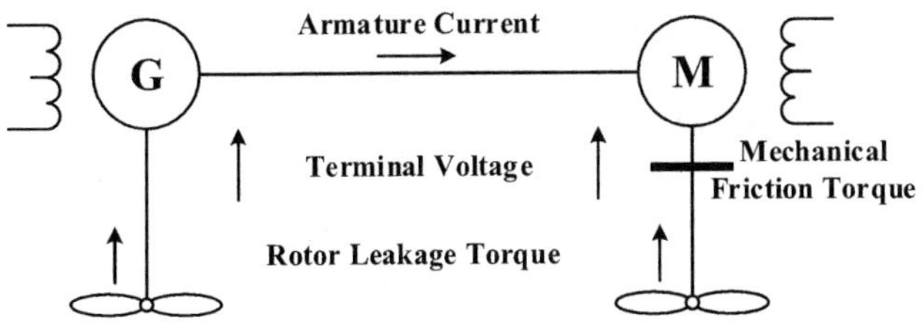

FIGURE 2.16 The process of back-to-back startup.

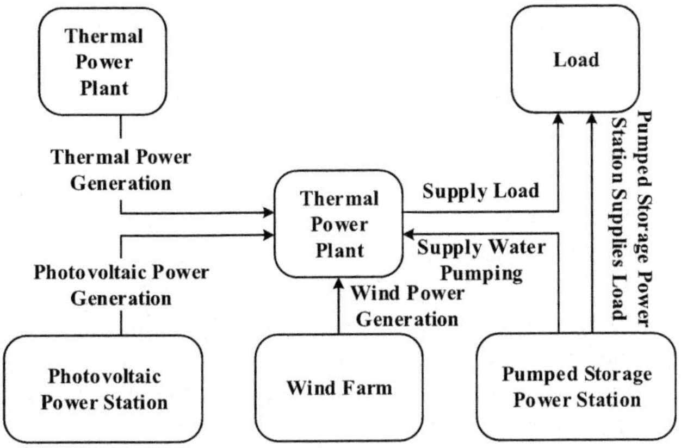

FIGURE 2.17 Structure of the pumped storage-thermal power joint scheduling system.

reached 900.7 billion kWh, accounting for approximately 20% of total power generation. Moreover, the total installed capacity of wind and photovoltaic power (1.28 billion kW) has surpassed that of coal-fired power (1.18 billion kW). Due to the intermittent, random, and fluctuating nature of wind and photovoltaic power generation, relying solely on thermal power units for scheduling has become insufficient to maintain stable grid operation. Therefore the joint scheduling of pumped storage and thermal power has become a key means of stabilizing grid net load fluctuations, see Fig. 2.17.

The formula for calculating net load is as follows:

$$P_{\text{eq},t} = P_{\text{L},t} - P_{\text{v},t} - P_{\text{w},t} \tag{2.1}$$

where $P_{\text{eq},t}$ represents the net load power; $P_{\text{L},t}$ represents the load demand; $P_{\text{V},t}$ represents the power of the photovoltaic power station; $P_{\text{W},t}$ represents the power of the wind power plant.

During the actual operation and scheduling of the power grid, PSUs follow unified grid scheduling instructions to pump water or generate power at fixed times throughout the day, or to set reference values for these actions. Essentially, this approach delineates the boundaries of the scheduling problem [7], as

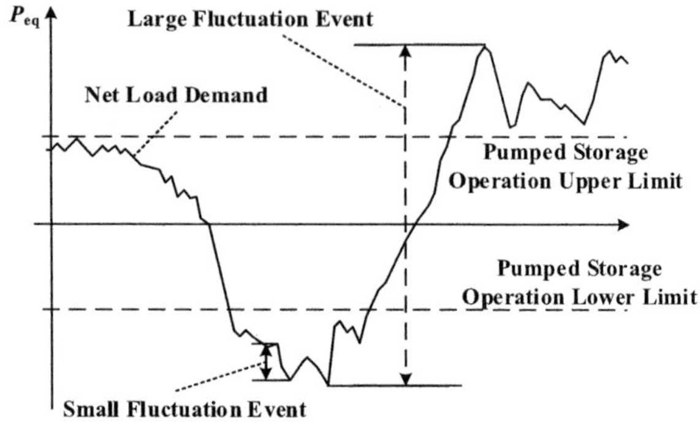

FIGURE 2.18 Schematic diagram of pumped storage unit operation.

illustrated in Fig. 2.18. When the grid's net load demand exceeds the reference value range, PSUs are activated to alleviate the pressure on thermal power units during peak and valley load periods. However, this means that PSUs are only activated during these peak and valley periods, and their ability to smooth out net load fluctuations during other periods is not fully utilized. The fixed-period scheduling mode struggles to fully leverage the operational efficiency of PSUs. As a flexible scheduling mode, the variable-period scheduling mode effectively promotes the absorption of high proportions of renewable energy and improves the operational efficiency of PSUs. Day-ahead optimal scheduling based on the variable-period scheduling mode has received widespread attention.

The day-ahead optimization mathematical model for the joint scheduling system aims to minimize the overall operational costs. It comprehensively considers the operational costs of thermal power units (including ramp-up/down costs, C_{RP}, and coal consumption costs C_{TH}), the operational costs of pumped storage stations C_{PS}, and the penalty costs for wind and solar power curtailment, as well as load shedding C_{V}. The objective function is formulated as follows:

$$\min C = C_{\mathrm{RP}} + C_{\mathrm{TH}} + C_{\mathrm{PS}} + C_{\mathrm{v}} \tag{2.2}$$

The operational costs of thermal power units encompass several components, which can be summarized as follows:

$$\begin{cases} C_{\mathrm{RP}} = \sum_{n=1}^{N} \left[\left(\mathrm{Coe}_n^{RU} + \mathrm{Coe}_n^{RJ} \right) \sum_{t=1}^{T} \max\left(P_{G,n,t} - P_{G,n,t-1}, 0 \right] \\ C_{\mathrm{TH}} = \sum_{t=1}^{T} \sum_{n=1}^{N} \left(a P_{\mathrm{G},n,t}^2 + b P_{\mathrm{G},n,t} + c \right) \end{cases} \tag{2.3}$$

where N represents the total number of thermal power units in the system; T represents the total scheduling period; Coe_n^{RU} and Coe_n^{RJ} represent the upramping and downramping cost coefficients of thermal power unit n, respectively; $P_{G,n,t}$ represents the active power output of thermal power unit n at time t; a, b, and c are all fuel cost coefficients of thermal power units.

The operating cost of pumped storage power stations is:

$$C_{PS} = \sum_{l=1}^{T} \sum_{h=1}^{H} \left(c_{PS}^{P} S\left(P_{PS,h,l}^{P}\right) + c_{PS}^{G} S\left(P_{PS,h,l}^{G}\right) \right) + \sum_{l=1}^{T-1} \left(c_{PS}^{PG} S_{PS,l,l+1}^{PG} \right) \tag{2.4}$$

where H represents the number of PSUs; c_{PS}^{P} and c_{PS}^{G} are the cost coefficients for the pumping state $S\left(P_{PS,h,l}^{P}\right)$ and generating state $S\left(P_{PS,h,l}^{G}\right)$ of the PSUs, respectively; $S\left(P_{PS,l,l+1}^{PG}\right)$ is the switching state variable for pumped storage and generation operations of the pumped storage unit at times t and $t+1$; c_{PS}^{PG} is the cost coefficient for $S_{PS,h,l+1}^{PG}$

The penalty cost for wind and solar power curtailment is:

$$C_V = \sum_{t=1}^{T} \left(\theta_W P_{W,t}^{ab} + \theta_V P_{V,t}^{ab} + \theta_L P_{L,t}^{ab} \right) \tag{2.5}$$

where θ_W represents the wind curtailment penalty coefficient; θ_V represents the solar curtailment penalty coefficient; θ_L represents the load shedding penalty coefficient; $P_{W,t}^{ab}$ is the wind curtailment power at time t; $P_{V,t}^{ab}$ is the load shedding power at time t; $P_{L,t}^{ab}$ represents the load shedding power at time t.

In terms of constraints, the most basic constraints considered in everyday-ahead scheduling optimization models include power balance constraints, upper and lower power limits for thermal power units, ramp rate constraints, start-stop constraints, spinning reserve constraints, as well as minimum and maximum output limits for PSUs in their pumping states. Depending on the specific focus of the problem, additional constraints may be imposed on top of these basic constraints. For instance, when aiming to reduce the number of thermal power unit startups and adjustments during valley periods of net load when PSUs are pumping, and to alleviate pressure on thermal power units during peak periods of net load when PSUs are generating electricity, constraints on the maximum number of unit startups and shutdowns, as well as constraints on the number of operating units, can be introduced. When dealing with different research objects, different constraint relationships are added to the model to make it more targeted.

In general, the joint scheduling sequence and strategies for pumped storage and thermal power can be divided into three categories: synchronous dispatch of pumped storage and thermal power, timed pumping and generating dispatch of pumped storage, and variable-time pumping and generating dispatch of pumped storage. The synchronous dispatch strategy optimizes the output of PSUs and thermal power units in the system simultaneously, but this strategy does not fully

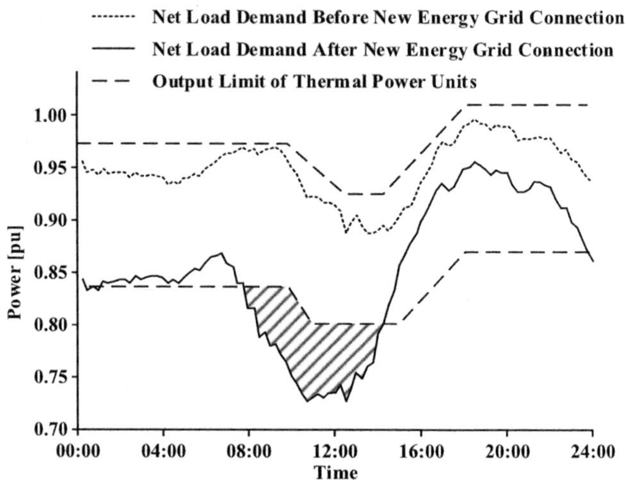

FIGURE 2.19 Graph of net load curve changes before and after new energy grid integration.

consider the efficiency loss of pumped storage during pumping and generating, which can easily lead to frequent startups and shutdowns of units, making it difficult to guarantee dispatch efficiency. The timed pumping and generating dispatch strategy can achieve good results when the penetration rates of wind and photovoltaic power generation are low, as the peak-to-valley fluctuations of daily system load are relatively stable and load differences are small at this time. However, as the penetration rate of renewable energy generation increases, as shown in Fig. 2.19, the peak and valley periods of the daily net load curve shift, the peak-to-valley difference increases, and the degree of fluctuation intensifies. As a result, the timed pumping-and-generating dispatch strategy gradually becomes inapplicable. In contrast, the variable-time pumping and generating dispatch strategy can effectively leverage the role of pumped storage, reduce the peaking burden on thermal power units, and strongly promote the absorption of wind and photovoltaic power. It is the best choice for addressing the aforementioned challenges.

2.2.4 The integrated application demonstration of pumped storage

Variable-speed pumped storage technology is currently a key research focus, and the following will focus on the integrated application demonstration of variable-speed PSUs [8]. Early variable-speed units adopted segmented pole-changing methods, typically with two speed levels. However, due to the limited number of speed levels, relatively low economic benefits, and high manufacturing requirements for the motors, their development was greatly restricted. In 1935, German engineer Tuxen proposed the concept of dual-axis excitation, initiating research into modern AC flux technology. Based on this,

TABLE 2.4 Examples of variable-speed pumped storage power stations.

Power Station Names	Rated Capacity/MVA	Speed Range/rpm	Excitation Device
Goldkronach Pumped Storage Power Station	340.4	300–346.6	CYC
AVCE Pumped Storage Power Station	195	576–626.4	CYC
Ōkawa Pumped Storage Power Station	395	330–390	CYC
Komaru Pumped Storage Power Station	345	576–624	CYC
Yagikizawa Pumped Storage Power Station #2	85	130–156	CYC
Shiobara Pumped Storage Power Station #3	360	345–405	CYC
Okuizumi Pumped Storage Power Station Phase II #2	345	407–450	GTO
Yamahara Seawater Pumped Storage	31.5	423–477	GTO
Azunogawa Pumped Storage Power Station	475	480–520	IEGT
Takami Pumped Storage Power Station	105	208–254	CYC

a significant amount of theoretical research and practical exploration on the asynchronous operation of synchronous motors was conducted internationally, accumulating valuable experience. Currently, some variable-speed pumped storage power stations that have been commissioned globally are shown in Table 2.4.

With advancements in power electronics and control technology, high-capacity AC excitation technology has developed significantly, leading to the emergence of variable-speed operating unit technology using three-phase AC excitation. The development of this technology provides new possibilities for the modernization and efficient operation of pumped storage power stations. Especially when dealing with the intermittency and volatility of renewable energy generation, variable-speed PSUs can offer more flexible regulation capabilities, effectively promoting the absorption of wind power, photovoltaics, and other renewable energy sources.

The research on AC excited motors and pumped storage technology in China began in the 1980s. After decades of development, China has made significant achievements in variable-speed and variable-pole technology. In 2016 the world's largest operational variable-speed and variable-pole unit was located at the Panjiakou Pumped Storage Power Station in Qianxi County, Hebei Province, see Fig. 2.20. This power station was equipped with three 90 MW variable-speed and variable-pole mixed-flow reversible pump-turbines and three conventional mixed-flow hydro-generators with a unit capacity of 150 MW each. Under pump conditions, when the head was between 65.1 and 85.7 m, the rotational speed was 142.86 r/min; when the head was between 36 and 66.4 m, the rotational speed was 125 r/min. Under turbine conditions, with a maximum head of 85 m, the output was 100 MW; at a rated head of 71.6 m, the output was 90 MW with a flow rate of 145.4 m^3/s; and at a minimum head of 36 m, the output was 26.95 MW.

In July 2024, the Fengning Pumped Storage Power Station in Hebei Province, operated by State Grid Xinyuan, broke the world record for the largest installed capacity of a pumped storage power station. It installed 12 PSUs with a unit capacity of 300,000 kW each, capable of storing nearly 40,000,000 kWh of energy at one time, ranking first in the world in energy storage capacity. This significantly promoted the progress of variable-speed PSUs based on AC excitation in China.

2.3 Compressed air energy storage

Compressed Air Energy Storage (CAES) is widely regarded as a technology suitable for GW-scale large-scale electric power energy storage following pumped storage, see Fig. 2.21. Compared to pumped storage power stations, CAES power stations offer greater flexibility in site selection, being able to utilize containers such as abandoned mines, salt caverns, and gas tanks as energy storage units. They can be constructed and installed in most regions of China, possessing broad application prospects [9]. For instance, the "Energy Storage No. 1" project in Yingcheng, Hubei, utilizes underground salt caverns for gas storage, with a capacity of 700,000 m^3, equivalent to 260 standard swimming pools.

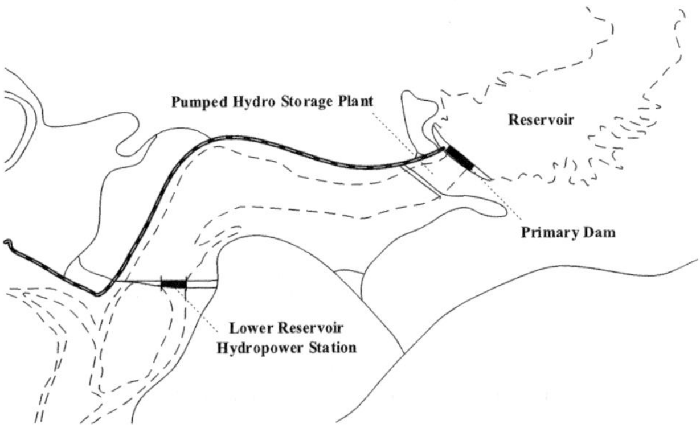

(a) Geographical Location of Panjiakou Pumped Storage Power Station

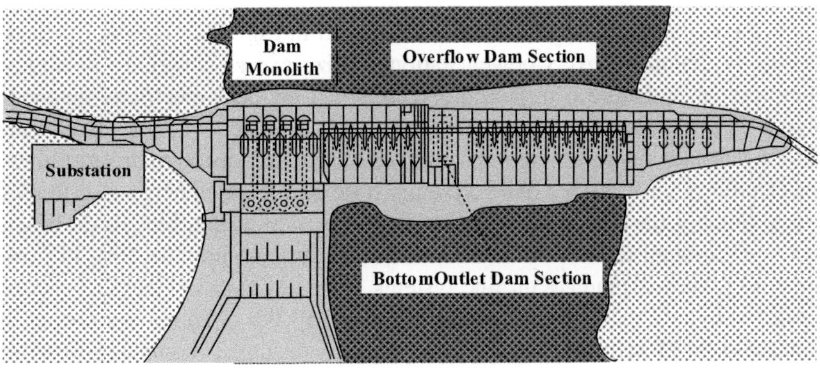

(b) Layout of the Panjiakou Pumped Storage Power Station Hub

FIGURE 2.20 Engineering layout drawing of Panjiakou Pumped Storage Power Station.

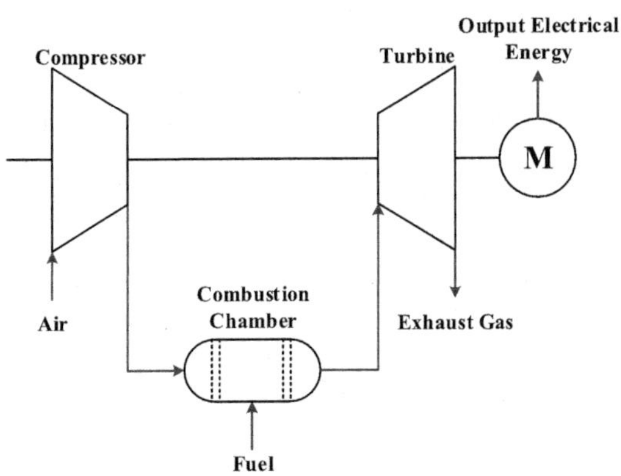

FIGURE 2.21 Schematic diagram of gas turbine system.

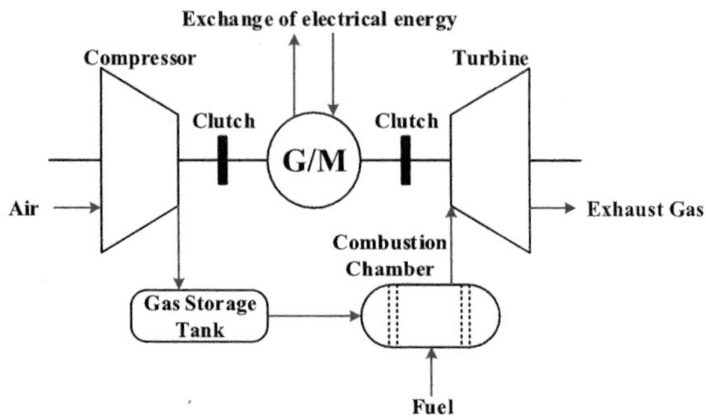

FIGURE 2.22 Schematic diagram of a traditional combustion-supplemented CAES system.

The specific structure of CAES is illustrated in Fig. 2.22. During off-peak electricity demand, CAES power stations utilize excess electricity to highly compress air and store it in pressure vessels. During peak electricity demand, high-pressure air is released into the combustion chamber, where it mixes with natural gas and undergoes vigorous combustion and expansion to drive turbines for electricity generation. CAES systems store and release energy at different times. Notably, the compressor does not dissipate additional turbine work. When the same amount of fuel is consumed, CAES systems provide 20% to 60% more useful power than gas turbine systems. For every kWh of electrical energy produced by a CAES system, approximately 0.75 kWh of compression energy and about 4420 kJ of fuel internal energy are consumed (compared to 6700 to 9400 kJ/kWh for gas turbines). This efficient energy conversion and storage method gives CAES technology significant advantages and potential for large-scale energy storage.

To enhance the performance of CAES systems and expand their application scenarios, many scholars have optimized the thermodynamic cycles of traditional systems, proposing advanced CAES systems such as Advanced Adiabatic Compressed Air Energy Storage (AA-CAES, see Fig. 2.23). and Liquid Air Energy Storage (LAES). AA-CAES stores the heat generated during compression in a heat accumulator and uses the stored heat to preheat the high-pressure air entering the expander during power generation. This entire process does not require fossil fuel combustion, significantly improving the system's cycle efficiency. LAES, building upon AA-CAES, liquefies the high-pressure air and stores it in tanks. During power generation, liquid air is pressurized and heated to vaporize, thereby eliminating the need for large gas storage chambers, significantly increasing energy density, and enhancing the feasibility of mobile energy storage.

Given the excellent performance of advanced CAES systems, China has included advanced CAES technology in its 14th Five-Year Plan. Overall, this technology is still transitioning from engineering demonstration to commercial

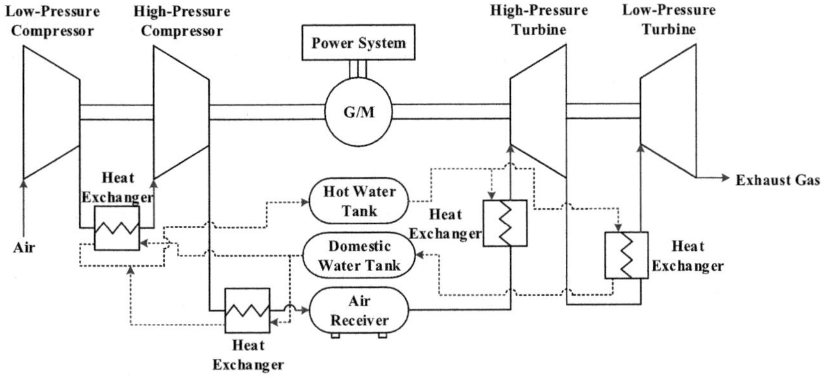

FIGURE 2.23 Schematic diagram of AA-CAES system.

application, and a fully operational and management system has not yet been established. Therefore there is an urgent need to study the optimal design and operational planning methods for advanced CAES systems.

2.3.1 Compressor and expander technology

Compressors and expanders, as the critical components of CAES systems, function during off-peak hours at night. Surplus electrical energy drives compressors to compress air, while interstage heat exchangers absorb and store the "compression heat" in high-temperature storage tanks. Subsequently, stored cold energy is utilized to cool the compressed air, liquefying it. During daytime peak hours, the liquefied air is first pressurized; as it vaporizes, it absorbs heat, generating substantial cold energy. The vaporized air then recovers cold energy from the vaporization process in heat exchangers. It simultaneously absorbs "compression heat" from the high-temperature storage tanks through interstage reheaters, driving the expander to produce work.

Compressors can be classified into two major categories based on their working principles: volumetric (piston compressors) and turbomachinery (centrifugal compressors and axial compressors). Piston compressors are suitable for various discharge pressure conditions with medium to small flow rates. However, when the flow rate demand exceeds 100 kNm^3/h, the size of a single unit becomes excessively large, often surpassing the limits of conventional materials. Axial compressors offer high throughput capacity and efficiency but have relatively low single-stage pressure ratios. To achieve higher discharge pressures, multiple stages are typically used, which reduces economic efficiency. Centrifugal compressors are renowned for their high single-stage pressure ratios, compact structures, and wide operating ranges, making them particularly suitable for CAES systems with large flow rates and high discharge pressures. In CAES systems, the required pressure ratio for compressors is often 40–80. Due to the limited pressure rise per stage of a single impeller, multistage compression is commonly employed to meet these demands in practical engineering applications.

Multistage centrifugal compressors can be divided into single-shaft and multishaft types. Among them, multishaft centrifugal compressors are more suitable for large-scale CAES systems due to their high efficiency, compact structure, and ease of interstage heat recovery. Generally, the fixed volume of the gas storage container causes the exhaust pressure inside the centrifugal compressor to increase as the internal pressure rises. Efficient operation of centrifugal compressors under continuously varying conditions over a wide pressure-ratio range is a key technology in CAES systems. Currently, the commonly used methods for regulating the operating conditions of centrifugal compressors in engineering are shown in Table 2.5.

Turbine expanders are classified into axial flow, radial inflow, and radial-axial flow types based on the direction of gas flow in the impeller. They are further categorized into reaction turbines and impulse turbines depending on whether the gas continues to expand in the impeller. In CAES systems, the first few stages of turbines typically have small volumetric flow rates and high pressures at the inlet, so single-stage, high-expansion ratio, and compact radial inflow turbines are generally used. For the later stages, where the volumetric flow rates are larger and the expansion ratios are smaller, radial-axial flow turbines with strong flow capacity are often employed.

2.3.2 Thermal (cold) storage heat exchanger technology

As the core equipment for realizing energy decoupling and coupling in CAES systems, thermal (cold) storage heat exchangers play a crucial role in both traditional single CAES systems and hybrid systems coupled with other CAES systems. High-efficiency, low-cost heat exchange devices are key to improving the overall energy efficiency of these systems.

During the energy storage phase, the heat exchange medium in the heat exchanger on the compressor side absorbs the "compression heat" generated during the compression of air, which is then stored in the thermal storage unit after heating. The high-pressure, low-temperature gas after heat exchange is stored in the air storage unit. This phase achieves decoupling of internal energy into thermal energy and potential energy. During the energy release phase, the heat exchanger on the turbine expander side uses the high-temperature heat exchange medium from the thermal storage unit to heat the high-pressure gas released from the air storage unit. The high-pressure, high-temperature air then enters the turbine expander, and the heat exchange medium returns to the thermal storage unit after heat exchange. This phase completes the coupling of pressure potential energy and compressed thermal energy, see Fig. 2.24.

Heat exchangers come in various shapes and structures, which can be classified into shell-and-tube, plate, and special types, see Fig. 2.25. The

TABLE 2.5 Table of operating condition adjustment methods.

Operating Condition Adjustment Method	Basic Content	Advantages	Disadvantages
Throttling Regulation	By artificially altering the resistance characteristics of the piping network through adjustments to the throttle valve opening, the operating point of the combined compressor and piping network system can be changed.	Simple in structure, low in cost, and easy to adjust.	Additional throttling losses result in poorer economic performance.
Bypass Regulation	By diverting a portion of the compressor outlet airflow, either venting it or reducing its pressure through throttling, and then feeding it back into the compressor inlet, significant adjustment of the compressor's working flow rate can be achieved. In extreme cases, zero flow output is possible.	The adjustment range is larger than that of other adjustment methods.	Due to its poor economic performance, it cannot be used as a normal adjustment method for compressors.
Variable Speed Regulation	Adjust by altering the compressor's rotational speed to significantly change its operating range.	Adjustment without additional energy loss, resulting in good economic performance.	Multishaft centrifugal compressors have numerous shaft systems and complex rotor dynamics designs, which result in a very narrow safe operating speed range. Therefore it is not feasible to significantly alter the main shaft speed over a wide range.
Variable Geometry Component Regulation	Adjusting operating conditions by altering the geometric parameters (typically angles) of a certain component.	Flow losses are relatively small, leading to higher efficiency.	Requires complex mechanical control mechanisms.

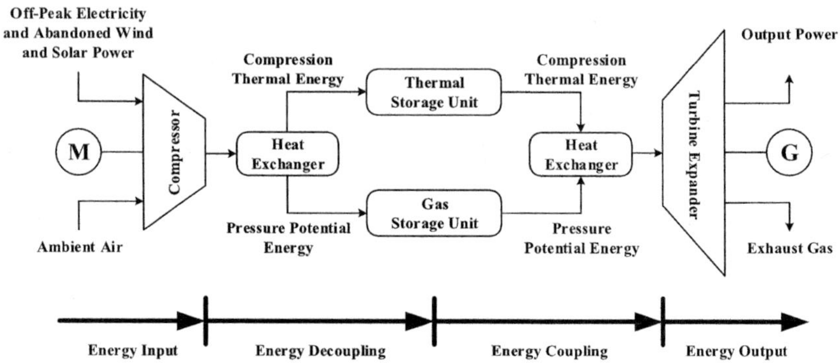

FIGURE 2.24 Diagram of energy exchange in CAES.

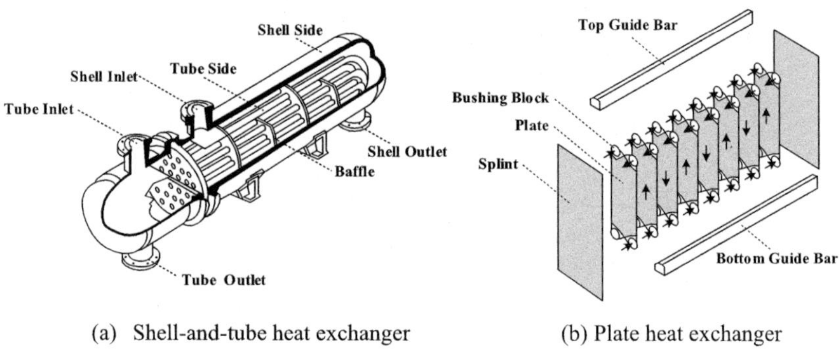

FIGURE 2.25 Typical heat exchange devices.

heat transfer coefficient (U) for plate heat exchangers and shell-and-tube heat exchangers ranges from 1.0 to 3.1 $W/m^2 \cdot K$, respectively. Plate heat exchangers have an advantage due to their high heat transfer coefficient and larger heat transfer area per unit volume, resulting in superior heat exchange performance compared to shell-and-tube heat exchangers. However, plate heat exchangers have poorer pressure-bearing capacity and temperature resistance, making them suitable for scenarios with smaller flow rates. For large-scale CAES power stations, considering technical feasibility and economics, shell-and-tube heat exchangers are often the preferred choice.

Currently, research on shell-and-tube heat exchangers at home and abroad has been extensive, primarily focusing on technical improvements on both the tube and shell sides. Shell-and-tube heat exchangers consist of a shell, a tube bundle, and baffles. The tube bundle is composed of many parallel tubes, and the shell forms a certain space around the tubes. Fluid flows between the tube and shell, and heat transfer is achieved through thermal conduction between them. Baffles are usually located inside the shell, and by setting salt-discharging holes

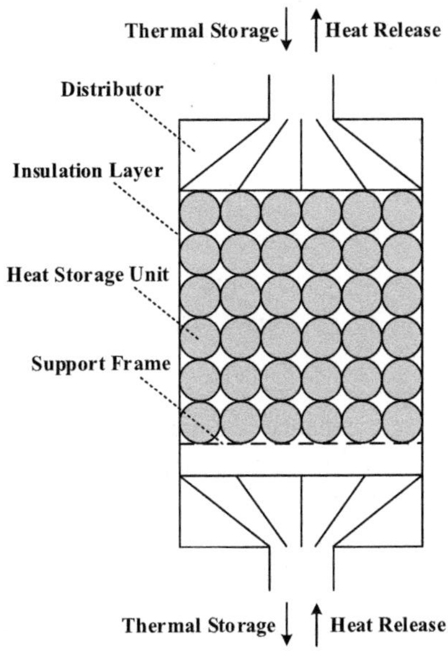

FIGURE 2.26 Diagram of packed bed structure.

in the baffles, the mixing degree of the fluid is increased, the thermal resistance during heat conduction is reduced, and the heat exchange efficiency is improved.

Taking the packed bed phase change heat storage exchanger in an adiabatic CAES system as an example, as shown in Fig. 2.26, it employs a packed bed heat storage system based on a shell-and-tube heat exchanger. The packed bed structure offers a larger fluid contact area, and the internal turbulent mixing significantly enhances heat transfer efficiency. Table 2.6 presents the properties of different heat transfer fluids. In terms of heat transfer fluid materials, liquid heat transfer fluids mostly use thermal oils or molten nitrate liquids. Their relatively high thermal capacity and thermal conductivity allow the heat exchanger to operate at low circulation speeds. Gas heat transfer fluids often use air or high-temperature flue gas. Although their heat transfer coefficients are limited, their advantages of high safety, low risk, and low cost enable the heat exchanger to maintain good performance over a wide range of applications without posing environmental issues or significant technical challenges. In recent years, nano-phase change materials (nano-PCM) have emerged as a new type of material with high heat storage density, stable operating temperatures, and low costs for large-scale implementation. They are used to improve the heat storage materials in large shell-and-tube heat exchangers. Among them, the geometric configuration of single-phase change material-packed beds and the proportional allocation of various phase change material-filled layers in cascaded arrangements of multiple phase change materials are important

TABLE 2.6 Parameters of heat transfer fluid.

Types of Heat Transfer Fluids	Specific Heat Capacity Cp/[J/(kg·K)]	Operating Temperature/K
Thermal oil	3212	623.15
Air	1124	923.15
Air/Oil/Lava		823.15/673.15/823.15
Liquid metal lead-bismuth	146	673.15
Oil-Al_2O_3 nanofluid	1961/2132	500/600

research areas. To further enhance the performance of heat exchangers, a more detailed analysis of their internal mechanisms is needed, and in-depth discussions on optimization measures for energy-saving technologies, fouling issues, and flow resistance are required.

2.3.3 Integrated application demonstration of CAES

CAES technology is divided into two categories based on whether a heat source is required: the combustion-supplemented type and the noncombustion-supplemented type. Currently, there are only two large-scale commercial CAES power stations globally: the Huntorf power station in Germany and the McIntosh power station in the United States, both of which use natural gas as a combustion-supplemented heat source. The Huntorf power station in Germany can generate electricity continuously for 2 hours after 8 hours of continuous compressor charging. However, due to heat loss from gas turbine exhaust, its average operating efficiency is only 42%. In contrast, the McIntosh power station has made technical improvements based on the Huntorf power station. By recovering waste heat from gas turbine exhaust and utilizing underground salt caverns for gas storage, it can compress air continuously for 41 hours and generate electricity continuously for 26 hours, with an average operating efficiency of the power station increased to 54%. To overcome the limitations of traditional combustion-supplemented CAES technology, researchers have developed emerging technologies such as adiabatic CAES, isothermal CAES, and liquid CAES. These technologies store and release energy more efficiently and are environmentally friendly, aiming to fully utilize the heat generated by the system and improve its overall efficiency. A summary of some of these emerging technologies is shown in Table 2.7.

China's research into CAES technology began relatively late, but in recent years, thanks to the concerted efforts of universities and enterprises, remarkable research successes and engineering achievements have been made. The research team at Tsinghua University, using high-temperature thermal oil as the heat storage medium and underground abandoned salt caverns as the gas storage containers, employed two-stage centrifugal compression and two-stage axial

TABLE 2.7 Partial summary of CAES power station projects.

Project Location	Technology Type	Operation Time	Gas Storage Container	Power MW	Efficiency %
Huntorf, Germany	Combustion-supplemented CAES	1978	Underground Salt Cavern	290	44~46
McIntosh, USA	Combustion-supplemented CAES	1991	Underground Salt Cavern	110	52~54
Ohio, USA	Noncombustion-supplemented CAES	2001	Underground Salt Cavern	2700	
Kamijyuna-gawa-cho, Japan	Noncombustion-supplemented CAES	2001	Underground Abandoned Mine Pit	2	<40
Adele, Germany	Adiabatic CAES	2010	Underground Salt Cavern	90	60~70
Jintan, Jiangsu	Thermal-storage CAES	2021	Underground Salt Cavern	60	61.2
Zhangjiakou, Hebei	Thermal-storage CAES	2022	Storage Tank	100	70.5
Yingcheng, Hubei	Thermal-storage CAES	Under Construction	Underground Salt Cavern	300	
Jiuquan, Gansu	Thermal-storage CAES	Under Construction	Cavern	300	
Feicheng, Shandong	Noncombustion-supplemented CAES	2024	Underground Salt Cavern	300	

flow expansion technologies. In 2021 they successfully grid-connected and commissioned a CAES power station in Jintan, Jiangsu Province. This power station can compress air for 8 hours a day and generate electricity for 5 hours, with a conversion efficiency of up to 61.2%, approximately 6% higher than that of the McIntosh power station in the United States. The team led by Chen Haisheng from the Institute of Engineering Thermophysics, Chinese Academy of Sciences, compressed air using multistage compression with interstage cooling, recovered heat generated during compression using a pressurized water heat storage system, and stored the compressed air in above-ground high-pressure tanks. Drawing on this experience, a 100 MW CAES power station was built and grid-connected in Zhangjiakou, Hebei Province, in 2022, achieving a conversion efficiency of 70.5%, setting a new high for conversion efficiency.

The industrialization of CAES power stations in China is still in its infancy and faces numerous bottlenecks and challenges. For instance, relevant standards for defining and calculating the system efficiency of energy storage power

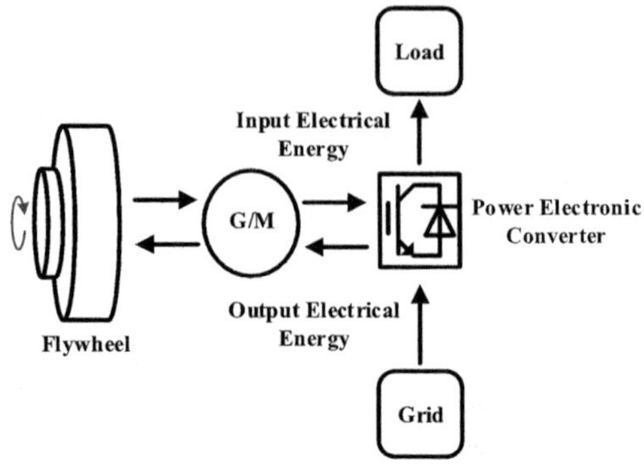

FIGURE 2.27 Working principle of flywheel energy storage system.

stations are still incomplete, and the construction of adiabatic CAES power stations on the new energy generation side still needs further development. Overall, the demonstration applications of CAES projects have laid a solid foundation for realizing a clean, sustainable, and efficient energy future. Continuous innovation and practice in this field will lead to a new round of development in China's energy technologies and make more significant contributions to building a green and low-carbon society.

2.4 Flywheel energy storage

Flywheel energy storage, as a type of power-oriented energy storage technology, boasts advantages such as long cycle life, high instantaneous power, low operational power consumption, and no geographical constraints. Unlike large-scale energy-oriented storage technologies such as pumped hydro storage and CAES, flywheel energy storage, despite its relatively small total energy storage capacity, features high energy density and rapid response times, enabling it to swiftly compensate for grid power imbalances. It has broad application prospects as a grid-forming energy storage technology and can provide stable control for power systems [10].

2.4.1 Working principle and key technologies of flywheel energy storage systems

Flywheel energy storage systems, also known as flywheel mechanical batteries, operate by storing kinetic energy in the form of high-speed rotation of a flywheel rotor when the grid load is low. During peak grid load times, the rotating flywheel rotor drives an electrical generator, and the power is then converted through a power converter to output the appropriate voltage and current for the load, see Fig. 2.27.

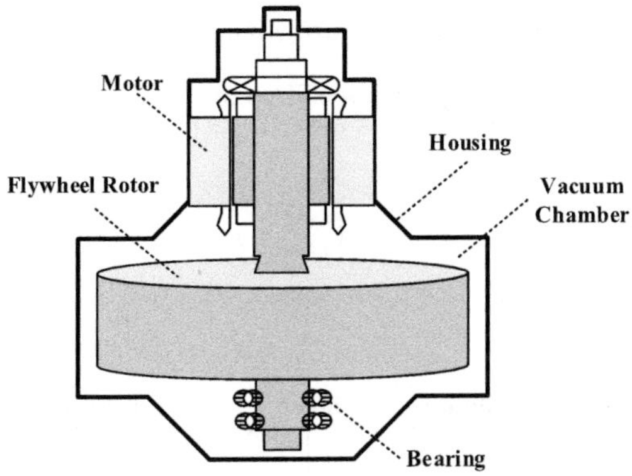

FIGURE 2.28 Diagram of flywheel energy storage system architecture.

The flywheel energy storage system adopts a modular structure, as shown in Fig. 2.28. The main components include the flywheel rotor, flywheel bearing, energy converter, and generator/motor. The key technologies of the flywheel energy storage system revolve around these four major parts.

The flywheel rotor is coaxially connected to the electric motor and serves as the main energy storage component of the flywheel energy storage system. In engineering, uniformly radially symmetric rotors are often used. The equivalent value of their moment of inertia is:

$$J = \int r^2 \mathrm{dm} = Kmr^2 \tag{2.6}$$

where m represents the mass of the rotor; r represents the radius of the rotor; K is a constant that is influenced by the mass distribution and geometric shape of the flywheel rotor.

The rotational kinetic energy E stored during high-speed rotation is given by:

$$E = \frac{1}{2} J\omega^2 \tag{2.7}$$

where J represents the rotational inertia of the shaft system with the flywheel rotor as the main component; ω is the instantaneous angular velocity of the flywheel rotor during rotation.

Considering the complexity of the winding manufacturing process, composite material flywheel rotors typically adopt a simple ring shape. The manufacturing process utilizes winding molding methods, among which the wet winding molding method is cost-effective and produces products with good airtightness. This method is frequently used when fabricating flywheel rotors the composite materials listed in the Table 2.8.

TABLE 2.8 Flywheel rotor material performance parameters.

Material Type	Density $\rho/(kg{\cdot}m^3)$	Tensile Strength σ/Mpa	Strength-to-Density Ratio $\sigma(Mpa{\cdot}m^3{\cdot}kg^{-1})/\rho$
Steel and Iron	7830	1300	0.17
Titanium Alloy	5100	1200	0.24
Glass Fiber	1900	1300	0.68
Carbon Fiber	1550	6300	4.06

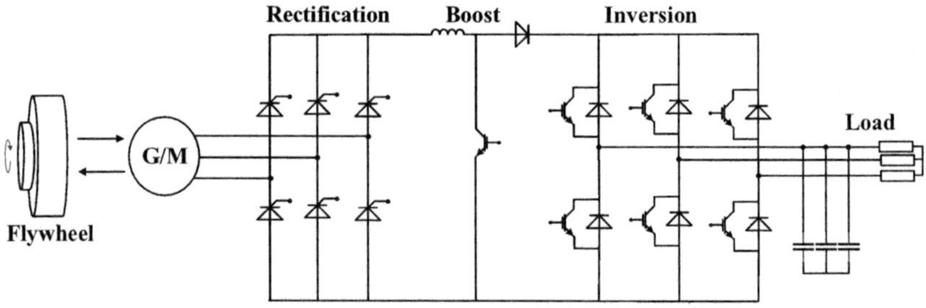

FIGURE 2.29 Energy converter for flywheel energy storage.

Flywheel bearings play a crucial role in supporting and stabilizing the flywheel rotor during high-speed operation. A reasonable bearing design can reduce friction loss in the flywheel rotor, ensuring that it remains within the designated operating position range and operates safely even under disturbance conditions, thereby guaranteeing the stable operation of the flywheel energy storage system.

The main types of bearings used in flywheel energy storage include mechanical bearings, permanent magnet bearings, high-temperature superconducting magnetic bearings, and combined bearings. The first three are single bearings and struggle to meet the development needs of modern flywheel energy storage systems. Consequently, an increasing number of researchers are turning their attention to the study of hybrid magnetic bearings, as shown in Table 2.9.

The power electronic converter serves as an energy converter in the flywheel energy storage system, adjusting the input and output of energy. As shown in Fig. 2.29, with the assistance of auxiliary equipment, the power electronic converter facilitates the energy conversion process from electrical energy to mechanical energy and back to electrical energy.

The energy conversion process can be divided into 3 stages:

1) Conversion of electrical energy to mechanical energy: The power electronic converter regulates the charging current, rectifying the AC from the grid into DC power, which drives the motor/generator to rotate, converting kinetic energy into mechanical energy of the flywheel rotor.

TABLE 2.9 Analysis of main types and characteristics of flywheel bearings.

Bearing Type	Working Characteristics
Mechanical Bearing	It belongs to the category of contact bearings, replacing sliding friction with rolling friction. It has the advantages of good working stability and simple control, but it has a short service life, high operating noise, and high friction loss.
Permanent Magnet Bearing	It is a noncontact bearing that utilizes magnetic force to levitate the rotor in the gap of a vacuum chamber. It has low friction loss, high charging and discharging efficiency, and can achieve high-speed operating conditions. However, it cannot achieve stability in all six degrees of freedom and needs to cooperate with other bearings to provide a stable levitation force for the flywheel rotor.
High-Temperature Superconducting Magnetic Levitation Bearing	It is a noncontact bearing that uses a high-temperature superconductor as the stator and a permanent magnet as the rotor. Leveraging the flux pinning effect of high-temperature superconducting magnets, it can achieve stable rotation of the flywheel without the need for control. However, high-temperature superconducting magnetic levitation bearings have high manufacturing` costs, low displacement stiffness, relatively small bearing capacity, and poor stability at high flywheel speeds.
Combined Magnetic Bearing	Combining several types of bearings for use, a bearing system dominated by permanent magnet bearings and high-temperature superconducting magnetic levitation bearings has the characteristics of longer life, lower noise, and lower loss compared to a bearing system dominated by mechanical bearings.

2) Storage of electrical energy in the form of mechanical energy: The flywheel rotor rotates at high speed, and the power electronic converter provides a low voltage to counteract rotor friction losses and maintain the rotational speed of the flywheel.

3) Conversion of mechanical energy back to electrical energy: The power electronic converter converts the mechanical energy released by the flywheel rotor into AC power with a frequency and phase that match the grid. As the flywheel continues to decelerate, the boosting stage responds to maintain a stable voltage.

In flywheel energy storage, the electric motor serves as an integrated component with independent input and output. It functions as a motor during Stage 1 for charging and as a generator during Stage 2 for discharging, which significantly reduces the system's volume and weight. In practical engineering, the following aspects need to be considered: (1) Under motor operating conditions, it should have high torque and output power to meet the system's energy storage efficiency requirements as quickly as possible; (2) High energy conversion efficiency and a wide speed regulation range; (3) Low no-load losses to meet long-term energy storage operation requirements and ensure a long service life.

2.4.2 Demonstration of integrated flywheel energy storage applications

Flywheel energy storage technology has a wide range of integrated application fields. With its advantages of long service life, rapid response, and high operational efficiency, it has achieved numerous applications in power system frequency regulation, distributed power generation system power regulation, and uninterrupted power supplies both domestically and internationally.

In 2011 Beacon Power Corporation of the United States completed the construction of the first 20 MW/5 MW·h flywheel energy storage frequency regulation power station in Stephentown, New York. In 2014 it built a second 20 MW frequency regulation power station in Pennsylvania. The Azores region of Portugal employs distributed power generation systems. Due to stability issues associated with wind and hydroelectric power generation, a 350 kW/5 kWh flywheel energy storage system was introduced and connected to the 400 V bus, significantly enhancing the stability of the regional grid frequency.

Since 2005, German company ATZ has been researching flywheel energy storage systems rated at 5 kWh/250 kW. In collaboration with L-3MM, they produced a 5 kWh/250 kW high-temperature superconducting magnetic-levitation flywheel energy storage system for use as an Uninterrupted Power Supply (UPS). In the event of a power outage, it can discharge at the millisecond level to stabilize the grid.

Active Power Corporation, based in the United States, has developed the Clean Source series of UPS flywheel energy storage systems. The 1500iC series has a single-unit rated capacity of 250 kV·A and supports up to six units operating in parallel, with a total capacity of up to 1.5 MV·A, suitable for reliable power supply to data centers and critical equipment.

In 2022 Beijing Qifeng Juneng collaborated with Huaneng Laiwu Power Plant to complete a flywheel integrated project for a 1-million-kilowatt ultra-supercritical thermal power unit. The project established a 6 MW/50 kW·h flywheel energy storage array comprising ten 600 kW units, demonstrating the application of advanced energy storage technology to enhance thermal power generation efficiency and reliability. In 2023 China National Power Investment Corporation designed a flywheel energy storage system with a power of 5 MW and an energy capacity of 175 kWh. The system consists of five flywheel units, each with a power of 1 MW and an energy capacity of 35 kWh. These units are first paralleled on the DC side before being connected to the AC grid. Each flywheel unit has a charge/discharge duration of 126 seconds, demonstrating strong charging and discharging capabilities that enhance the primary frequency regulation capability of the Shiziping Wind Farm, thus improving the stability and reliability of the power supply. In March 2024, Guoneng Ningxia Lingwu Power Generation Co., Ltd. equipped its 2×600 MW thermal power units with 22 MW/4.5 MWh flywheel energy storage for frequency regulation, enhancing the precision and flexibility of the thermal power units. By the end of October 2024, the first flywheel of the National Energy Administration's demonstration project – Penglai Company's 4 MW/1 MWh magnetic levitation flywheel energy storage project – was successfully lifted, marking the project's imminent entry into the commissioning phase. This project not only represents a significant breakthrough in domestic flywheel energy storage technology but is also the world's largest individual flywheel energy storage device. Its independent application in Automatic Generation Control (AGC) secondary frequency regulation is a first in China.

Overall, the technical characteristics of flywheel energy storage technology are highly compatible with the frequency regulation market, and its advantages in power-frequency regulation have been recognized. Meanwhile, it has also found widespread application in other fields. However, in China, this technology is still in the early stages of industrialization and faces challenges such as high costs and a lack of verified demonstration projects of sufficient maturity and scale.

2.5 Thermal and cold energy storage

Thermal and cold energy storage technology stores thermal or cold energy in the form of sensible heat, latent heat, or chemical thermal energy, and releases it in large quantities when needed [11]. Liquid-cooled energy storage technology involves connecting the energy storage device to a liquid-cooling system, where a high-thermal-conductivity liquid (such as ethylene glycol solution) serves as the heat transfer medium. During operation of the energy storage device, the coolant absorbs heat transferred through the liquid-cooled plate. Subsequently, the coolant circulates through the liquid cooling system, transporting the heat to a cooling tower or other heat-dissipation equipment for dissipation. After heat dissipation, the coolant is pumped back to the energy storage device, enabling continuous heat dissipation and temperature control. The advantages of low cost, high efficiency, and large scale make thermal and cold energy storage technology have a wide range of application scenarios in energy, industrial, and building sectors.

2.5.1 Sensible heat storage

Sensible heat storage involves storing or releasing heat based on the temperature change of the storage medium, with no phase change occurring during the process, see Fig. 2.30. The stored heat can be calculated using the following formula:

$$Q = M \cdot c_p \cdot \Delta T \tag{2.8}$$

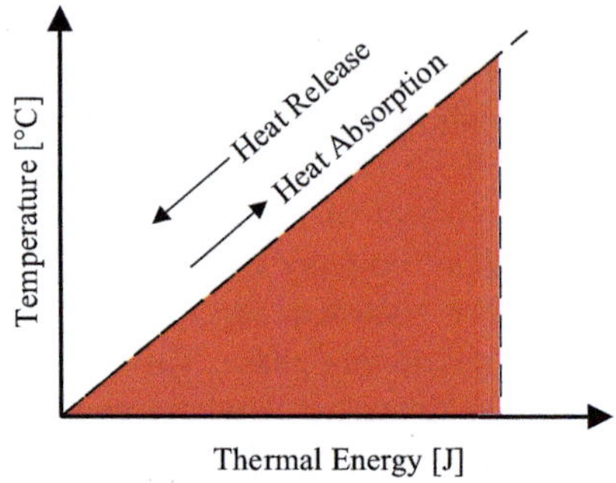

FIGURE 2.30 Curve for temperature and heat stored in sensible heat storage.

TABLE 2.10 Relevant properties of sensible heat storage media.

Type of Medium	Phase State	Temperature Range for Energy Storage °C	Density $\rho/(kg \cdot m^{-3})$	Thermal Conductivity $\lambda/W \cdot m^{-1} \cdot K^{-1}$	Specific Heat Capacity $c/kJ \cdot kg^{-1} \cdot °C^{-1}$
Aluminum	Solid	<650	2707	204	0.96
Cast Iron	Solid	<1100	837	29.3	0.46
Sand	Solid	<1500	1555	0.15–0.25	0.92
Concrete	Solid	<1000	2240	1.70	0.84
Water	Liquid	0–100	998	0.027	4.18
Heat Transfer Oil	Liquid	80–320	600–800	0.12	0.76
Molten Salt	Liquid	260–550	1700–2000	0.6–1.2	1.53

where M represents the mass of the storage medium; c_p is the specific heat capacity of the medium; ΔT is the change in temperature of the medium; Q represents the thermal energy stored in the medium.

In engineering applications, storage materials with high specific heat capacity, low cost, and high thermal stability during heat storage and release cycles are typically selected. Table 2.10 presents the key energy storage properties of commonly used solid and liquid storage media. Solid media have high density but low specific heat capacity. To store the same amount of heat energy, a larger mass and volume of solid media are required.

In the medium and low temperature ranges, such as those suitable for hot water, heating, and air conditioning systems, liquid materials with water perform best, while solid materials generally choose gravel and other similar materials. Fig. 2.31 shows a typical application scenario for solid sensible heat storage, where gravel is used to store heat and provide a heat source for indoor air

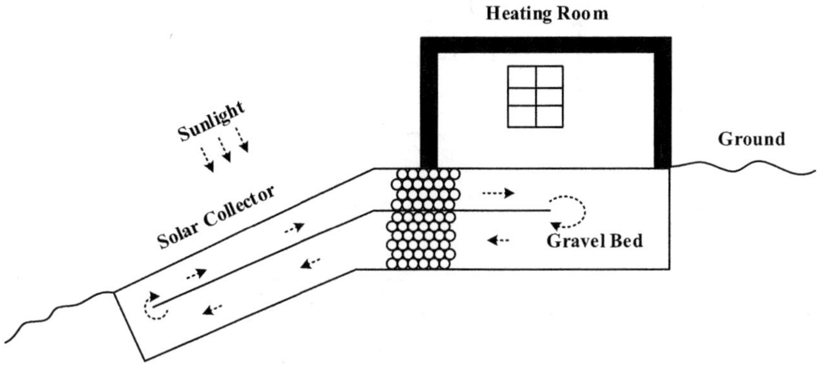

FIGURE 2.31 Diagram of solid gravel heat storage device.

heating systems. In the medium and high-temperature ranges, high-temperature concrete is suitable as a solid sensible heat storage medium. It typically uses a structure filled with granular beds to exchange heat with fluids and has been applied in solar thermal power generation. The aforementioned materials have high specific heat capacities, are abundant in supply, are low in price, are nontoxic and noncorrosive, and exhibit strong thermal stability.

2.5.2 Latent heat storage

Latent heat storage achieves thermal energy storage by undergoing phase changes in phase change materials during heat absorption and release processes. Compared to sensible heat storage, latent heat storage has a higher energy storage density and offers advantages in heat transfer optimization and hot spot temperature control [12], see Fig. 2.32.

The formula for calculating the heat stored in latent heat storage is as follows:

$$Q_{PCM} = \int_{T_i}^{T_m} mc_p dT + ma_m \Delta h_m + \int_{T_m}^{T_f} mc_p dT \tag{2.9}$$

where T_i and T_f are the initial and final temperatures during the heat storage process, respectively; m is the mass of the heat storage medium; a_m is the melting coefficient; c_p is the specific heat capacity; Δh_m is the latent heat per unit mass; T_m is the melting point of the medium.

Latent heat storage materials can be classified into four categories based on their phase change forms and processes: solid-gas, liquid-gas, solid-solid, and solid-liquid phase change storage materials. The advantages and disadvantages of these four types of materials are shown in Table 2.11.

Phase change materials are mainly classified into two categories: organic and inorganic heat storage materials. Organic heat storage materials typically have phase-change temperatures below 200 °C and are mainly used in medium- and

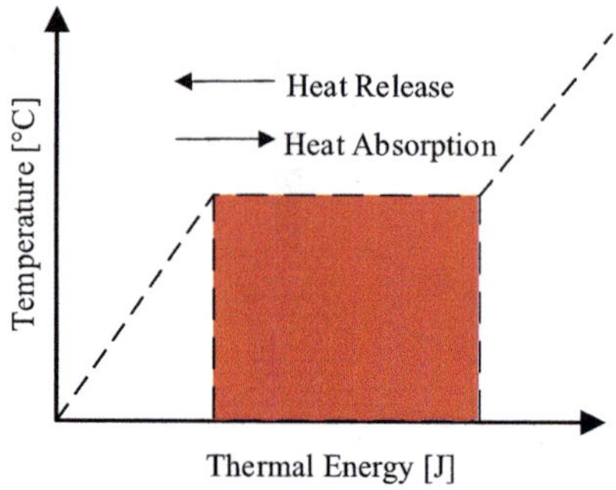

FIGURE 2.32 Curve representing the relationship between temperature and heat stored in latent heat storage.

TABLE 2.11 Classification and advantages/disadvantages of phase change energy storage materials.

Material Type	Advantage	Disadvantage
Solid-Gas Phase Change Energy Storage Material	The latent heat of phase change is large.	There is a gas release, with significant volume changes.
Liquid-Gas Phase Change Heat Storage Material		
Solid-Solid Phase Change Energy Storage Material	The phase change reversibility is good, with no issues of supercooling or phase separation.	The latent heat of phase change is small, the thermal conductivity is relatively low, and the price is high.
Solid-Liquid Phase Change Energy Storage Material	The energy storage density is high.	There are issues of supercooling and phase separation, and leakage is prone to occur.

low-temperature thermal energy applications. However, for high-temperature phase change heat storage applications such as solar thermal power generation, inorganic materials are more suitable due to their higher phase change temperatures, including inorganic salts, metals, and their alloys. Among inorganic salts, nitrates, chlorides, carbonates, and fluorides are widely used, with their phase change temperatures increasing in that order. When used as phase change heat storage media, metal and alloy materials exhibit significant advantages, including large latent heat of phase change and high energy storage density, which results in higher heat storage per unit cost. Additionally, they have extremely high thermal conductivity, allowing for rapid charging and discharging processes. Furthermore, metal and alloy materials are nontoxic and odorless. However, the melting points of metals and alloys are generally high (often exceeding 500 °C), limiting their application in high-temperature heat

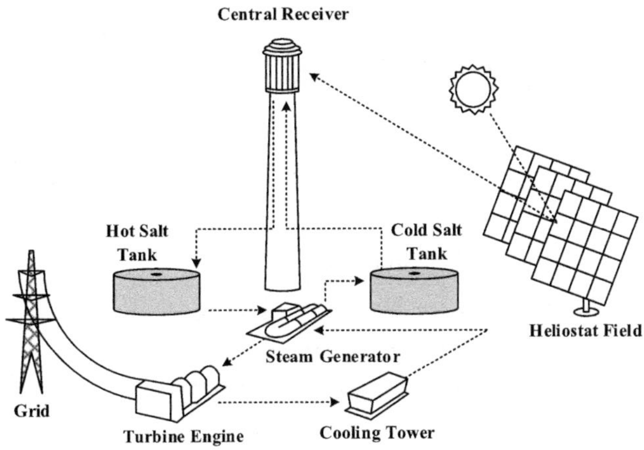

FIGURE 2.33 Molten salt phase change heat storage system for solar thermal power stations.

storage scenarios. Currently, aluminum-based alloys, particularly aluminum-silicon alloys, are research hotspots.

Fig. 2.33 illustrates a molten salt phase change heat storage system, and its operating principle is as follows: During sunny periods, the high-temperature heat transfer fluid generated by the solar heat absorber is divided into two streams. One part directly enters the steam generator to produce steam for power generation through thermal-to-work conversion. The other part is transported to the phase-change heat storage unit, where it heats the solid heat storage medium, causing it to melt and store thermal energy. When it is nighttime or solar energy is insufficient, the heat transfer fluid at a lower temperature in the steam generator enters the phase change heat storage unit. The heat storage medium solidifies and releases heat, heating the low-temperature heat transfer fluid to a high temperature before returning it to the steam generator. The process continues to heat water to produce steam that drives the generator, thereby achieving continuous and stable power output.

2.5.3 Thermochemical energy storage

Thermochemical energy storage primarily relies on a reversible thermochemical reaction ($A + \Delta B \rightleftharpoons C + D$), where the storage material A absorbs energy such as solar energy ΔB, and converts it into C and D for storage. When energy is needed, C and D are fully contacted to react and produce A, releasing the stored chemical energy. Generally, the thermochemical energy storage process can be divided into three stages: the heat storage stage, the storage stage, and the heat release stage. These three stages can achieve a closed cycle by properly storing the energy storage medium, thereby enabling a longer energy storage duration [13], see Fig. 2.34.

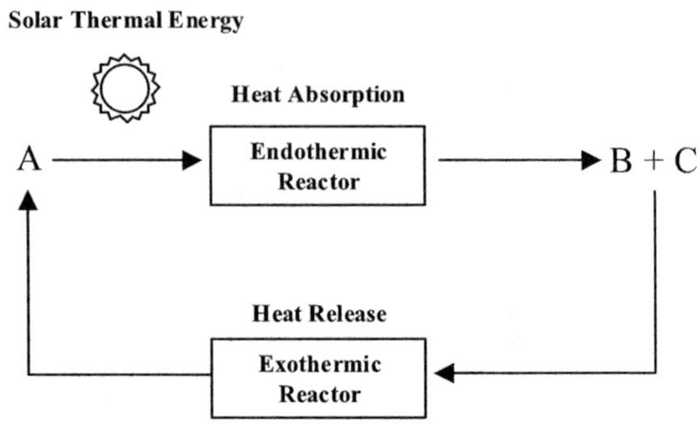

FIGURE 2.34 Schematic diagram of thermochemical energy storage.

TABLE 2.12 Examples of thermochemical energy storage systems.

Thermo-chemical Energy Storage Materials	Reaction Formula	Energy Storage Density ρ/kJ·kg^{-1}	Reaction Temperature/°C
$MgSO_4 \cdot 6H_2O$	$MgSO_4 \cdot 6H_2O(s) + heat \rightleftharpoons MgSO_4 \cdot 0.1H_2O(s) + 5.9H_2O(s)$	611	150
NH_3	$NH_3 + \Delta H \rightleftharpoons 1/2N_2 + 3/2H_2$	745	400~700
$Ca(OH)_2$	$Ca(OH)_2 \rightleftharpoons CaO + H_2O$	437	350~900
MgH_2	$MgH_2 \rightleftharpoons Mg + H_2$	580	250~500
$CaCO_3$	$CaCO_3 \rightleftharpoons CaO + CO_2$	692	700~1000

The energy storage density of thermochemical energy storage materials typically ranges from 0.5 to 3 GJ/m^3, which can be 8 to 10 times that of sensible heat materials and more than twice that of latent heat materials. Additionally, they exhibit low thermal losses during long-term storage and are considered one of the most promising heat storage materials. Table 2.12 presents the performance and reaction temperatures of common thermochemical energy storage systems. Thermochemical heat storage materials can be divided into low-temperature and medium-to-high-temperature categories based on their operating temperature range. Low-temperature thermochemical heat storage materials are mainly hydrated salts, widely used in building heating applications. Medium-to-high-temperature thermochemical heat storage materials include metal hydroxide systems, ammonia decomposition systems, and carbonate systems. However, these medium-to-high-temperature systems face challenges such as harsh reaction conditions, high reactant costs, high toxicity, catalyst deactivation, inadequate storage safety, and short cycle life. Currently, they are still in the basic research stage.

2.5.4 Demonstration of integrated application of thermal energy storage and cold storage technologies

Currently, approximately 85% of global thermal energy storage technologies are applied in building heating and district heating systems. Most of the installed capacity of thermal energy storage facilities originates from Nordic countries such as Denmark, Germany, and Sweden. Roughly 70% of large-scale solar district heating plants have been constructed in Denmark, making it a global leader in terms of installed capacity for solar district heating systems.

The Høje Taastrup pit thermal energy storage project is located in the western suburbs of Copenhagen, Denmark, and is jointly owned by Høje Taastrup District Heating and VEKS (a heat transmission company in the western suburbs of Copenhagen). This 70,000 m^3 pit thermal energy storage facility is used for short-term thermal energy storage, primarily serving peak load reduction in the local district heating network. It optimizes electricity production in the Copenhagen area, significantly reducing fossil fuel consumption and carbon dioxide emissions. The Vojens solar cross-seasonal thermal energy storage district heating system, located in Vojens, Denmark, is currently the largest solar seasonal heat storage heating project in the world. Owned by Vojens Fjernvarme, it has a solar collector area of approximately 70,000 m^2 and a cross-seasonal heat storage tank with a volume of 200,000 m^3. The Bornholm Island molten salt thermal energy storage demonstration project is located on Marstal Island in Denmark and is owned by Hyme Energy and Bornholms Energi & Forsyning (BEOF). Using proprietary corrosion control technology, it is the world's first thermal energy storage system to deploy molten hydroxide salts, with a storage capacity of 1 MW/20 MWh. In China significant progress has been made in the application of molten salt thermal energy storage technology.

In terms of Concentrated Solar Power (CSP) and molten salt thermal energy storage, the Dunhuang 100 MW molten salt tower CSP demonstration project was independently developed and funded by Beijing Shouhang AIQIWEI Energy Saving Technology Co., Ltd., with full intellectual property rights. The project's heliostat field has a total reflective area of over 1.4 million square meters. The heliostats concentrate sunlight onto the top of the absorber tower to heat the molten salt. The molten salt then enters the steam generator system to produce superheated steam, driving the turbine generator set to generate electricity. Meanwhile, some of the heat is stored in molten salt tanks to accumulate energy for full-load power generation after sunset. The Yumen "CSP + Photovoltaic + Wind Power" demonstration project, located in Yumen City, Gansu Province, has a total installed capacity of 700 MW and a heat-collecting area of 1.3 million square meters. The project integrates CSP, photovoltaic, and wind power, with an annual electricity generation capacity of up to 1.75 billion kWh, saving 521,000 tons of coal and reducing emissions by 1.353 million tons annually.

FIGURE 2.35 Scene of Dunhuang's 100 MW molten salt tower CSP station.

In terms of thermal energy storage for peak shaving and frequency regulation in thermal power plants using molten salt, the Huaneng Dezhou Power Plant's modular molten salt thermal energy storage project for peak shaving and frequency regulation has a total thermal energy storage capacity of 180,000 kWh, consisting of 12 thermal energy storage units. It has an efficient regulation capacity of 50,000 kW for 4 hours, solving the problems associated with the application of traditional technologies in thermal power plants. The Guoneng Suzhou Thermal Power's 1000 MWh molten salt thermal energy storage project is China's first GWh-scale molten salt thermal energy storage coupling project for decoupling thermal power units. By using molten salt thermal energy storage technology, it achieves deep peak shaving in thermal power units, greatly improving the flexibility of thermal power plants, see Fig. 2.35.

References

[1] Digambar M. Tagare, Hydroelectric generation—pumped storage, minor hydroelectric, and Oceanic-based systems, in: Electricity Power Generation: The Changing Dimensions, IEEE, 2011.

[2] S. Li, W. Li, L. Chen, et al., Real-time robust state estimation for large-scale low-observability power-transportation system based on meta physics-informed graph TimesNet, IEEE Transactions on Smart Grid (2024) 5500–5513.

[3] Alejandro Garces, Hydrothermal scheduling, in: Mathematical Programming for Power Systems Operation: From Theory to Applications in Python, IEEE, 2022.

[4] A. Faraji, S.A. Sadrossadat, W. Na, F. Feng, Q.-J. Zhang, A new macromodeling method based on deep gated recurrent unit regularized with Gaussian dropout for nonlinear circuits, IEEE Transactions on Circuits and Systems. I, Regular Papers (2023) 2904–2915.

[5] Jicheng Li, Functional characteristics of excitation control and starting system of reversible pumped storage unit, in: Design and Application of Modern Synchronous Generator Excitation Systems, IEEE, 2019.

[6] P. Wang, Q. Mou, X. Liu, W. Gu, X. Chen, Start-up control of a synchronous condenser integrated HVDC system with power electronics based static frequency converter, IEEE Access (2019) 146914–146921.

[7] S. Xia, Z. Ding, T. Du, D. Zhang, M. Shahidehpour, T. Ding, Multitime scale coordinated scheduling for the combined system of wind power, photovoltaic, thermal generator, hydro pumped storage, and batteries, IEEE Transactions on Industry Applications (2020) 2227–2237.

[8] M. Valavi, A. Nysveen, Variable-speed operation of hydropower plants: a look at the past, present, and future, IEEE Industry Applications Magazine (2018) 18–27.
[9] S. Mei, R. Li, L. Chen, X. Xue, An overview and outlook on advanced adiabatic compressed air energy storage technique, in: Proceedings of the CSEE, 2018, pp. 2893–2907+3140.
[10] O. Bamisile, Z. Zheng, H. Adun, D. Cai, N. Ting, Q. Huang, Development and prospect of flywheel energy storage technology: a citespace-based visual analysis, Energy Reports (2023) 494–505.
[11] U. Herrmann, D.W. Kearney, Survey of thermal energy storage for parabolic trough power plants, Journal of Solar Energy Engineering (2002) 145–152.
[12] H. Chen, Y. Wang, J. Li, Experimental research on a solar air-source heat pump system with phase change energy storage, Energy and Buildings (2020) 228–237.
[13] F. Schaube, A. WoRner, R. Tamme, High temperature thermochemical heat storage for concentrated solar power using gas–solid reactions, Journal of Solar Energy Engineering (2011) 1006–1012.

Chapter 3

Electrochemical energy storage batteries and materials

Gui Chen, Zihan Xiong, Yuhao Xia, Haotian Liang, Hui Wang, and Renzong Hu
School of Materials Science and Engineering, South China University of Technology, Guangdong Provincial Key Laboratory of Advanced Energy Storage Materials, Guangzhou, China

Contents

Modern Energy Storage. https://doi.org/10.1016/B978-0-32-395419-8.00009-X

3.1 Electrochemical energy storage

Electrochemical energy storage systems rely on the conversion between electrical and chemical energy [1]. These systems store and release energy through electrochemical reactions during the charge and discharge cycles of the battery. Commonly used electrochemical energy storage technologies include aqueous batteries, organic lithium/sodium-ion batteries, flow batteries, and high-temperature sodium-sulfur (NaS) batteries [2]. Specifically, aqueous batteries encompass lead-acid and nickel-metal hydride (NiMH) types.

3.1.1 Lead-acid battery

Since its invention by Planté in 1859, the lead-acid battery has been in use for over 160 years and remains a well-established technology. It is the most widely deployed electrochemical power source worldwide. Lead-acid batteries offer numerous benefits, including stable voltage, a broad range of operating temperatures and currents, high cycle endurance, reliable storage performance, and low manufacturing costs. However, they also have several drawbacks, such as large size, low energy density, susceptibility to environmental corrosion, limited cycle life, and significant self-discharge [3].

In the last decade, research on lead-acid batteries has increasingly centered on lead-carbon technology. By incorporating highly active carbon materials into the negative electrode, the rapid capacity loss due to sulfation during charging is effectively mitigated. This advancement can notably extend the battery cycle life and enhance its fast charge and discharge performance.

3.1.2 Nickel-metal hydride battery

The NiMH battery is an alkaline aqueous rechargeable battery that utilizes a concentrated potassium hydroxide (KOH) solution as the electrolyte. Developed in the late 1980s, it has gained widespread use. During charging, the $Ni(OH)_2$ at the positive electrode undergoes oxidation to form NiOOH by losing electrons. Meanwhile, at the negative electrode, water in the alkaline electrolyte accepts electrons, leading to the release of hydrogen atoms, which are then absorbed by the hydrogen storage alloy to form metal hydride. The discharge process is essentially the reverse of this reaction.

NiMH batteries are characterized by their high energy density and lack of memory effect. They can also tolerate a certain degree of overcharging and deep discharging. Environmentally friendly and possessing high recycling value, NiMH batteries are widely considered a green and eco-friendly type of secondary battery [4].

3.1.3 Lithium-ion battery

Lithium-ion batteries are known for their high energy density, efficient charge and discharge cycles, and quick response times, making them the fastest-growing energy storage technology. These batteries operate through the movement of lithium ions between the positive and negative electrodes. During charging, lithium ions are released from the positive electrode, move through the electrolyte, and intercalate into the negative electrode, which becomes enriched with lithium. The discharge process reverses this movement. Key components of a lithium-ion battery include the positive and negative electrode materials, electrolyte, and separator. The positive electrode material plays a crucial role in determining the battery's core performance, such as capacity and lifespan, and typically accounts for about 40% of the total battery cost.

Currently, lithium-ion batteries are the most widely used energy storage technology, second only to pumped storage in terms of installed capacity. Additionally, lithium-ion batteries offer a higher working voltage of up to 3.6 V, which is three times greater than that of nickel-cadmium (NiCd) or nickel-metal hydride (NiMH) batteries. They also do not contain toxic cadmium, making them easier to dispose of after use compared to NiCd batteries. However, safety concerns surrounding lithium-ion batteries have posed challenges to their further development [5]. Lithium-ion batteries are susceptible to overheating, and when exposed to high voltages, they can be damaged, leading to thermal runaway and even combustion. To enhance safety, these batteries are equipped with mechanisms that regulate voltage and internal pressure, though these features can add weight and, in some cases, limit performance. Furthermore, lithium-ion batteries undergo aging during use, resulting in a gradual decrease in capacity, and they typically fail after a few years of operation.

3.1.4 Redox flow battery

The Redox Flow Battery (RFB) is regarded as one of the most promising technologies for large- and medium-scale energy storage [6]. It features a distinct design in which energy is stored in electrolyte-filled reservoirs. Typically, the redox materials for the positive and negative electrodes are dissolved in the electrolytes, which are separated by a membrane. Unlike conventional electrochemical energy storage systems, flow batteries involve redox reactions between the electrolyte solutions and the electrodes during charging and discharging. In this setup, the electrolyte materials not only conduct charged particles but also store energy, while the electrodes serve primarily as interfaces for energy conversion.

As a result, because the energy conversion reactions in an RFB typically occur only at the surface of the electrodes and do not damage the internal structure, the integrity of the internal components is preserved, giving flow batteries a longer cycle life. Additionally, design flexibility is another major advantage. The effective decoupling of energy and power in flow batteries allows for flexible adjustments to meet user needs, making them highly attractive for large-scale energy storage applications.

3.2 The development and current status of electrochemical energy storage

The origins of electrochemical energy storage technology date back to the 19th century. In 1800, Alessandro Volta introduced the Voltaic Pile, marking the advent of electrochemical battery storage. By the late 20th century, lithium-ion batteries, developed by Sony, rapidly gained market dominance and found widespread use in portable electronics, power tools, electric vehicles, and other applications. In the 21st century, emerging technologies such as sodium-sulfur and flow batteries have been developed, making them viable for large-scale energy storage systems. Today, governments around the world are enacting policies to foster the growth of electrochemical energy storage, supporting the adoption of electric vehicles and renewable energy. This surge in demand has driven rapid advancements and mass production in energy storage technologies, resulting in a continuous reduction in costs. Looking ahead, electrochemical energy storage technology is expected to evolve towards higher energy density, improved safety, longer lifespan, and lower costs, forming a strong technological basis for the global energy transition and sustainable development [7].

3.2.1 The development history of electrochemical energy storage batteries

The battery, often referred to as a chemical power source, is the core element of electrochemical energy storage and is essential for driving the development and application of these technologies [8]. Chemical power sources are categorized into primary batteries, fuel cells, and secondary batteries based on their applications. Primary batteries, or dry cells, are consumed during discharge and cannot be recharged. They mainly include alkaline and acid-based batteries. Fuel cells, in contrast, do not rely on active materials but generate electricity by catalytically reacting fuel with an oxidizer, directly converting chemical energy into electrical energy. Secondary batteries, or rechargeable batteries,

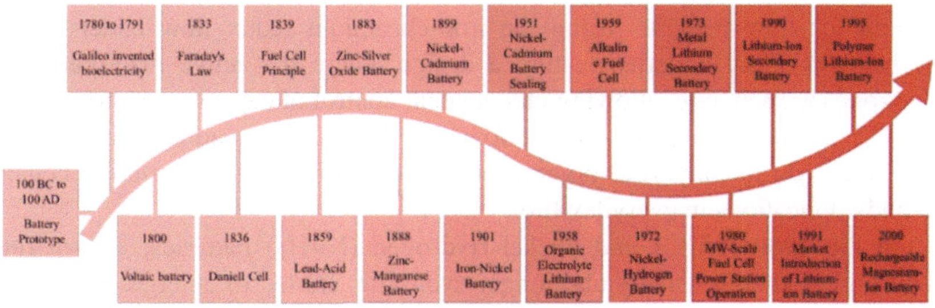

FIGURE 3.1 Timeline of electrochemical battery development.

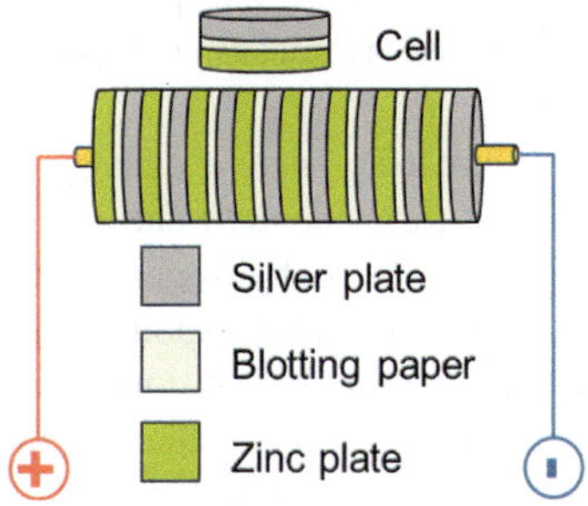

FIGURE 3.2 Schematic diagram of voltaic cell structure.

are designed for reversible discharge and charge cycles, allowing the active materials to be reused. Examples include lead-acid, NiMH, and lithium-ion batteries.

The origins of energy storage batteries date back to the late 18th century. Italian scientist Alessandro Volta first described the phenomenon of "Voltaic electricity," where a pair of silver and iron plates, separated by saltwater-soaked paper and connected by external wires, produced a brief electrical spark. This discovery laid the foundation for chemical power sources. In 1859 French scientist Gaston Planté discovered that a pair of lead plates soaked in sulfuric acid could produce a discharge current after removing the charging current, leading to the invention of the lead-acid battery, as shown in Fig. 3.1. The lead-acid battery has been in use since its invention. In 1838, French scientist Henri Andre invented the silver-zinc secondary battery based on the lead-acid battery, which was commercialized 50 years later.

The flow battery is another key large-scale, high-efficiency electrochemical energy storage system, as shown in Fig. 3.2. The concept of the flow battery was introduced in 1974, and it entered the industrialization phase by the end of the 20th century. In 1971, Ashimura and Miyake studied polarization in the positive electrode of a redox-type fuel cell flowing through a porous carbon electrode, laying the groundwork for modern flow batteries. Three years later, Thaller defined the basic structure, operating principles, and material selection criteria for flow batteries using the iron-chromium redox flow battery (Fe-Cr battery). Flow batteries feature a design where the active materials and electrodes are separate, allowing independent control of capacity and power. As research advanced, various types of flow batteries, including aqueous inorganic, inorganic hybrid, anhydrous organic, and aqueous organic flow batteries, began to find applications in everyday life.

In the 1980s Ni-Cd and Nickel-Hydrogen (Ni-H_2) batteries, known for their higher energy density and longer cycle life, started to emerge. However, due to the environmental and health concerns associated with cadmium, Ni-H_2 batteries gradually replaced Ni-Cd batteries and became widely adopted in electric vehicles, portable electronics, and power systems. Since the 1970s, Ni-H_2 batteries have also served as specialized energy storage solutions for satellites and space stations.

To address the low volumetric energy density of Ni-H_2 batteries, the Brookhaven and Philips laboratories studied and reported a hydrogen storage alloy called $LaNi_5$, which can store hydrogen within its lattice structure. This alloy behaves like a "water-absorbing sponge" and can store and release hydrogen under certain electrochemical conditions. In 1995 the Ni-MH battery, based on AB_5 hydrogen storage alloy as the negative electrode, was first industrialized by Sanyo Corporation in Japan and entered the market, quickly gaining attention. Over the following decades, it has been widely researched. The Ni-MH battery, also known as low-pressure Ni-H_2 batteries, offers a 166% increase in energy density compared to traditional Ni-H_2 batteries and has become the mainstream Ni-H_2 battery currently under study. In this book, unless otherwise specified, the Ni-H_2 battery refers to the Ni-MH battery.

The lithium-ion battery, developed after the Lead-acid and Ni-H_2 batteries, represents the next generation of secondary batteries. Thanks to the low oxidation-reduction potential of lithium metal (−3.045 V vs. RHE) and its low density (0.534 g cm^{-3}), lithium-ion batteries offer higher energy density and a wider operating voltage range. In 1913, researchers first measured the exact electrode potential of lithium metal (−3.3044 V vs. SCE). In 1973, M. Stanley Whittingham discovered that lithium ions could reversibly deintercalate from the TiS_2 positive electrode, paving the way for the lithium-metal secondary battery prototype using titanium sulfide as the positive electrode and metallic lithium as the negative electrode. The first commercial Li||MoS_2 secondary battery was introduced in Canada in 1988, but it was later discontinued due to internal short-circuiting that caused severe explosions.

Despite this setback, the attempt played a pivotal role in the eventual development of Lithium-ion batteries. To this day, enhancing the safety of Lithium-ion secondary batteries continues to be a major research priority.

In the 1980s Michel Armand introduced the concept of the "rocking-chair battery." That same year, John B. Goodenough investigated various LMO_2 (M = V, Cr, Co, Ni) layered-structure materials for the positive electrode, ultimately discovering Lithium Cobalt Oxide ($LiCoO_2$, LCO), which offered

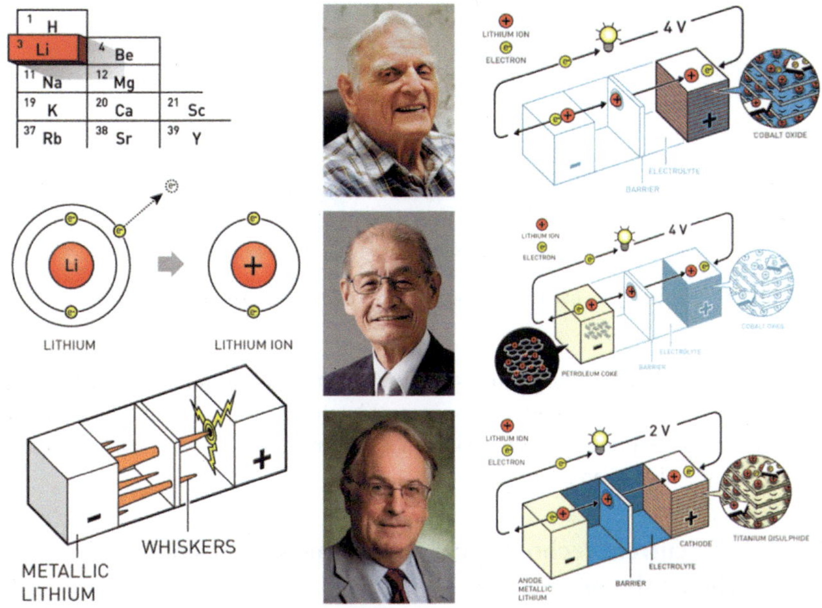

FIGURE 3.3 Lithium-ion batteries and the 2019 Nobel Prize laureates in chemistry [9].

a high voltage and stability in air. This discovery provided an ideal material for the rocking-chair lithium-ion battery. Later, Akira Yoshino achieved reversible lithium-ion intercalation and deintercalation in petroleum coke using an organic electrolyte. In 1991 in collaboration with Sony Corporation, he launched the first commercial lithium-ion battery (Coke||$LiPF_6$+PC+DEC||$LiCoO_2$) and popularized the concept of lithium-ion batteries. Over the next two decades, Lithium-ion batteries rapidly evolved and became widely adopted in mobile electronics, electric vehicles, and energy storage systems, making them the dominant technology for electrochemical energy storage.

On October 9, 2019, the Royal Swedish Academy of Sciences awarded the Nobel Prize in Chemistry to John B. Goodenough, M. Stanley Whittingham and Akira Yoshino, in recognition of their outstanding contributions to the development of Lithium-ion batteries. The three scientists and their key discoveries are shown in Fig. 3.3.

With the rise of lithium-ion battery technology, sodium-ion and potassium-ion batteries have also been explored. Among them, sodium-ion batteries offer a significant cost advantage in the long term and are considered a promising alternative to lithium-ion batteries, especially as lithium resources become scarcer. Currently, sodium-ion battery development remains in the early stages worldwide. In addition to these traditional new energy storage technologies, advances in science and technology have led to increased research into high-energy-density batteries such as lithium-metal and sodium-

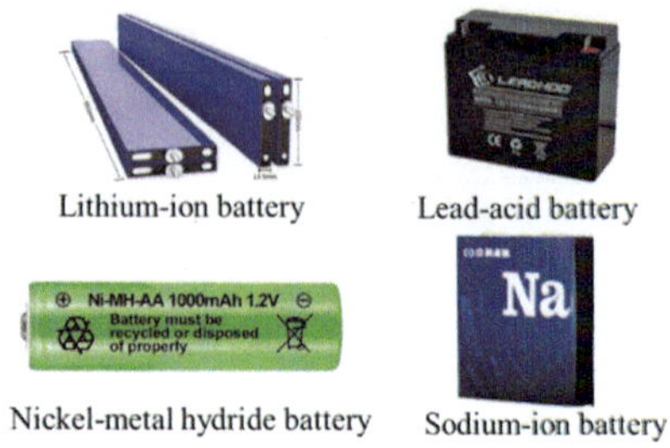

FIGURE 3.4 Various types of energy storage batteries sold on the market.

metal batteries, including lithium-sulfur (Li-S) and sodium-sulfur batteries. Furthermore, innovative battery concepts like lithium-air (Li-air), sodium-air (Na-air), magnesium-ion (Mg-ion), aqueous zinc-ion (Zn-ion), and dual-ion batteries have also been proposed.

3.2.2 The current technological development status of electrochemical energy storage

In recent years, energy and environmental challenges have become critical issues for the sustainable development of nations and economies. According to the International Energy Agency's "World Energy Outlook 2022," the exploration and adoption of clean energy are central to addressing global energy challenges in the long term [10]. Increasingly, countries and regions are setting renewable energy targets, driving energy transitions, and accelerating structural shifts in the global energy economy. Clean energy research and industries are entering a phase of rapid growth. Given the high global energy consumption, developing clean energy technologies, particularly electrochemical energy storage, is vital. The demand for electrochemical energy storage is growing quickly, with higher expectations for both energy density and power density. Secondary batteries, known for their safety, reliability, and recyclability, are widely used in consumer electronics, electric vehicles, and modern agriculture, positioning them as a key technology for the energy sector. Currently, secondary batteries in large-scale industrial production include lead-acid, Ni-H_2, flow, lithium-ion, and sodium-ion batteries, as shown in Fig. 3.4.

Lead-acid batteries are the earliest developed and industrialized secondary batteries. Their main advantages include high safety, mature technology, simple construction, and low cost. Additionally, lead-acid batteries can maintain stable cycling within a wide temperature range of 0–45 °C. However, their energy density is relatively low, ranging from 30 to 60 Wh kg^{-1}, limiting their ability to perform deep discharge and high-power discharge during operation. Currently, lead-acid battery production accounts for more than 85% of global lead usage and is widely applied in fields such as electric motorcycles, low-speed vehicles, emergency backup power sources, and wind energy systems. However, the use of lead raises significant safety and environmental concerns, and the recycling and disposal of used lead-acid batteries have garnered considerable public attention.

Hydrogen energy, as one of the most sustainable and cost-effective renewable resources, offers key advantages for energy storage systems. Ni-MH batteries began being used in 1983 on communication satellites in the international geosynchronous orbit (GEO), where these Ni-H_2 batteries, based on platinum-catalyzed electrodes, demonstrated excellent cycling stability with over 40,000 cycles. In recent years, to develop new Ni-MH battery anode materials, researchers have focused on perovskite-type oxides. Perovskite-type oxides have a distinctive crystal structure, offering significant potential for energy conversion and storage. $LaFeO_3$, in particular, is seen as a promising material for Ni-MH battery anodes due to its low cost, environmental benefits, and strong high-temperature stability. Since the 1970s, Ni-MH batteries with impressive lifespans exceeding 40 years have been primarily used in satellites and highly specialized energy storage systems for aerospace stations.

Flow batteries play a key role in electrochemical energy storage, with several types, including traditional Vanadium Redox Flow Batteries (VRFB), Zinc-Bromine Flow Batteries (Zn-Br), emerging organic flow batteries, and semisolid flow batteries. Vanadium redox and zinc-bromine flow batteries have already been commercialized, and research mainly focuses on improving materials such as electrolytes, membranes, and flow channel structures in various flow battery systems. Beijing Baineng Huitong was the first to develop China's first zinc-bromine flow energy storage battery, as well as key materials such as battery membranes, electrodes, and electrolytes. Flow batteries have a high power density, reaching the megawatt scale, and possess a long lifespan of 15–20 years, as well as deep-discharge capability. However, flow batteries suffer from low energy density, slow charge/discharge rates, and inefficient energy conversion. They are mainly used in ultra-large-scale power station energy storage, with around 5% of large-scale energy storage projects adopting flow batteries in the past three years.

Lithium-ion batteries, with an energy density of 200–250 Wh kg^{-1}, are widely used in mobile devices, electric vehicles, energy storage systems, and other industries. They offer advantages such as high working voltage, energy efficiency, fast response speed, long lifespan, and environmental friendliness. However, safety issues, particularly the risk of explosion during charging, remain a major concern. Additionally, high production costs and inconsistency issues hinder further development. Despite this, lithium-ion batteries remain the dominant technology in electrochemical energy storage [11]. The primary types of lithium-ion batteries in industrial use are based on lithium iron phosphate, ternary cathodes, and silicon-carbon anodes. Although a recycling industrial chain for retired power batteries has been established, the recycling process remains limited by traditional hydrometallurgical and pyrometallurgical methods, and its economic efficiency needs improvement.

Research on sodium-ion batteries began in the 1980s. Compared to lithium-ion batteries, sodium-ion batteries are more abundant and cost-effective, with good electrochemical performance and thermal stability, making them suitable for certain large-scale power applications. However, due to sodium's higher reduction potential and larger ionic radius, it cannot easily intercalate into the graphite layers used in lithium-ion batteries, necessitating the search for new anode materials. Sodium-ion batteries also have lower energy density, ranging from 100 to 180 Wh kg^{-1}, which makes them better suited for large-scale energy storage. In 2019 Zhongke Haina launched China's first sodium-ion battery energy storage station, and CATL laid out a comprehensive industrial chain for sodium-ion batteries by 2023. However, due to technological limitations, the application of sodium-ion batteries is still restricted to demonstration projects and has not yet been widely adopted.

Compared to other energy storage technologies, high-temperature sodium-sulfur batteries offer advantages in capacity and lifespan, and many countries are actively investing in sodium-sulfur battery technologies. Since 1992, high-temperature sodium-sulfur batteries have been in practical use for nearly 30 years. These high-power NaS batteries are primarily used for peak shaving, wind power generation, emergency power supplies, etc., to enhance power reliability. However, the brittle ceramic separator material adds manufacturing complexity and safety concerns. Additionally, the high operating temperature of 300–350 °C presents unresolved safety issues, which have hindered its large-scale application.

3.3 Main types and evaluation factors of electrochemical energy storage batteries

Electrochemical energy storage systems primarily include the battery pack, Battery Management System (BMS), Energy Management System (EMS), Power Conversion System (PCS), and additional components such as the thermal management system. Among these, the battery makes up 60% of the total cost and is the most crucial component. The electrochemical energy storage battery consists mainly of positive and negative electrode materials, separators, electrolytes, casings, and protection circuits. Key performance evaluation factors for these batteries include voltage, capacity, charge/discharge efficiency, energy density, and electrochemical impedance [12]. This chapter provides a brief introduction to the composition, working principle, evaluation factors, and testing techniques of mainstream electrochemical energy storage batteries.

3.3.1 Composition and basic principles of electrochemical energy storage batteries

In electrochemical energy storage batteries, chemical energy is converted into electrical energy through redox reactions at electrodes with different potentials in a specific electrolyte environment [13]. The anode, with a more negative potential, and the cathode, with a more positive potential, are key to the battery's function. The energy density depends on the battery's operating voltage and the reversible capacity of the electrode materials, with the operating voltage determined by the potential difference between the anode and cathode. When the external circuit is open, there is no current, and although the electrodes have a potential difference (open-circuit voltage), no energy conversion occurs. When the external circuit is closed, a current flows due to the potential difference, triggering redox reactions at the electrodes. This results in charge transfer, reactant/product transfer, and ion migration. The charging and discharging processes involve opposite charge and mass transfer. Therefore, the reversible redox reactions at the electrodes are crucial for energy conversion. This section covers the composition and basic principles of various electrochemical energy storage batteries, including aqueous batteries, organic alkali-metal-ion batteries, and flow batteries.

3.4 Key materials of electrochemical energy storage batteries

The performance of electrochemical batteries depends largely on the properties of key materials, such as the cathode, anode, and electrolyte. For example, the transition from lithium to lithium-ion batteries was enabled by advancements in materials like titanium disulfide cathodes, LCO cathodes, and graphite anodes. Therefore, the research, development, and application of these materials are vital for the growth of the battery industry. This chapter focuses on the key materials used in widely adopted energy storage systems (lead-acid, lithium-ion, Ni-MH, and flow batteries) and emerging technologies (sodium-ion and solid-state batteries) [14]. It will explore the physical and chemical properties of these materials, their performance characteristics, and relevant national and industry standards, including specifications for commercialized systems.

3.4.1 Lead-acid batteries

3.4.1.1 Working principle and battery structure of lead-acid batteries

Lead-acid batteries consist of components such as positive and negative active materials, electrolytes, grids, separators, and casing [15]. The positive electrode is typically lead dioxide (PbO_2), the negative electrode is spongy lead metal (Pb), and the electrolyte is a sulfuric acid solution with a mass fraction of 37%–42%. The grid serves as the current collector material for both the positive and negative electrodes, supporting and securing the active materials while also transmitting electrons between the active substances and the terminals. Separators are often made of porous polyethylene material, which isolates the positive and negative plates to prevent short circuits, while the porous structure

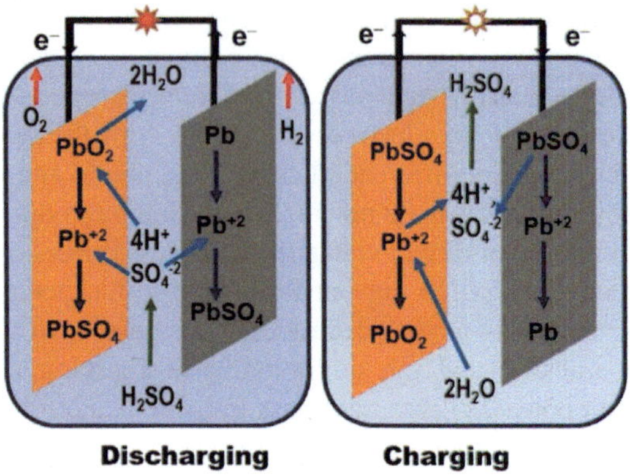

FIGURE 3.5 Schematic diagram of the working principle of lead-acid batteries [16].

facilitates the movement of the electrolyte. As the earliest commercialized energy storage battery, lead-acid batteries have the following main advantages:

(1) The longest commercialization time, stable performance, and high applicability.

(2) The electrolyte is nonflammable, ensuring higher safety, with good high-rate discharge performance, making it suitable for engine starting.

(3) Excellent performance across a wide temperature range, with the ability to function between −40 °C and +60 °C.

The lead-acid battery operates based on redox reactions between the various valence states of lead, as depicted in the schematic diagram in Fig. 3.5. The charge and discharge processes are explained by the reactions outlined in Eqs. (3.1) to (3.6):

$$\text{Discharge:}\quad \text{Anode: } Pb + 2H_2SO_4 \rightarrow PbSO_4 + 2H^+ + 2e^- \tag{3.1}$$

$$\text{Cathode: } PbO_2 + 2H_2SO_4 + 2H^+ + 2e^- \rightarrow PbSO_4 + 2H_2O \tag{3.2}$$

$$\text{Overall Reaction: } PbO_2 + 2H_2SO_4 + Pb \rightarrow 2PbSO_4 + 2H_2O \tag{3.3}$$

$$\text{Charging:}\quad \text{Anode: } PbSO_4 + 2H^+ + 2e^- \rightarrow Pb + H_2SO_4 \tag{3.4}$$

$$\text{Cathode: } PbSO_4 + 2H_2O \rightarrow PbO_2 + H_2SO_4 + 2e^- \tag{3.5}$$

$$\text{Overall Reaction: } 2Pb_2SO_4 + 2H_2O \rightarrow PbO_2 + 2H_2SO_4 + Pb \tag{3.6}$$

During discharge, lead atoms at the negative plate lose two electrons and undergo an oxidation reaction with SO_4^{2-} from the electrolyte to form $PbSO_4$. The electrons produced at the negative terminal flow through the load to the positive terminal, generating a current. At the positive terminal, Pb^{4+} in lead dioxide undergoes a reduction reaction to become Pb^{2+}, which combines with

SO_4^{2-} from the electrolyte to form lead sulfate ($PbSO_4$). At the same time, the H^+ ions produced from hydrolysis at the positive terminal also react with free OH^- ions in the electrolyte, resulting in the discharge product, H_2O. During charging, under the influence of external current, $PbSO_4$ at the positive terminal decomposes to produce Pb^{2+}. The free Pb^{2+} loses two electrons and undergoes an oxidation reaction with H_2O in the electrolyte to form PbO_2. At the negative terminal, $PbSO_4$ also dissociates under the external current, generating Pb^{2+}, which gains electrons and undergoes a reduction reaction, forming spongy Pb that attaches to the surface of the negative electrode.

During charging, sulfate ions (SO_4^{2-}) in the electrolyte move toward the positive terminal, while hydrogen ions (H^+) head toward the negative terminal. This results in a decrease in sulfuric acid concentration. The nonconductive discharge product, $PbSO_4$, accumulates at both terminals, raising the internal resistance of the battery. Moreover, as the electrolyte's pH drops, the electromotive force (EMF) of the lead-acid battery is reduced, causing electrochemical polarization. The capacity loss over extended storage is mainly due to side reactions during the later stages of charging. These reactions include the electrolysis of water, which releases gases and reduces the electrolyte's moisture, increasing the electrolyte's concentration and internal resistance, ultimately lowering battery capacity [17].

3.4.1.2 Key materials of lead-acid batteries

As the most established secondary battery, the performance and lifespan of lead-acid batteries are largely influenced by their key materials [18]. These batteries typically feature PbO_2 as the positive electrode, spongy Pb as the negative electrode, and a diluted sulfuric acid electrolyte. Presently, lead-acid batteries are extensively used in automotive starting, backup power, and power systems, making them a vital energy storage option in various critical applications.

(1) Lead Dioxide (PbO_2) Positive Electrode Material

The PbO_2 positive electrode plate is an essential part of lead-acid batteries [19]. The positive plate consists of a layer of PbO_2 active material coated onto a lead or lead alloy grid. Depending on the method of integration with the grid, it is categorized into flat and tubular structures, with the flat structure being more commonly used in the market. In terms of physical properties, PbO_2 has a relatively high density and hardness, which enhances the structural stability and mechanical strength of the electrode. Chemically, PbO_2 is highly stable in strong acids and has good ionic conductivity, which aids in the swift ion transport within the battery, enhancing charge/discharge efficiency.

Fig. 3.6 illustrates the crystal structure of the PbO_2 positive electrode material in lead-acid batteries, featuring both the orthorhombic α-PbO_2 and tetragonal β-PbO_2 phases. Among them, α-PbO_2 has smaller particle sizes, with good contact between particles, a denser structure, and excellent mechanical properties, but it has poor conductivity (resistivity around 10^{-3} Ω cm^{-1}). On

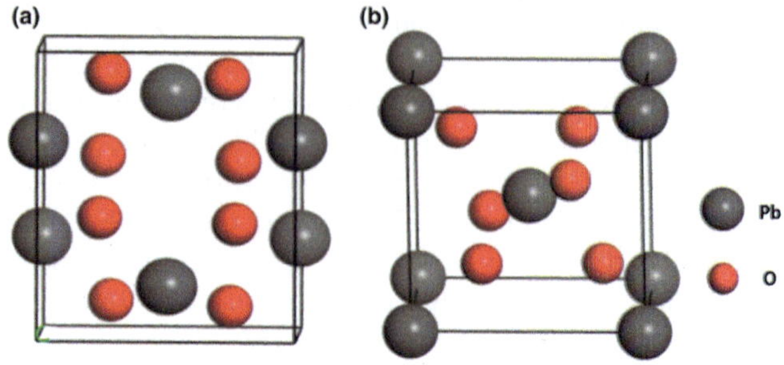

FIGURE 3.6 Crystal structure of PbO_2 positive electrode material: (a) α-PbO_2 and (b) β-PbO_2 [20].

the other hand, β-PbO_2 has larger particles and exhibits better catalytic activity, structural stability, and conductivity (10^{-4} Ω cm^{-1}) in acidic electrolytes. Therefore, in the design of the positive plate structure, dense α-PbO_2 serves as the structural framework of the positive electrode material, while β-PbO_2 acts as the active substance in the positive electrode material.

However, in practical operation, the PbO_2 positive electrode often experiences structural damage, leading to phenomena such as softening and material shedding, which reduce the battery's lifespan. Research on enhancing PbO_2-positive electrodes is currently concentrated on the following key areas:

(1) Doping Modification: By doping other elements (such as Sn, Sb, etc.) into PbO_2, its crystal structure and electronic structure are altered, thereby improving its electrochemical performance.

(2) Nanoparticle Formation: By preparing PbO_2 into nanoscale materials, its specific surface area, ion transport rate, porosity, and overall reactivity can be significantly increased, thereby enhancing its electrochemical performance.

(3) Multiphase compositing involves combining PbO_2 with materials like carbon and conductive polymers to create composite materials that enhance electrochemical performance through synergistic effects.

(2) Lead Negative Electrode Material

The negative plate of a lead-acid battery consists of a layer of lead (Pb) coated on an alloy grid. During charging, the Pb layer forms two distinct structures: a spongy framework made up of interconnected Pb crystals and additional Pb crystals deposited on this framework. The composition, quality, structure, and manufacturing process of the Pb negative electrode have a significant impact on the performance and longevity of the lead-acid battery. During charge and discharge cycles, the Pb negative electrode experiences "sulfation," leading to the formation of $PbSO_4$. This poorly reversible $PbSO_4$ limits the active material at the negative electrode from participating in the charge and discharge process, causing an increase in internal resistance and a reduction in cycle life [21]. To mitigate sulfation, researchers have explored the

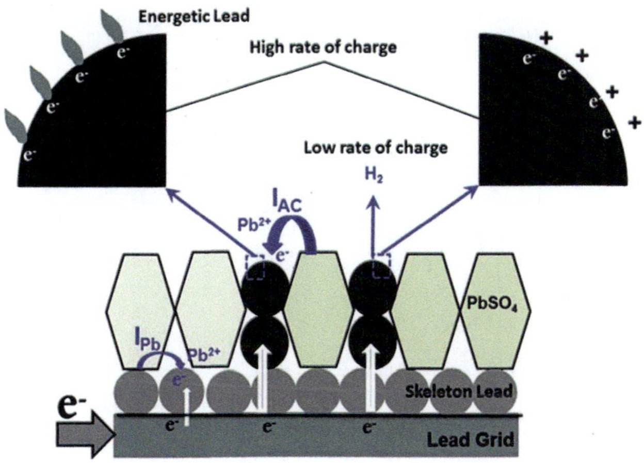

FIGURE 3.7 The suggested charging mechanism for the lead-carbon electrode [22].

use of inorganic and organic additives, including various alloys such as Pb-Sb, Pb-Ca, Pb-Sn, Pb-Ca-Sn, and elements like Al and Se.

(3) Lead-Carbon Battery

To enhance the lifespan of the lead-acid battery's negative electrode, researchers have integrated supercapacitors, known for their high specific power, with lead-acid batteries, which offer high specific capacity. This integration forms the lead-carbon battery, a highly efficient new energy storage device. The charge mechanism of the lead-carbon electrode is illustrated in Fig. 3.7. Carbon-based materials with a high specific surface area, such as graphite carbon, activated carbon, carbon nanotubes, or carbon aerogels, are added to the lead negative electrode. This results in the formation of a stable Pb-C alloy-like compound that greatly improves the performance of the negative electrode.

Lead-carbon batteries offer notable advantages over traditional lead-acid batteries, such as higher power output, greater charge acceptance, longer cycle life, enhanced recyclability, and improved safety. However, several technical and process challenges remain, including the low hydrogen evolution overpotential of carbon-based materials, alterations in charge/discharge reactions due to carbon doping, reduced battery reaction activity, and the low viscosity of carbon-based materials, all of which require further attention.

3.4.2 Flow battery

3.4.2.1 Working principle and battery structure of flow batteries

Flow batteries, or redox flow batteries, utilize liquid-state redox pairs for both the positive and negative electrodes. These batteries feature a distinctive design,

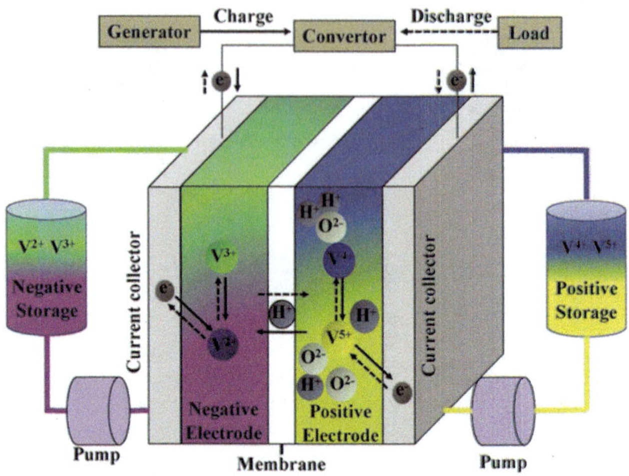

FIGURE 3.8 Diagram illustrating the structure of an all-vanadium redox flow battery system [23].

with energy stored in electrolyte tanks, making them a promising solution for medium- and large-scale energy storage. Traditional metal or inorganic aqueous flow batteries have shown dependable electrochemical performance and have been commercially proven for large-scale applications.

A flow battery consists of several components, including the stack, electrolyte, electrolyte storage tanks, membrane, circulation system, current collectors, and flow field plates. The stack, where the electrochemical reactions occur, is formed by multiple cells connected either in series or in parallel. Fig. 3.8 shows the schematic of an all-vanadium redox flow battery system, highlighting the key components of the flow battery stack unit, such as the bipolar plates, electrode frame, electrodes, and ion-conducting membrane. The bipolar plates play a critical role in connecting cells in series and separating the electrolytes to prevent self-discharge. The electrode frame and sealing gasket create a sealed space for the electrochemical reaction, helping to fix the electrodes and prevent electrolyte leakage. The electrodes function as the reaction interface for electrochemical processes and are closely tied to the power density and polarization losses in flow batteries.

Flow battery electrodes are typically made from carbon-based materials like carbon felt and graphite felt, known for their porous structure, high surface area, and numerous reaction sites. These materials also offer good conductivity, reducing internal resistance losses. The ion-conducting membrane selectively allows electrolyte ions to pass, completing the current circuit and preventing the mixing of positive and negative electrolyte solutions, which helps avoid internal short circuits.

The primary distinction between flow batteries and traditional batteries is their energy storage mechanism. In flow batteries, the electrolyte material

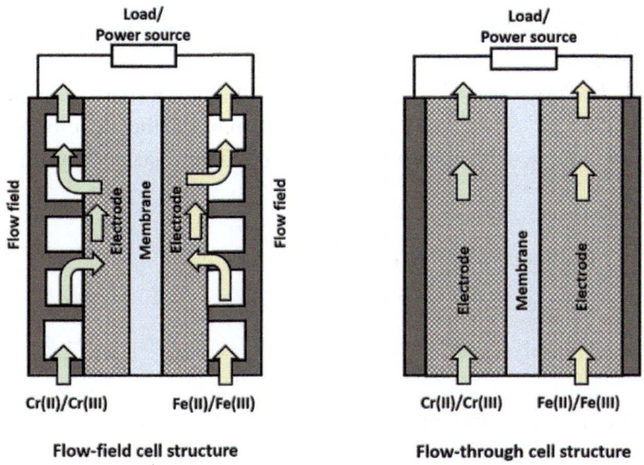

FIGURE 3.9 Iron-chromium flow battery operating principle diagram [24].

serves both as a medium for conducting charged particles and as the energy storage medium, while the electrode materials mainly act as the interface for chemical reactions, without storing energy themselves. The schematic diagram in Fig. 3.9 demonstrates the working principle of an iron-chromium flow battery. As shown, the liquid electrolyte is stored in external tanks. When activated, circulating pumps move the positive and negative electrolyte solutions through the reaction stack and back to the storage tanks. As the redox-active species in the electrolyte pass over the electrodes, they undergo oxidation/reduction reactions, converting chemical energy into electrical energy. Simultaneously, electrons flow through the external circuit, while charged particles from the electrolyte solutions migrate across the membrane due to the electric field generated by the electrodes, completing the battery circuit. The discharge reactions in a flow battery are described by Eqs. (3.7) to (3.9):

$$\text{Cathode: } M_1^{a+} + ne^- \rightarrow M_1^{a-n} \quad (3.7)$$

$$\text{Anode: } M_2^{b+} - ne^- \rightarrow M_2^{b+n} \quad (3.8)$$

$$\text{Overall Reaction: } M_1^{a+} + M_2^{b+} \rightarrow M_1^{a-n} + M_2^{b+n} \quad (3.9)$$

The discharge process in flow batteries is thermodynamically spontaneous, and the maximum number of electrons (n) that a single M_1/M_2 pair can absorb or release determines the molar energy density of the electrolyte. To enhance the energy density of flow batteries, one approach is to select ions with higher oxidation states. Additionally, the electrolytes in the storage tanks can be pumped to multiple reaction stacks, enabling the battery's output power to be adjusted by modifying the number and size of the stacks. Currently, the most commonly used flow batteries include iron-chromium, zinc-bromine, and all-vanadium types.

In 1974 NASA created the iron-chromium flow battery, utilizing the Fe^{3+}/Fe^{2+} and Cr^{3+}/Cr^{2+} redox pairs. This marked the development of the first true flow battery, with its working principle demonstrated by Eqs. (3.10) to (3.12):

$$\text{Cathode: } Fe^{3+} + e^- \rightleftharpoons Fe^{2+} \tag{3.10}$$

$$\text{Anode: } Cr^{2+} - e^- \rightleftharpoons Cr^{3+} \tag{3.11}$$

$$\text{Overall Reaction: } Fe^{3+} + Cr^{2+} \rightleftharpoons Fe^{2+} + Cr^{3+} \tag{3.12}$$

The iron-chromium flow battery uses the Fe^{2+}/Fe^{3+} ion pair in the positive electrolyte and the Cr^{2+}/Cr^{3+} ion pair in the negative electrolyte as the energy carriers. However, this system has a working voltage of only 1.18 V, offering limited voltage range advantages. Moreover, issues such as low redox activity of the Cr^{2+}/Cr^{3+} pair, hydrogen evolution during cycling, high operating temperatures (around 65 °C), and capacity degradation caused by cross-contamination between the positive and negative electrolytes significantly hinder the practical use of iron-chromium flow batteries.

In the 1970s the zinc-bromine flow battery, which uses Zn/Zn^{2+} and Br_2/Br^- as redox pairs, was introduced. Unlike the iron-chromium flow battery, which involves a single-phase transition, the zinc-bromine flow battery operates as a liquid flow system based on a gas/liquid phase transition, making it a two-phase hybrid power system. The working principle of this battery is illustrated in Eqs. (3.13) to (3.15):

$$\text{Cathode: } Br_2 + 2e^- \rightleftharpoons 2Br^- \tag{3.13}$$

$$\text{Anode: } Zn - 2e^- \rightleftharpoons Zn^{2+} \tag{3.14}$$

$$\text{Overall Reaction: } Br_2 + Zn \rightleftharpoons Zn^{2+} + 2Br^- \tag{3.15}$$

In the 1980s researchers from the University of New South Wales developed and successfully demonstrated the first all-vanadium flow battery. Since then, extensive research has been conducted globally on various components of the all-vanadium flow battery, including electrode materials, electrolytes, membranes, and flow channel designs. Today, it stands as the most advanced flow battery technology, with its specific reaction process outlined in Eqs. (3.16) to (3.18):

$$\text{Cathode: } O_2^+ + 2H^+ + e^- \rightleftharpoons VO^{2+} + H_2O \tag{3.16}$$

$$\text{Anode: } V^{2+} - e^- \rightleftharpoons V^{3+} \tag{3.17}$$

$$\text{Overall Reaction: } VO_2^+ + 2H^+ + V^{2+} \rightleftharpoons VO^{2+} + H_2O + V^{3+} \tag{3.18}$$

From the working principles of the flow batteries described above, the following characteristics can be inferred:

(1) Flow batteries offer independent control of output power and capacity, enabling separate optimization. This flexibility allows for modular design and eliminates the usual capacity-power trade-off seen in traditional batteries.

(2) In flow batteries, energy conversion reactions take place solely at the electrode surface. Unlike traditional lithium-ion batteries, where particles are inserted or extracted from the electrode, the electrochemical reaction in flow batteries does not involve this process. During discharge, only the concentration of active species in the electrolyte changes, and deep discharge does not cause structural degradation of the electrodes. As a result, flow batteries experience minimal wear during cycling, contributing to their long cycle life and ability to handle prolonged deep discharges.

(3) In a flow battery, the positive and negative electrolyte solutions are kept in separate storage tanks, which helps to prevent self-discharge reactions when the system is not in use. When activation is required, the electrolyte solution can be rapidly circulated for discharge by simply starting the circulation pump.

(4) Aqueous flow batteries utilize nonflammable energy storage media, providing greater safety and reducing the likelihood of incidents. In contrast to lithium-ion batteries that rely on organic solvent electrolytes, flow batteries offer enhanced safety.

(5) Flow batteries offer a unique advantage in that the liquid electrolyte in the storage tank can be replaced, allowing for rapid "charging," similar to refueling a car. This process separates charging and discharging, with energy being accumulated during the liquid transfer. The replaced electrolyte can be recharged in an electrolyzer at a station, making flow batteries highly suitable for large-scale industrial energy storage.

3.4.2.2 Key materials in flow batteries

Flow batteries, a versatile and adaptable electrochemical energy storage technology, store active materials in liquid form within the electrolyte. This provides significant advantages in energy density, cycle life, and scalability. Flow batteries are categorized into types like vanadium flow, zinc-bromine flow, and iron-chromium flow, each offering distinct chemical properties and application benefits, making them ideal for various power system regulations, energy storage demands, and renewable energy integration scenarios.

(1) Iron-Chromium Flow Battery

Recently, as renewable energy sources like wind and solar power are increasingly integrated into the grid, the need for energy storage with longer durations has become more critical. The iron-chromium flow battery has attracted attention as a potential solution due to its long cycle life, environmental benefits, high efficiency, strong adaptability, and low cost [24]. It is expected to become a mainstream technology in flow batteries and a strong competitor in the long-duration energy storage field.

Currently, iron-chromium flow batteries primarily use carbon-based materials for their electrodes, with three-dimensional mesh-structured graphite felt being the most commonly used. This is due to its high electrical conductivity, large surface area, good acid resistance, and wide electrochemical window. To improve the reaction activity of the Cr^{2+}/Cr^{3+} redox pair and reduce hydrogen evolution, various electrode modifications have been explored. These include surface treatments such as soaking in strong oxidants or refining carbonization processes to improve the electrodes' surface and electrochemical properties. Additionally, researchers have enhanced the specific surface area and active sites of the electrodes by adding a SiO_2 layer to the graphite felt, which boosts the reactivity of chromium ions. Initially, the electrolytes used in these batteries were $FeCl_2 + HCl$ for the positive electrode and $CrCl_3 + HCl$ for the negative electrode. However, performance degradation occurs over time due to the aging of the negative electrode electrolyte. To counteract this, researchers have introduced additives that significantly improve the electrochemical reactivity of Cr^{3+} ions and suppress the hydrogen evolution reaction.

Increasing the operating temperature of the battery can enhance the activity of the negative electrode reactions. However, higher temperatures can reduce the ion selectivity of the proton exchange membrane, causing active ions from the positive and negative electrolytes to penetrate into the opposite side. This leads to a significant decrease in the effective charge of the electrolyte and reduces its lifespan. To address this issue, a common approach in the iron-chromium battery system is to premix the positive and negative electrolytes. This reduces the concentration difference between the iron and chromium ions, effectively slowing the ion crossover rate. Although this approach decreases the utilization efficiency of active ions, it does not impose additional costs, as iron and chromium are abundant and inexpensive elements.

(2) Zinc-Bromine Flow Battery

Zinc-bromine flow batteries have a standard cell voltage of up to 1.85 V and a theoretical energy density of up to 440 Wh kg^{-1}, which surpasses that of other flow battery systems. While $ZnBr_2$ is highly soluble in water, Br_2 has poor solubility, leading to an actual energy density of approximately 60–85 Wh kg^{-1}, much lower than its theoretical potential. The slightly water-soluble Br_2 can also react with Br^- in the solution to form Br^{3-}, which can easily damage the ion-conducting membrane and cause capacity degradation in the battery. Therefore, bromine scavengers are typically added to the electrolyte of zinc-bromine flow batteries to inhibit membrane penetration. However, the addition of these scavengers can lead to phase separation in the solution, reducing the cycle stability.

Additionally, during cycling, zinc undergoes phase transitions between its solid and ionic forms, with irregular ion deposition on the zinc electrode, resembling dendrite formation observed in lithium metal batteries. This reduces the cycle life of zinc-bromine flow batteries. The toxicity of Br_3^- ions and the formation of zinc dendrites further impede the commercialization of this flow battery system [25].

The zinc-bromine flow battery consists primarily of bipolar plates, positive and negative electrodes, membranes, and electrolytes. Future development of zinc-bromine batteries focuses on the following key aspects:

(1) In terms of electrode materials, the current focus is mainly on developing catalytically active substances that can be efficiently plated on the electrodes to significantly enhance the activity of the electrodes for the bromine redox reaction.

(2) The focus regarding the electrolyte is on discovering additives that enhance conductivity, while also boosting the battery's power output and cycle life;

(3) Researchers are exploring polymer composite membranes that are easier to manufacture and more cost-effective than Nafion proton exchange membranes, while maintaining the battery's performance.

(3) Vanadium Redox Flow Battery

The standard electrode potential of a VRFB is 1.26 V. The redox reactions in the battery involve only four oxidation states of vanadium ions, so the cross-contamination of active substances has minimal impact on capacity loss. If the positive and negative electrolytes mix, the oxidation states of vanadium can be restored via a "rebalancing" process. VRFBs have a longer cycle life, reaching up to 20,000 charge/discharge cycles. To better meet practical needs, research has focused on increasing their volumetric energy density, which is tied to the solubility limits of vanadium ions in the electrolyte [26].

The electrolyte in a VRFB plays a crucial role in determining the battery's performance. Its composition directly impacts both the performance and safety of the VRFB.

Therefore standardizing the technical conditions for the composition of the electrolyte is extremely important. In practical production and everyday use, VRFBs may encounter issues such as electrolyte leakage, short circuits, and overcharging or overdischarging, which could lead to accidents, causing personal injuries and economic losses.

Due to its inherent advantages, the VRFB is highly suitable for large-scale energy storage. However, at present, the development of flow batteries still requires significant progress, particularly in key materials. Fundamental theoretical research is urgently needed, especially in critical areas such as electrolyte, electrode, and Proton Exchange Membrane (PEM) materials, to enable large-scale commercialization.

3.4.3 Lithium-ion batteries

3.4.3.1 Working principle and structure of lithium-ion batteries

Lithium-ion batteries are composed of several key components, including cathode and anode materials, electrolytes, separators, and outer shells. Common cathode materials include layered lithium cobalt oxide, spinel manganese-based materials, Ni-rich cathode materials, and polyanion-type materials. Common anode materials include graphite and lithium titanate. The liquid electrolyte is typically an organic solution formed by dissolving lithium salts in ester solvents. Common lithium salts include lithium hexafluorophosphate ($LiPF_6$), and the ester solvents are usually a mixture of one or more of Ethyl Methyl Carbonate (EMC), Ethylene Carbonate (EC), and Propylene Carbonate (PC). Sometimes, to meet various performance requirements, several additives are introduced into the liquid electrolyte. The current use of organic electrolytes raises safety issues because of their low boiling temperatures and flammability, which may pose risks. Consequently, replacing liquid electrolytes with solid electrolytes has emerged as a crucial focus for the advancement of lithium-ion batteries. The separators are typically microporous films made of PolyPropylene (PP) or PolyEthylene (PE), and their role is to isolate the cathode and anode to prevent short circuits. To enhance battery safety, some strategies are used, such as coating the separators with oxide ceramic coatings and ceramic electrolytes to improve the thermal stability of polymer separators. Lithium-ion batteries are currently available in various shapes, including cylindrical, button, pouch, square, and blade configurations.

The lithium-ion battery is essentially a concentration gradient battery. Its charge and discharge processes are primarily based on the migration of Li^+ between the positive materials and negative materials inside the battery. When charging and discharging, Li^+ is extracted from the crystal lattices of the cathode materials and inserted into the crystal lattices of the anode materials, while electrons work in the external circuit, thereby achieving the energy conversion process.

Taking the classic LCO||graphite battery as an example, as shown in Fig. 3.10, during the charging process, Li^+ is extracted from the LCO cathode, and then migrates through the electrolyte to the anode, where they are inserted into graphite, forming LiC_6 after several transformation steps. In this process, electrons move from the anode through the external circuit to the cathode to balance the total charge. This results in a decrease in the potential of the graphite anode and an increase in the potential of the LCO cathode [27]. In the subsequent discharging process, Li^+ is released from LiC_6 at the anode, goes through the liquid electrolyte back to the cathode, and then reinserts into the LCO cathode. During this process, electrons flow through the external circuit, from the cathode to the anode, causing the potential of the graphite anode to increase and the potential of the LCO cathode to decrease.

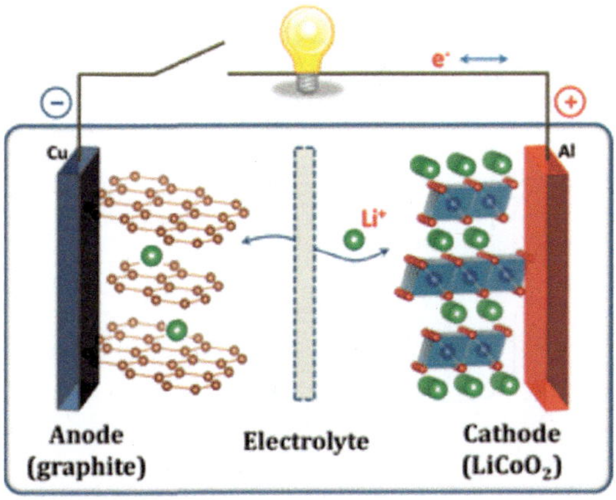

FIGURE 3.10 Schematic diagram of the operation of a lithium cobalt oxide||graphite lithium-ion battery [28].

$$\text{Charge:} \quad \text{Positive: } LiCoO_2 - xLi^+ - xe^- \rightleftharpoons Li_{1-x}CoO_2 \tag{3.19}$$

$$\text{Negative: } 6C + xLi^+ + xe^- \rightleftharpoons Li_xC_6 \tag{3.20}$$

$$\text{Discharge:} \quad \text{Positive: } Li_{1-x}CoO_2 + xLi^+ + xe^- \rightleftharpoons LiCoO_2 \tag{3.21}$$

$$\text{Negative: } Li_xC_6 - xLi^+ - xe^- \rightleftharpoons 6C \tag{3.22}$$

$$\text{Total reaction:} \quad LiCoO_2 + 6C \rightleftharpoons Li_{1-x}CoO_2 + Li_xC_6 \tag{3.23}$$

The specific electrochemical reactions are shown in Eqs. (3.19)–(3.23). The delithiation/lithiation capacity of the cathode and anode materials determines the overall capacity. Thus, advancing electrode materials with a high capacity for lithium storage is essential for the development of lithium-ion batteries with high energy density. In previous studies, lithium-ion battery technology has made significant breakthroughs and has achieved successful commercialization. Compared to the previously mentioned lead-acid batteries and nickel-hydride batteries, lithium-ion batteries have these advantages:

(1) Higher working voltage: Lithium-ion batteries have a higher working voltage, typically ranging from 3.2 V to 3.7 V, which is three times that of nickel-hydride batteries and 1.8 times greater than lead-acid batteries.

(2) Greater energy density: Lithium-ion batteries offer a significantly higher energy density. For example, the Lithium Iron (Fe) Phosphate (LFP) blade battery released by BYD in 2021 has a single-cell capacity of 202 Ah and an energy density of up to 185 Wh kg^{-1}. Meanwhile, the ternary lithium-ion battery developed by CATL can exceed 220 Wh kg^{-1}, far surpassing lead-acid and nickel-hydride batteries.

(3) Lower self-discharge rate: Lithium-ion batteries exhibit a relatively low self-discharge rate, only 3%–9%, while nickel-cadmium batteries have a self-discharge rate of around 20%, and nickel-hydride batteries range from 20% to

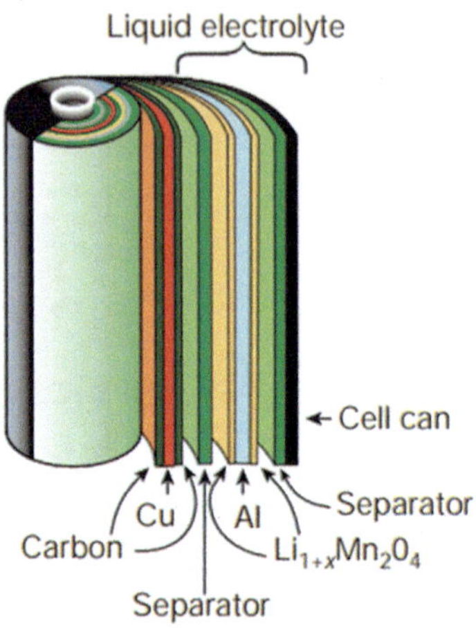

FIGURE 3.11 Material composition of a typical lithium-ion battery [29].

30%. As a result, lithium-ion batteries can retain charge for a longer time under the same conditions.

(4) Safe and fast charging and discharging: Lithium-ion batteries allow for safe and rapid charging and discharging, with high charging efficiency and large average output power.

(5) Environmentally friendly: The materials employed in lithium-ion batteries are free from toxic elements like cadmium, lead, or mercury, making them a relatively environmentally friendly and clean chemical energy source.

3.4.3.2 Key materials for lithium-ion batteries

Lithium-ion batteries are currently widely employed in instantaneous high-power output and energy storage applications. The composition of materials in commercially available lithium-ion batteries, as shown in Fig. 3.11 primarily includes cathode and anode materials, electrolyte, separator, and current collector.

As the key component responsible for supplying Li^+ in lithium-ion batteries, the production cost of cathode materials constitutes over 30% of the total cost. Moreover, the significantly lower specific discharge capacity of cathode materials in comparison to anode materials has hindered further advancements in the development of lithium-ion batteries. The development of cathode materials with high energy density and low cost has been a key research priority in the lithium-ion battery industry. During operation, the cathode plays two critical roles. First, it supplies Li^+ for insertion and extraction within the battery. Second, it provides the necessary Li^+ for the formation of the solid electrolyte interface (CEI/SEI) at the interface between the electrodes and the electrolyte.

Cathode materials can be classified into two types based on their reaction mechanisms: solid solution reaction types and multiphase conversion reaction types. In general, cathode materials for lithium-ion batteries must fulfill the following requirements:

(1) High redox potential with a flat and long voltage platform to ensure a high and stable output voltage for the full cell.

(2) Abundant lithium storage sites allow significant Li^+ insertion and extraction, providing the material with high specific discharge capacity.

(3) Fast ion diffusion channels facilitate the complete insertion and extraction of Li^+, ensuring the reversibility of electrochemical processes at the electrode.

(4) High electrical conductivity and ion diffusion rate to ensure rapid transport of Li^+ and electrons, thus giving the battery good rate performance.

(5) Good structural stability, so that Li^+ insertion and extraction do not affect the material structure, and it does not react with the electrolyte within the electrochemical window.

(6) Simple preparation process, abundant resources, and environmental friendliness.

The anode is another core component of lithium-ion batteries, with the structure and performance of anode materials determining the battery's electrochemical properties. Unlike the wide variety of commercialized cathode materials, current commercialized anode materials mainly include carbon-based materials like graphite, silicon-based materials, and lithium titanate, with graphite being the most widely used. However, graphite has limitations, such as low specific capacity and poor rate performance. These drawbacks hinder further advancements in the energy density and charging power of lithium-ion batteries. Therefore, the advancement of anode materials remains a key research focus in the field.

Based on different lithium storage mechanisms, anode materials can be categorized into insertion-type anode materials (such as graphite, TiO_2, $Li_4Ti_5O_{12}$, Nb_2O_5, etc.), alloying anode materials (such as Si, Sn, Sb, Ge, Al, and Bi), and conversion reaction anode materials (such as transition metal oxides, sulfides, phosphides, and nitrides). Additionally, some metal compounds (such as SnO_2, SnS, Sb_2S_3, SnSe, SiO_x, etc.) store lithium through both conversion reactions and alloying reactions. Currently, anode materials for lithium-ion batteries generally need to meet the following requirements:

(1) Low operating potential, ensuring that, when matched with the cathode material, the system has a sufficiently high output voltage.

(2) Good structural and chemical stability.

(3) Good electrical and ionic conductivity, allowing Li^+ to easily insert and extract.

(4) Small voltage variation during the insertion and extraction of Li^+ in the anode material, ensuring a more stable output voltage.

(5) Abundant resources and low cost, being environmentally friendly and safe for human use during production and operation.

The electrolyte, which acts as the "blood" for internal material transport within the battery, is just as important as the electrode materials in determining the battery's energy density, power density, cycle life, and safety performance. The electrolyte primarily consists of a solvent, additives, and lithium salts. Currently, the most widely used electrolyte is liquid. The battery's anode, cathode, and separator are all immersed in the liquid electrolyte. The electrolyte serves as the medium for Li^+ transport during charge and discharge cycles. It provides active Li^+ as conductive ions and creates ion channels that help the Li^+ move freely between the anode and cathode. Consequently, the following specifications must typically be fulfilled by lithium-ion battery electrolytes:

(1) Good ionic conductivity and electronic insulation, allowing for efficient ion transport, minimizing self-discharge. At room temperature, the ionic conductivity of the liquid electrolyte in lithium secondary batteries should be greater than 10^{-3} S cm^{-1}.

(2) A broad electrochemical window, ensuring that the electrolyte does not decompose within the operating voltage range of both the cathode and anode.

(3) The electrolyte must be chemically inert toward battery parts, such as separators, electrode substrates, and battery packaging materials, ensuring good safety.

The separator is also an indispensable part of battery design. It is typically a thin, porous membrane that provides electrical insulation between the electrodes, while also facilitating ion transport and stabilizing the CEI film on the positive materials and the SEI film on the negative materials. An ideal separator needs to have great mechanical properties, and chemical, electrochemical, and thermal stability. To determine if a membrane material is suitable as a separator, factors such as the high wettability of the electrolyte, good permeability, and high porosity for effective ion transport should be considered. The porous nature of the separator needs to be optimized to avoid internal short circuits and ensure full penetration of the electrolyte. Meanwhile, an appropriate thickness must be maintained for the separator to balance its mechanical performance and Li^+ transport capability. Additionally, the separator should be able to prevent the diffusion of electrode components or cathode products when the temperature rises. Furthermore, mechanical properties, ionic conductivity, and flexibility also influence the performance of the separator.

3.4.3.3 Lithium cobalt oxide cathode material

In 1980 John B. Goodenough first researched and reported the $LiCoO_2$ cathode material. Due to its simple preparation process, long cycle life, high operating voltage, and good rate performance, LCO became one of the first commercialized cathode materials and is still one of the most widely studied and applied cathode materials today.

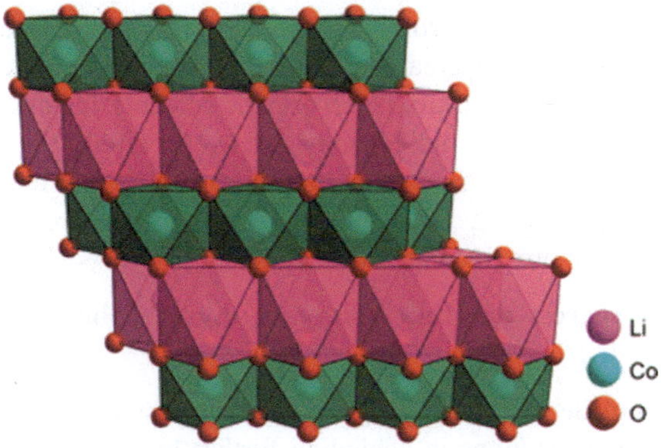

FIGURE 3.12 Crystal structure diagram of the layered LCO cathode material [30].

(1) Structure of Lithium Cobalt Oxide Cathode Material

The layered structure of LCO is the earliest commercialized cathode material system for lithium-ion batteries. Its crystal structure, shown in Fig. 3.12, adopts an α-$NaFeO_2$-type structure, belonging to the hexagonal crystal system with a space group of R-3m. In this structure, oxygen atoms (O) are arranged in a face-centered cubic close-packing at the 6c site, forming the structural framework. Lithium (Li) and cobalt (Co) atoms occupy the 3a and 3b sites on the (111) plane in an ordered manner, alternating and occupying octahedral interstices on both sides of the oxygen layers. The LiO_6 and CoO_6 octahedra are formed through ionic and covalent bonds with adjacent oxygen atoms. The arrangement of this "–O–Li–O–Co–O–Li–O–Co–" structure allows Li^+ to migrate between the CoO_2 layers, creating a two-dimensional Li^+ diffusion pathway.

(2) Electrochemical Performance of Lithium Cobalt Oxide Cathode Material

The charge and discharge processes of LCO cathodes are driven by the redox reaction between Co^{2+}/Co^{3+}. The potential plateau for the oxidation and reduction of Li^+ during its insertion and extraction is around 4 V. During charging, Li^+ is extracted from the cathode material, losing one electron in the oxidation reaction. During discharging, Li^+ gains one electron in the reduction reaction as it is reinserted into the cathode material. LCO has a theoretical specific capacity of 274 mA h g^{-1}, according to calculations. The reaction equations for the charge and discharge processes are as follows:

$$\text{Charging reaction:} \quad LiCoO_2 - xLi^+ - xe^- \rightarrow Li_{1-x}CoO_2 \tag{3.24}$$

$$\text{Discharging reaction:} \quad Li_{1-x}CoO_2 + xLi^+ + xe^- \rightarrow LiCoO_2 \tag{3.25}$$

During the charging process, as the charging voltage increases, more Li^+ is extracted from LCO. However, LCO experiences a phase change when the quantity of lithium extraction grows, and the particles' internal stress and strain rise as well, leading to a decrease in structural stability. Furthermore, when $x \leq 0.5$, Li_xCoO_2 starts to destabilize in organic solvents, leading to an oxygen loss reaction that causes a rapid capacity decay. Therefore, the commercially available LCO cathode typically extracts only up to half of the Li^+, and the charging termination voltage is set at 4.2 V. The actual reversible specific capacity utilized is approximately 140–150 mA h g^{-1}.

(3) Research Progress on High Voltage Lithium Cobalt Oxide Cathode Materials

$LiCoO_2$ has made significant progress over the past 40 years, but its energy density still faces limitations in achieving further breakthroughs. To improve its energy density, researchers have increased the charging cutoff voltage to 4.6–4.8 V, allowing for more Li^+ extraction and insertion, effectively enhancing its energy density. Therefore, in the research on $LiCoO_2$ cathodes, enhancing the working lifespan at high operating voltages is currently a mainstream research direction.

High operating voltages (greater than 4.5 V) not only cause significant volumetric changes in $LiCoO_2$, leading to irreversible phase transitions, but they also trigger issues such as oxygen evolution and severe interface degradation. At high voltages, the surface lattice oxygen of $LiCoO_2$ is easily oxidized to the peroxide state (O^-), which not only damages the CEI film but also transforms into O_2, leading to severe gas generation. At the same time, Co^{4+} is prone to reacting with carbonate-based electrolytes under high voltage, producing CO_2 and other gases.

To enable $LiCoO_2$ to function effectively under high voltage, current research strategies focus on modifying the structure/interface of $LiCoO_2$, optimizing and modifying electrolytes and separators, and exploring new binder materials. Among these, the primary methods for modifying $LiCoO_2$ itself include bulk doping and surface modification.

3.4.3.4 Lithium iron phosphate cathode material

In 1997 John B. Goodenough and others reported a series of polyanion-type cathode materials, $LiMPO_4$ (where M = Ni, Co, Mn, Fe, etc.). Among them, LFP stands out for its excellent conductivity, high thermal safety, and high reversibility, making it the preferred cathode material for high-power, high-capacity, and large-size lithium-ion batteries. Currently, LFP has achieved large-scale commercialization and is widely used in some applications, such as energy storage power stations and vehicle power batteries.

(1) Structure of Lithium Iron Phosphate Cathode Material

The crystal structure of the LFP cathode material is an ordered olivine structure, as shown in Fig. 3.13, with a space group of Pnma. In this structure, the oxygen atoms are slightly distorted in a hexagonal close-packed

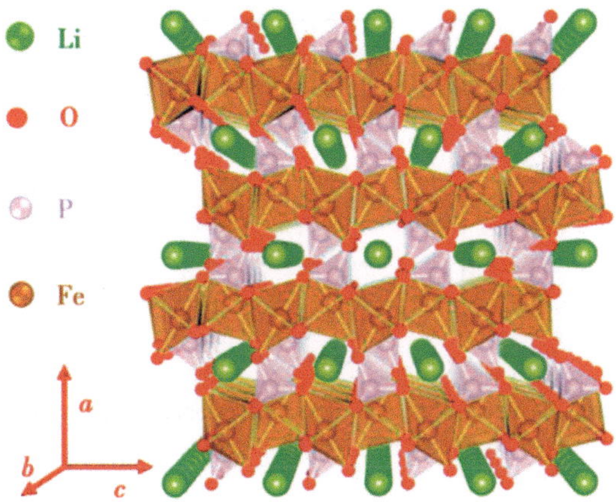

FIGURE 3.13 Crystal structure diagram of the LFP cathode [31].

arrangement, forming the framework, with phosphorus (P) atoms occupying the tetrahedral interstitial sites. Iron (Fe) atoms occupy the octahedral interstitial sites at the 4a position, while lithium (Li) atoms occupy the octahedral interstitial sites at the 4c position. Each FeO_6 octahedron is connected to four other FeO_6 octahedra via common angles on the b- and c-planes, forming a Z-plane. The LiO_6 octahedra are linked into a linear chain along the b-axis, sharing edges. The FeO_6 octahedra share edges with two LiO_6 octahedra, and the PO_4 tetrahedra share one edge with the FeO_6 octahedron and two edges with the LiO_6 octahedra. This structure provides LFP with a one-dimensional Li^+ diffusion pathway.

The structure of $FePO_4$ is like that of LFP, with the main framework of $FePO_4$ slightly deformed into an orthogonal symmetry. As a result, the olivine structure is stable when LFP delithiates and converts into $FePO_4$, with only minor variations in volume and density (volume drops by 6.81%, density rises by 2.59%). This prevents significant volume changes in the cathode during charging and discharging, and the slightly reduced volume of the de-lithiated cathode helps compensate for the volume increase in the carbon-based anode. Additionally, the high bond energy of the P–O covalent bonds in the PO_4^{3-} group contributes to the stability of oxygen atoms, preventing oxygen loss. As a result, both LFP and $FePO_4$ exhibit excellent thermal stability in N_2 or O_2 atmospheres, making them safe and reliable cathode materials for lithium-ion batteries.

(2) Electrochemical Performance of $LiFePO_4$ Cathode Materials

The initial LFP charge and discharge curve at different rates are displayed in Fig. 3.14. The charge and discharge process of LFP involves the redox reaction between Fe^{2+} and Fe^{3+}, with the first charging plateau appearing around 3.5 V.

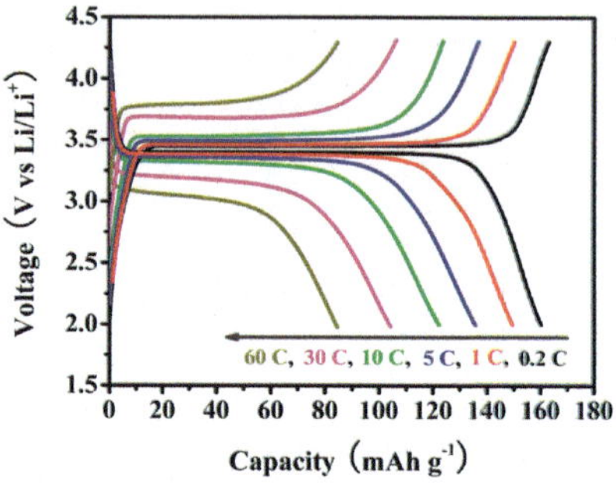

FIGURE 3.14 Charge and discharge curves of the LFP cathode at different rates [32].

After delithiation, the product of LFP is $FePO_4$, so the actual charge-discharge process exists in a state where $FePO_4$ and LFP coexist in two phases.

During charging, Li^+ is extracted from the FeO_6 layer, and the Fe^{2+} undergoes oxidation to Fe^{3+}. During discharge, the process reverses, with Li^+ being reinserted into the FeO_6 layer, and Fe^{3+} is reduced back to Fe^{2+}. The reaction of the charge-discharge process is as follows:

$$\text{Charging reaction: } LiFePO_4 - xLi^+ - xe^- \rightarrow xFePO_4 + (1-x)LiFePO_4 \tag{3.26}$$

$$\text{Discharging reaction: } FePO_4 + xLi^+ + xe^- \rightarrow xLiFePO_4 + (1-x)FePO_4 \tag{3.27}$$

LFP is a mixed conductor of both electronic and ionic conductivity, but its conductivity is relatively low, around 9–10 S cm^{-1}. In the LFP crystal structure, the PO_4 tetrahedra create certain restrictions for lithium-ion diffusion, allowing it to diffuse only along the *c*-axis, resulting in a lower lithium-ion diffusion coefficient. This restricts the material's rate performance and low-temperature behavior. As a result, LFP is better suited for use under low current densities, with more significant capacity loss occurring under high current cycling conditions.

At high current cycling rates, as Li^+ is embedded into the material, the interface between LFP and $FePO_4$ gradually moves inward, causing the interface area to shrink. When the interface area becomes small enough (critical area), the rate of lithium-ion migration within LFP cannot match the rate of electron migration (current), causing the charging process to stop. The same occurs during discharging. Therefore, when LFP is discharged at high rates, a larger critical area is required due to the high current density, which leads to fewer Li^+ being available for migration, causing a reduction in discharge capacity under high-rate conditions.

(3) Research Progress of Lithium Iron Phosphate Cathode Materials

The development and use of $LiFePO_4$ materials are severely constrained by their low electronic conductivity and Li^+ diffusion coefficient, which lead to a relatively low actual specific capacity during the charge and discharge processes, particularly at low temperatures and high charge/discharge rates. To overcome these shortcomings, researchers have focused on surface coating and doping methods to enhance the electrochemical characteristics of LFP materials.

Using materials with excellent conductivity for coating layers not only enhances ion mobility and surface conductivity but also helps suppress excessive particle growth of LFP, shortening the path for Li^+ migration, and improving reaction kinetics. Coating materials commonly utilized encompass metal compounds, conductive polymers, and carbon-based coatings.

In addition, reducing the particle size increases the material's specific surface area. Therefore, synthesizing micron-sized and nano-sized $LiFePO_4$ particles can significantly shorten the Li^+ diffusion path, improve Li^+ migration efficiency, and enhance the material's conductivity. Currently, the primary techniques for reducing particle size involve mechanical ball milling, adjusting calcination temperatures, using ultrafine conductive particles with a high specific surface area as nucleation promoters, selecting uniform precursors during synthesis, and applying carbon coatings to restrict crystal growth.

To improve the intrinsic electrochemical activity of $LiFePO_4$, ion doping can be used to distort its lattice, thereby reducing electrochemical polarization and charge transfer resistance, and significantly enhancing the electronic conductivity and improving the Li^+ diffusion rate of $LiFePO_4$. Introducing a small amount of doping into Li sites creates broader Li^+ diffusion channels and reduces charge-transfer resistance, significantly enhancing the reversibility and kinetics of lithium intercalation and deintercalation. Doping with Fe sites can weaken the Li-O bond interactions, thereby accelerating the diffusion rate of Li^+. When anion doping occupies the O sites, it can effectively suppress the antisite defects of $LiFePO_4$, improve its conductivity, and increase Li^+ mobility, offering significant practical value.

(4) New Lithium Manganese Iron Phosphate ($LiMn_xFe_{1-x}PO_4$) Cathode Materials

Currently, $LiFePO_4$ has been widely used in energy storage stations and electric vehicles, and its theoretical specific capacity (180 Wh kg^{-1}) has almost been fully realized in actual production, making it difficult to further improve its energy density. The limited redox reaction potential of $LiFePO_4$ is the main reason for its restricted energy density. However, lithium manganese iron phosphate ($LiMn_xFe_{1-x}PO_4$) exhibits a redox potential platform at 4.1 V, providing a higher theoretical energy density. When the Mn content is too high ($x > 0.8$), the Jahn-Teller effect is activated during cycling, causing lattice distortion and impairing the material's structural stability. Nevertheless, its energy density rises with the Mn level. Research has shown that when the Mn

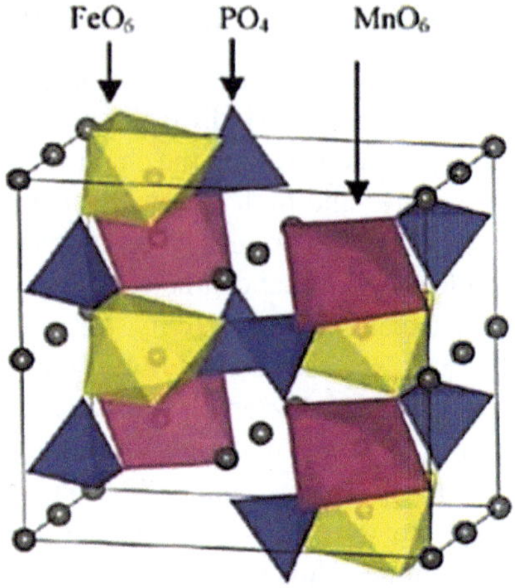

FIGURE 3.15 Crystal structure diagram of $LiFe_{0.5}Mn_{0.5}PO_4$ [33].

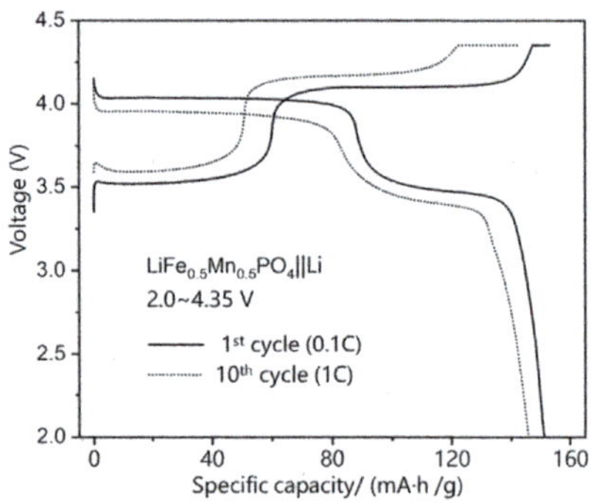

FIGURE 3.16 Charge and discharge curves of $LiFe_{0.6}Mn_{0.4}PO_4$.

content is between 0.6 and 0.8, $LiMn_xFe_{1-x}PO_4$ is more suitable as a high-capacity cathode material.

The crystal structure schematic of $LiMn_xFe_{1-x}PO_4$ is displayed in Fig. 3.15. As shown, $LiMn_xFe_{1-x}PO_4$ generally forms a uniform solid solution structure composed of FeO_6, MnO_6, and PO_4, and has a similar olivine structure to LFP. Because Mn has a larger atomic radius than Fe, it results in a larger unit cell volume compared to LFP.

The charge and discharge curves of $LiMn_{0.6}Fe_{0.4}PO_4$ at 0.1 C and 1 C rates in the 2.5~4.35 V range are shown in Fig. 3.16. $LiMn_{0.6}Fe_{0.4}PO_4$ exhibits two distinct charge and discharge platforms located at 3.4 V and 4.1 V, corresponding to the redox transitions of Fe^{2+}/Fe^{3+} and Mn^{2+}/Mn^{3+}, respectively. The 4.1-V charge/discharge platform increases the average operating voltage, thereby enhancing the energy density of the material. Additionally, the capacity contribution from the 4.1-V charge platform is notably higher than that from the

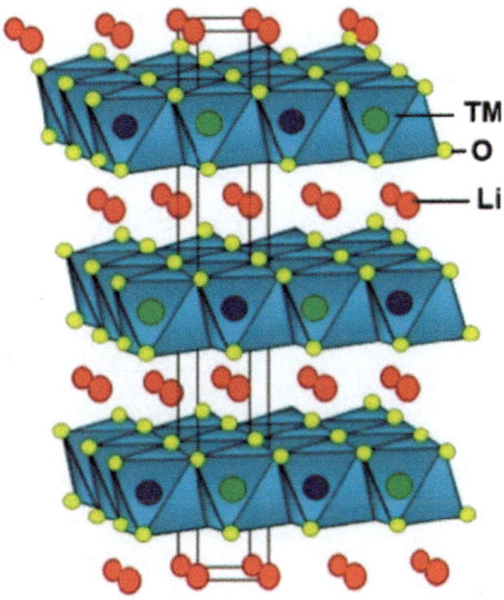

FIGURE 3.17 Crystal structure diagram of nickel-cobalt-manganese ternary layered cathode material [34].

3.4 V platform, which is due to the higher Mn content compared to Fe in the material.

3.4.3.5 Nickel-rich ternary cathode materials

In research on unit-layered materials such as $LiCoO_2$, efforts have been made to reduce production costs and improve safety by substituting and doping elements with similar chemical properties to cobalt, resulting in cathode materials with higher energy density. Among these, the nickel-cobalt-manganese lithium ternary layered cathode materials $LiNi_xCo_yMn_{1-x-y}O_2$ (NCM) is one of the most important categories. Ternary layered cathode materials combine the advantages of $LiCoO_2$, $LiNiO_2$, and $LiMnO_2$, demonstrating excellent electrochemical performance. Based on the different contents of Ni, Co, and Mn, the developed ternary cathode materials are named NCM333 ($LiNi_{1/3}Co_{1/3}Mn_{1/3}O_2$), NCM523 ($LiNi_{0.5}Co_{0.2}Mn_{0.3}O_2$), NCM622 ($LiNi_{0.6}Co_{0.2}Mn_{0.2}O_2$), NCM811 ($LiNi_{0.8}Co_{0.1}Mn_{0.1}O_2$), and NCM955 ($LiNi_{0.90}Co_{0.05}Mn_{0.05}O_2$), among others.

(1) Structure and Physical Properties of Nickel-Rich Ternary Layered Cathode Materials

A schematic diagram of the structure of ternary-layered cathode materials, as shown in Fig. 3.17, ternary cathode materials have the same R-3m space group as LCO and belong to the hexagonal crystal system, with an α-$NaFeO_2$-type layered salt structure. Lithium occupies the 3a position, while transition metals Ni, Co, and Mn occupy the 3b position, and oxygen occupies the 6c position in a cubic close-packed arrangement. Each transition metal atom is surrounded by six oxygen atoms, forming an MO_6 octahedral structure. In the c-axis direction,

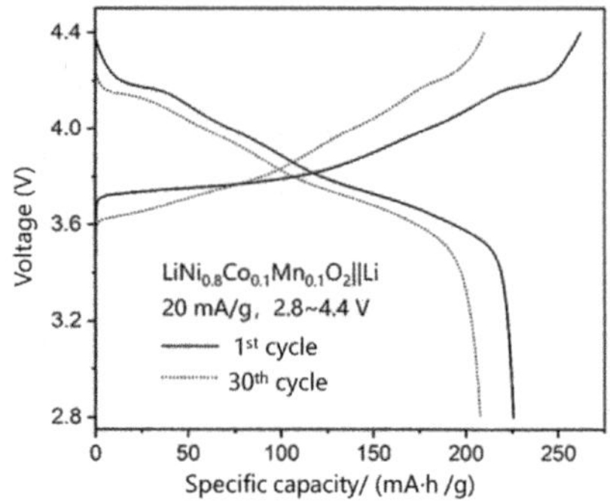

FIGURE 3.18 First and 50th charge and discharge curves of $LiNi_{0.8}Co_{0.1}Mn_{0.1}O_2$ cathode.

the transition metal layers alternate with lithium layers, which allows Li^+ to migrate between the layers, creating a two-dimensional Li^+ diffusion channel.

The electrochemical and physical properties of ternary layered cathode materials are typically affected by the relative proportions of Ni, Co, and Mn. Raising the Ni content tends to significantly boost the discharge capacity of the cathode material. This is because a higher Ni volume causes the crystal lattice to expand, which increases Li^+ free insertion in the layered structure and raises the battery's discharge capacity. However, since the radius of Ni^{2+} (0.069 nm) is similar to that of Li^+ (0.076 nm), a small amount of Ni^{2+} may occupy the Li positions during the delithiation process, resulting in cation mixing at the lattice sites. Cation mixing can influence both the discharge process and the cycling stability of the active material.

(2) Electrochemical Performance of Nickel-rich Ternary Layered Cathode Materials

Nickel-rich ternary layered cathode materials, which have a nickel content of 60% or higher, exhibit a high theoretical specific capacity (278 mA h g^{-1}) and a practical discharge capacity greater than 200 mA h g^{-1}. Taking NCM811 as an example, the charge-discharge curves for the first cycle and the 30th cycle are shown in Fig. 3.18. During the cycling process, the transition metal elements in the nickel-rich ternary layered cathode undergo a series of redox reactions, which can be divided into three stages:

(1) When the amount of delithiation is less than or equal to the proportion of Ni in the transition metals, Ni^{2+} is oxidized to Ni^{3+}.

(2) When the amount of delithiation is greater than the proportion of Ni but less than twice the proportion of Ni, Ni^{3+} is oxidized to Ni^{4+}.

(3) When the amount of delithiation is greater than twice the proportion of Ni and the sum of the proportion of Co, Co^{3+} is oxidized to Co^{4+}.

The specific reaction equations during the charge-discharge process of nickel-rich ternary layered cathode materials are as follows:

Charging process:

$$LiNi_xCo_yMn_{1-x-y}O_2 - nLi^+ - ne^- \rightarrow Li_{1-n}Ni_xCo_yMn_{1-x-y}O_2$$

Discharging process:

$$Li_{1-n}Ni_xCo_yMn_{1-x-y}O_2 + nLi^+ + ne^- \rightarrow LiNi_xCo_yMn_{1-x-y}O_2$$

During the charging process, due to the oxidation of Ni and Co, the unit cell parameter a initially decreases and then increases as Li^+ continues to deintercalate. Before the voltage reaches 4.4 V, the unit cell parameter c increases due to the repulsive forces between adjacent O layers as Li^+ are removed. However, after the charging voltage exceeds 4.4 V, the unit cell parameter c decreases because the increase in Ni–O bonds reduces the interlayer electronic repulsion. During the discharge process, the reduction in a and the increase in c lead to a decrease in the unit cell volume, with a shrinkage value of about 2%. This slight volume change indicates that the nickel-cobalt-manganese ternary composite cathode material exhibits good electrochemical performance.

(3) Research Progress of High-Nickel Ternary Layered Cathode Materials

To obtain cathode materials with higher energy density, ternary layered cathode materials are gradually developing towards higher nickel content and cobalt-free compositions. Currently, research on ternary materials at room temperature has been relatively perfected, and commercial application has been initially realized in the electric vehicle industry in China. However, to achieve large-scale commercial use of nickel-rich ternary layered cathode materials, it is still necessary to address the issue of poor electrochemical performance in low-temperature environments.

Nickel-rich ternary layered cathode materials experience a significant decline in performance at low temperatures. This is primarily due to the reduction in intrinsic lithium-ion diffusion kinetics as the temperature drops. As a result, the Li^+ diffusion rate decreases within the bulk of the cathode material, at the solid-liquid interface between the cathode and the electrolyte, and within the electrolyte itself. This leads to significant electrochemical polarization. Furthermore, low temperatures can cause the cathode-electrolyte interface to thicken, further impairing the battery's performance. Besides the decrease in lithium-ion diffusion kinetics, low temperatures also hinder the phase transition process of the cathode material, which presents a serious challenge to the material's kinetic performance.

In addition to the effects of low temperature, the intrinsic defects of the materials themselves are also significant factors affecting their low-temperature electrochemical performance. These issues include cation mixing, microcracks, and surface residual lithium, all of which can degrade the

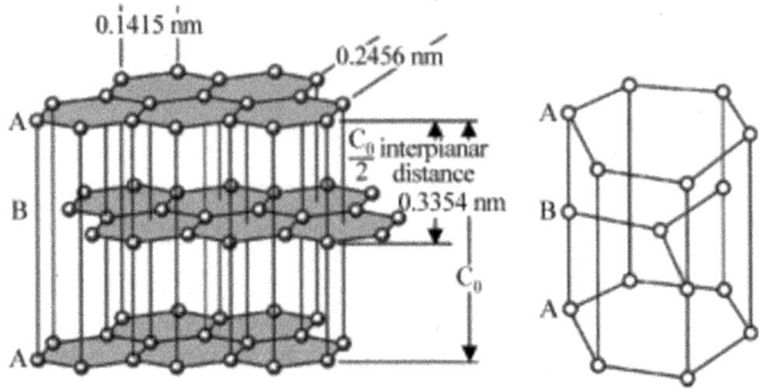

FIGURE 3.19 Structural diagram of natural graphite anode [35].

material's performance. To address these challenges and achieve the large-scale commercialization of nickel-rich cathode materials for both room and low-temperature applications, modification methods mainly include surface modification and element doping.

3.4.3.6 *Graphite-based anode materials*

Currently, the most widely used anode materials in commercial lithium-ion batteries are graphite-based. Depending on their composition and structure, graphite-based anode materials are mainly classified into Natural Graphite (NG), Artificial Graphite (AG), and Composite Graphite (CG). Among them, artificial graphite is further divided into several types, including Carbon MicroBead artificial graphite (CMB), Needle-coke Artificial Graphite (NAG), and Common Petroleum-coke Artificial Graphite (CPAG). Composite graphite refers to materials that contain at least two components, including both natural graphite and artificial graphite.

(1) Structural Characteristics of Graphite-Based Anodes

Graphite-based materials are typical intercalation-type anode materials based on the reversible intercalation and deintercalation reaction mechanism of Li^+. The classic natural graphite anode has a two-dimensional layered structure, where carbon layers are composed of sp^2 hybridized carbon atoms arranged in a hexagonal pattern, extending in two dimensions. The layers are connected by van der Waals forces, forming a two-dimensional layered structure.

The schematic diagram of the crystal structure of natural graphite is shown in Fig. 3.19. The graphite anode features a typical hexagonal layered structure, which includes the more stable hexagonal ABA (2H) sequence and the less stable rhombohedral ABC (3R) sequence. The layers are arranged in an AB stacking pattern, with an interlayer spacing of about 0.335 nm, held together by weak van der Waals forces arising from delocalized π-orbitals. Natural graphite exhibits a flaky morphology with two distinct surface types: the basal plane and the edge plane, the latter also referred to as the prismatic surface. On the basal

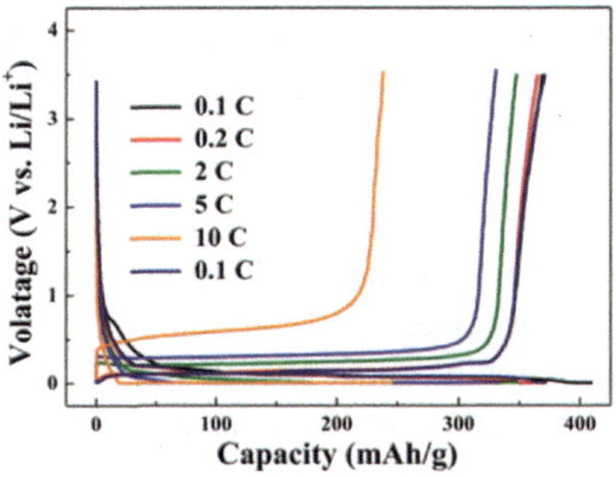

FIGURE 3.20 First and 50th charge and discharge curves of the graphite anode at 0.1 A g^{-1} [36].

plane, each carbon atom forms sp^2 hybridized bonds with three neighboring carbon atoms, forming C–C covalent bonds with a bond length of 0.142 nm and a bond angle of 120° between adjacent atoms. The edge planes can be categorized into "zigzag faces" and "chair faces" based on different stacking sequences. Generally, prismatic surfaces have higher surface energy and thus exhibit higher surface reactivity than the basal planes. As a result, the layered graphite materials often show anisotropy in their electronic, mechanical, and physicochemical properties.

(2) The Electrochemical Performance Indicators of Graphite-Based Anode Materials

The charge-discharge curves in the figure elucidate the rate capability of a graphite anode across various current rates (0.1 C to 10 C) in lithium-ion batteries. At low current rates (e.g., 0.1 C, 0.2 C), the graphite anode achieves a high discharge capacity (close to 400 mAh g^{-1}) with smooth voltage profiles (see Fig. 3.20), as the kinetics of Li^+ intercalation/deintercalation, including solid-phase diffusion within graphite and charge transfer at the Solid-Electrolyte Interface (SEI), are well synchronized with the current demand. Notably, the stable SEI film formed in the first cycle further facilitates uniform Li^+ transport at low rates. In stark contrast, at high current rates (e.g., 5 C, 10 C), the discharge capacity diminishes drastically, and the voltage curves show pronounced polarization. This stems from the exacerbated kinetic limitations, where Li^+ diffusion and charge-transfer processes cannot keep pace with the rapid current, leading to incomplete electrochemical reactions and even potential SEI degradation. Such results underscore that the graphite anode's rate performance is governed by kinetic processes and provide critical insights for optimizing battery designs targeting either high-energy or high-power applications. The overall electrochemical process is as follows:

$$6C + xLi^+ + xe^- \rightleftharpoons Li_xC_6 \qquad (3.28)$$

(3) Progress in Fast-Charging Research of Graphite Anodes

Graphite anodes are the most widely used anode material in commercial lithium-ion batteries. To further promote the widespread adoption of lithium-ion batteries, challenges such as capacity degradation and safety risks during fast charging must be addressed. As mentioned earlier, the deintercalation mechanism of Li^+ in graphite anodes is closely related to their ideal two-dimensional layered structure. The main issues limiting their fast-charging applications include:

(1) The diffusion dynamics in the two-dimensional layered structure are slow, with a long diffusion path from the edge of the layers to the interior of the material, and the structural stability of the graphene layers is poor.

(2) The kinetic processes occurring at the interface between the graphite anode and the electrolyte still need further clarification.

(3) During fast charging, surface lithium deposition and the formation of lithium dendrites can occur on the graphite anode, posing safety risks such as internal short circuits.

The structural design of graphite anodes is currently the main approach to improving their fast lithium storage performance. This includes four key aspects: crystal structure design, morphology growth regulation, interface engineering modification, and mixed-phase preparation.

The crystal structure design of graphite anodes can effectively accelerate the lithium-ion transport process, thereby enhancing the fast-charging performance of graphite-based anode materials. Techniques such as increasing the interlayer spacing of graphite, creating porous structures, doping the bulk phase, and introducing defects are commonly used. Increasing the interlayer distance of graphite can reduce the diffusion resistance of Li^+ and prevent the destruction of the two-dimensional structure during the intercalation and deintercalation processes, thus improving its fast-charging capability. Additionally, the porous structure in graphite anodes can effectively shorten the diffusion path for Li^+ embedding into the anode, thereby enhancing both charging and discharging performance.

In contrast to the conventional intercalation-based lithium storage mechanism in graphite anodes, the pseudocapacitive charge storage process offers both high energy and power density. Materials with a high density of defects are more likely to exhibit pseudocapacitive behavior. Therefore, enhancing the defect structures in graphite is an effective method to optimize its fast-charging performance, such as through bulk-phase doping and acid treatment.

Controlling the particle size and spherical morphology of graphite anodes can also effectively accelerate the kinetics of the process by shortening the diffusion path of Li^+. Adjusting the particle size of graphite anodes can increase their active sites, specific surface area, and Li^+ transport paths. Additionally, smaller particles of graphite can reduce the extent of side reactions with the electrolyte. This can improve the initial Coulombic efficiency and enhance the fast-charging ability.

Interface engineering can optimize the interface structure between the graphite anode and the commercial liquid electrolyte, effectively improving the diffusion rate of Li^+ at the interface and the electrochemical performance. This can be achieved through surface coating and interface modification. The preparation of graphite-based composites can take advantage of the enhanced synergy between the graphite material and the secondary phase, achieving a good combination of material properties and improving their electrochemical performance, such as alloying, heterojunction formation, and compound preparation.

3.4.3.7 Silicon-carbon composite materials

Silicon-carbon composite materials, as a key component of lithium-ion battery anodes, mark an important step forward in battery technology, aiming for higher energy density and longer cycle life. Silicon, as one of the main components, has an extremely high theoretical specific capacity, far exceeding traditional carbon materials. However, when silicon undergoes lithiation (the electrochemical alloying process), it faces a significant volume expansion issue, which, combined with low intrinsic electronic conductivity, hurts the battery's stability and durability [37].

To address this challenge, researchers have developed silicon-carbon composites, combining silicon with highly conductive carbon materials to provide a stable electrode structure and good electronic conductivity. The battery's cycle performance and safety are greatly enhanced by this composite material, which also successfully inhibits silicon's volume growth and lessens problems with the solid electrolyte interface. Silicon-carbon composites have gained significant attention in current lithium-ion battery research due to their promising potential in energy density, cycle life, and cost-effectiveness. These materials are anticipated to play a crucial role in the future of commercial applications.

Silicon-carbon anode materials broadly include two types:

(1) Silicon-carbon blend materials, which are obtained by physically mixing carbon materials with Si-based materials.

(2) Silicon-carbon composite materials are synthesized chemically, resulting in a silicon-carbon compound.

Among these, silicon-carbon composite materials have various structures due to the different synthesis methods, and different manufacturers design silicon-carbon materials in their unique ways.

Silicon-carbon anode materials are composed of a mix or composite of carbon materials and Si-based materials. The Si-based materials can either be pure silicon (Si) or silicon oxide (SiO_x), with differing crystal structures. Additionally, pure silicon exists in both crystalline and amorphous forms. The carbon materials can include both graphite and amorphous carbon.

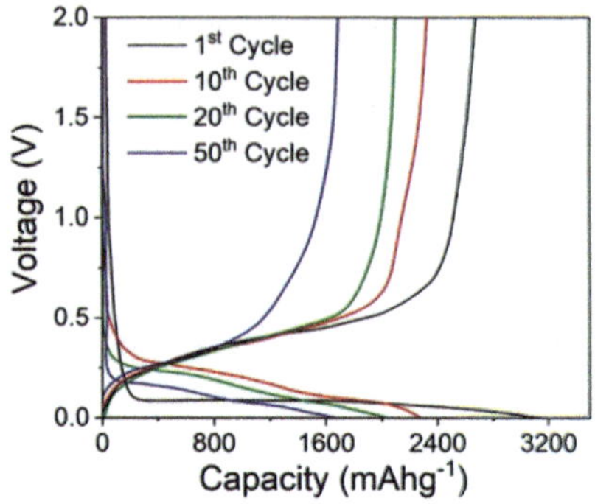

FIGURE 3.21 First and 50th charge and discharge curves of the Si anode at 0.1 A g^{-1}.

(1) Electrochemical Performance Indicators of Silicon-Carbon Anode Materials

Due to the diversity and complexity in the structure and composition design of silicon-carbon anode materials. The controversy surrounding the lithium storage mechanism of amorphous carbon in carbon materials, and the lithium storage characteristics of different silicon-carbon anodes vary. However, regardless of how the carbon materials and Si-based materials are selected, the lithium storage characteristics of the Si-based material will inevitably be reflected in the lithium storage process of the silicon-carbon anode. The charge and discharge curves of a silicon (Si) anode at 0.1 A g^{-1} for different cycle numbers are shown in Fig. 3.21 [38].

During the first discharge process, as the potential continuously decreases, Li+ diffuses and penetrates the Si lattice. When the potential reaches about 0.1 V, a phase transition begins, forming an amorphous Li_xSi alloy, which continues to form the $Li_{15}Si_4$ phase. This process corresponds to the long discharge plateau in the discharge curve. In the subsequent charging process, as the potential increases, the crystalline $Li_{15}Si_4$ undergoes delithiation and decomposes to form the amorphous Li_xSi alloy. The formed Li_xSi phase continues to undergo delithiation and decomposition to form amorphous Si. This process corresponds to the charging plateau located around 0.4 V in the charge curve. The electrochemical process is as follows:

Lithiation (Charge): $\text{Si(crystalline)} + x\text{Li}^+ + x\text{e}^- \rightarrow \text{Li}_x\text{Si(amorphous)}$

$$\text{Li}_x\text{Si(amorphous)} + (3.75 - x)\text{Li}^+ + (3.75 - x)\text{e}^- \rightarrow \text{Li}_{15}\text{Si}_4\text{(crystalline)}$$

Delithiation (Discharge):

$$\text{Li}_{15}\text{Si}_4 \rightarrow \text{Li}_x\text{Si(amorphous)} + y\text{Li}^+ + y\text{e}^- + \text{Li}_{15}\text{Si}_4\text{(crystalline, residual)}$$

$$\text{Li}_x\text{Si(amorphous)} \rightarrow \text{Si(amorphous)} + x\text{Li}^+ + x\text{e}^-$$

In comparison to pure Si materials, SiO_x materials form additional irreversible phases such as Li_2O and lithium silicates during the first discharge process. The subsequent processes, however, are similar to those of pure Si material lithiation. The formation of irreversible phases leads to a greater loss of active Li, reducing the initial Coulombic efficiency of the anode. However, the presence of these irreversible phases improves the stability and conductivity of the anode. As a result, SiO_x-based anodes exhibit lower initial Coulombic efficiency but better cycling stability compared to pure Si anodes. This is a crucial consideration when selecting Si-based materials for Si-C anodes and also dictates the direction for modifying such Si-C anodes.

(2) Structural Design of Silicon-Carbon Anodes

Compared to pure Si/SiO_X materials, although the volume change of Si-C anode materials is somewhat improved, the reason lies in the relatively lower Si content in Si-C materials, which results in smaller volume changes. As lithium-ion batteries continue to evolve towards higher energy densities, the silicon content in Si-C anodes will gradually increase. The lithiation/delithiation process of alloy-type anodes leads to unavoidable volume changes. To address this, it is essential to use effective structural design to mitigate the stress and strain caused by these volume changes. This approach is crucial to maintaining the stability of Si-based anode materials.

An ideal structure for Si-C composite anode materials should consist of a dense soft carbon outer layer and a composite internal structure (hard carbon/graphite/carbon fibers/nano-Si). In this structure, nanoscale Si particles are evenly distributed within the carbon matrix, forming micron-sized secondary particles. The relatively low absolute volumetric strain of nano-Si during the lithiation and delithiation processes reduces the electrode's volume expansion and pulverization. The carbon matrix further helps by buffering the volume expansion of the Si material. This matrix also provides an effective ionic and electronic conduction network, improving the material's kinetics. Micron-sized secondary particles contribute to the reduction of the specific surface area, improving the initial coulombic efficiency. Additionally, these particles increase the tap density, enhancing the overall volumetric energy density of the material.

However, factors such as the Si content and distribution state in the composite system, the porosity of the carbon matrix, surface characteristics, and doping of heteroatoms all have significant impacts on the anode's charge/discharge capacity, stability, and Coulombic efficiency.

On the path to the industrialization of silicon-carbon anode materials, in addition to excellent structural design, the preparation process and cost must also be considered. The advancement and commercial application of silicon-carbon anode materials will benefit from the continuous improvement of new materials and related technologies. With breakthroughs and innovations in technology, the electrochemical performance of silicon-carbon materials is expected to improve further, and production costs are likely to decrease gradually, thereby promoting their widespread adoption and application in the market. This will help meet the growing demand for high-energy-density lithium-ion batteries.

3.4.4 Sodium-ion batteries

3.4.4.1 *Working principle and battery structure of sodium-ion batteries*

Similar to lithium-ion batteries, the main components of sodium-ion batteries also include a positive electrode, a negative electrode, a separator, an electrolyte, a battery case, etc. The materials of the positive and negative electrodes are the most crucial parts of sodium-ion batteries, which determine the cycle capacity, cycle life, rate performance, and energy density of sodium-ion batteries. Common positive electrode materials include oxides, Prussian blue analogs, polyanion-based materials, and organic positive electrodes. Due to the relatively large ionic radius of Na^+, fewer negative electrode materials can realize the insertion/extraction of Na^+. Currently, hard carbon-negative electrode materials are widely used. Carbonates, as common solvents for the organic electrolyte of sodium-ion batteries, have a strong salt-dissolving ability [39]. The commonly used carbonate solvents are the same as those used in lithium-ion batteries, mainly including ethylene carbonate, propylene carbonate, diethyl carbonate, DiMethyl Carbonate (DMC), etc.

The working principle of sodium-ion batteries is similar to that of lithium-ion batteries. When the battery is in the charging state, Na^+ is extracted from the cathode material, transported through the electrolyte, passes through the separator, and moves to the anode end to be inserted into the anode material. In the external circuit, the anode releases electrons, and these electrons flow back to the anode through the external circuit to complete the charging process [40]. Subsequently, in the discharging state, Na^+ is extracted from the anode material under the action of the external electric field, transported through the electrolyte, and re-inserted into the cathode material. Meanwhile, electrons migrate from the negative electrode to the positive electrode through the external circuit to maintain charge conservation. Under ideal charging and discharging conditions, the extraction and insertion of Na^+ between the cathode and anode materials will not damage the crystal structure of the materials. Taking the full-cell system composed of the layered metal oxide cathode material Na_xMO_2 and the hard carbon anode material as an example, its charging and discharging reaction formulas are shown in Eqs. (3.29)–(3.30) as follows:

$$\text{Cathode:} \quad Na_xMO_2 - xLi^+ + xe^- \rightleftharpoons Na_{1-x}MO_2 \tag{3.29}$$

$$\text{Anode:} \quad C + xNa^+ + xe^- \rightleftharpoons Na_xC_n \tag{3.30}$$

$$\text{Total:} \quad Na_xMO_2 + nC \rightleftharpoons Na_{1-x}MO_2 + Na_xC_n \tag{3.31}$$

Compared with lithium-ion batteries, sodium-ion batteries possess the following characteristics:

(1) Sodium has relatively large mineral reserves, and its production and manufacturing costs are lower, which is conducive to large-scale promotion.

(2) Sodium ions have a relatively large ionic radius, avoiding the difficulty in the desolvation process in the electrolyte caused by an overly small radius, and accelerating their diffusion kinetic process, which provides more possibilities for the promotion of Na^+ batteries in cold regions.

(3) Sodium-ion batteries possess a relatively wide voltage window, which also offers opportunities for the further development of high-energy density sodium/potassium-ion batteries.

3.4.4.2 *Key materials of sodium-ion batteries*

Sodium-ion batteries, as an eye-catching energy storage solution, are gradually demonstrating their potential for large-scale applications. Compared with lithium-ion batteries, sodium-ion batteries have more abundant sodium resources and lower production costs. In addition, similar to lithium-ion batteries, the core components of sodium-ion batteries also include cathode, anode, separator, electrolyte, current collector, etc. The cathode and anode are the most critical parts of sodium-ion batteries, which largely determine the main performances of sodium-ion batteries, such as cycle life, rate performance, and energy density.

In recent years, a series of cathode materials for sodium-ion batteries have received extensive attention, such as oxides, Prussian blue analogs, and polyanion-based materials. Among them, oxides mainly include layered oxides and tunnel-structured oxides. Polyanion-based materials include phosphates, fluorophosphate, pyrophosphate, and sulfates. Layered oxides have a periodic layered structure and a relatively high energy density, but most of them are prone to absorbing water or reacting with air. Tunnel-structured oxides have a unique S-shaped channel in their crystal structure, possess good rate performance, and are highly stable in air and water, yet their specific capacity is relatively small. Polyanion-based materials feature an open three-dimensional framework and exhibit favorable rate performance. Nevertheless, their electrical conductivity is suboptimal. To enhance both their electronic and ionic conductivities, carbon coating and doping techniques must be implemented. Nevertheless, the application of these methods inevitably results in a decline in their volumetric energy density. Prussian blue analog materials possess an open three-dimensional channel structure, which enables the rapid migration of Na^+ within the channel. They demonstrate good structural stability and rate performance. However, challenges exist, including the arduous task of eliminating crystal water and the dissolution of transition metal ions [41].

Ideal positive electrode materials for sodium-ion batteries have the following properties:

(1) High redox potential, which is convenient for the full cell to obtain a higher working voltage and improve the energy density of the battery.

(2) High mass-specific capacity and volume-specific capacity.

(3) Good structural stability in the electrolyte to ensure a longer cycle life of the battery.

(4) High electronic conductivity and ionic conductivity, with appropriate sodium ion diffusion channels and a low ion migration barrier.

(5) High air-structural stability to avoid property deterioration problems caused by storage.

(6) Safe, nontoxic, low-cost raw materials, and easy to prepare.

In the industry, currently, various technical paths of sodium-ion battery-positive electrode materials have been laid out by enterprises. The layered oxide technical path is relatively close to the existing ternary lithium battery system, has a balanced comprehensive performance, and will take the lead in industrialization. The Prussian blue technical path has a high energy density, and with the promotion of companies such as CATL, it is expected to be applied to energy storage scenarios with higher energy density requirements. However, the polyanion path has a relatively low maturity and a relatively slow development speed.

Currently, the widely studied negative electrode materials for sodium-ion batteries can be mainly divided into five categories: carbon-based materials, alloy-based materials, metal oxides and sulfides based on conversion reactions, titanium-based composite materials with an insertion mechanism, and organic composite negative electrode materials. Among them, carbon-based materials, due to their excellent cycle stability and small volume expansion, are currently the most promising candidate negative electrode materials for commercialization. Therefore, as a negative electrode material for sodium-ion batteries, it should preferably meet the following requirements [42]:

(1) Have more sodium storage sites and a relatively high sodium storage-specific capacity.

(2) During the sodium ion insertion and extraction process, the structural change should be as small as possible to ensure good cycle performance.

(3) Have good compatibility with the electrolyte, and no side reactions occur.

(4) Have relatively high ion mobility, electronic conductivity, good chemical stability, and thermodynamic stability.

(5) Be able to form a good SEI film with the electrolyte and stably cycle within a wide voltage window.

(6) Be environmentally friendly and economical.

The electrolyte is the medium located between the cathode and the anode. As a bridge between the cathode and anode, it undertakes the role of transporting Na^+ between the cathode and anode, allowing Na^+ to move during the charging and discharging process. The electrolyte of sodium-ion batteries can be divided into liquid electrolytes and solid electrolytes. Currently, the commonly used electrolyte for sodium-ion batteries is the carbonate electrolyte. Carbonates, a common type of organic electrolyte solvent for sodium-ion batteries, generally have a strong salt-dissolving ability. Common carbonate solvents for sodium-ion batteries include ethylene carbonate, propylene carbonate, diethyl carbonate, and dimethyl carbonate.

The electrolyte is the medium located between the positive and negative electrodes. As a bridge between the positive and negative electrodes, it undertakes the role of transporting Na^+ between the positive and negative electrodes, allowing Na^+ to move during the charging and discharging process. The electrolyte of sodium-ion batteries can be divided into liquid electrolytes and solid electrolytes. Currently, the commonly used electrolyte for sodium-ion batteries is the carbonate electrolyte. Carbonates, as a common type of organic electrolyte solvent for sodium-ion batteries, generally have a strong salt-dissolving ability. The commonly used carbonate solvents for sodium-ion batteries mainly include ethylene carbonate, propylene carbonate, diethyl carbonate, and dimethyl carbonate.

3.4.4.3 *Layered oxide-based positive electrode materials*

With the iteration of sodium-ion battery technology, the transformation from laboratory research to market applications has achieved certain results, and it is currently in a critical incubation period for industrialization. Looking at global research institutions, laboratories, and industrial enterprises, each sodium-ion battery material system has its own characteristics. Among them, the layered oxide positive electrode materials are regarded as candidate positive electrode materials for next-generation high-performance sodium-ion batteries.

(1) Structure and Industry Standards of Layered Oxide-Based Positive Electrodes

In 1980 Delmas et al. in France first reported the sodium storage capacity of the Na_xCoO_2 layered oxide positive electrode material. The general structural formula is Na_xTMO_2, where the transition metal positions can be occupied by common transition metal ions (such as Ti, V, Cr, Mn, Fe, Co, Ni, Cu, Zn, etc.) or alkaline earth metal ions (Li, Na, Mg, etc).

Based on the coordination configuration of Na^+ in TMO_6 and the stacking mode of oxygen, layered oxides are classified into different structures such as O3, O2, P3, and P2. Here, O and P respectively represent different polyhedral coordination environments of Na^+. O indicates that Na^+ is located at the center of the octahedral coordination, while P indicates that Na^+ is located at the center of the trigonal prism coordination. The numbers represent the smallest repeating units of TMO_6 in the stacking of oxygen ions. The crystal structure of the classic O3-$NaNi_{1/3}Fe_{1/3}Mn_{1/3}O_2$ is schematically depicted in Fig. 3.22. It shares the same crystal structure as α-$NaFeO_2$, and the corresponding space group is R-3m [44].

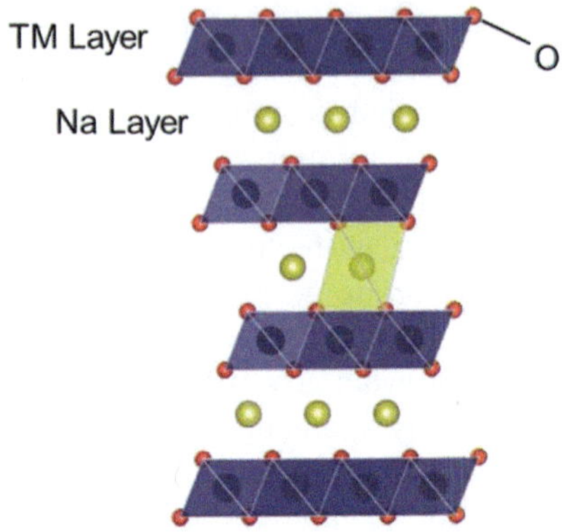

FIGURE 3.22 Schematic diagram of the crystal structure of the O3-$NaNi_{1/3}Fe_{1/3}Mn_{1/3}O_2$ positive electrode [43].

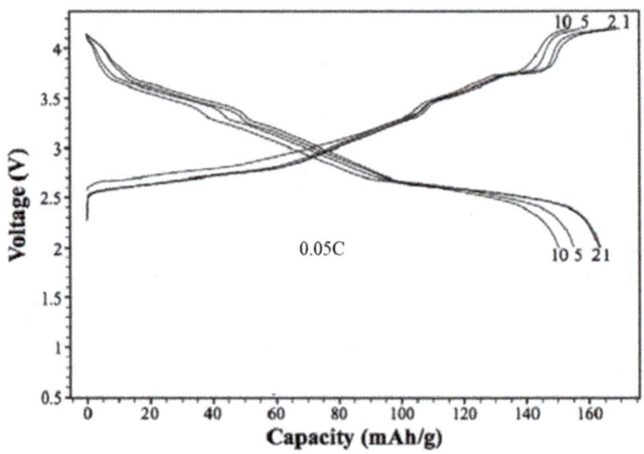

FIGURE 3.23 Charge and discharge curves of O3-$NaNi_{1/3}Fe_{1/3}Co_{1/3}O_2$ cathode at 0.05C [43].

(2) Electrochemical Behavior of Layered Oxide-Based Materials

Taking the classic O3-structured layered oxide $NaNi_{1/3}Fe_{1/3}Co_{1/3}O_2$ as an example, Fig. 3.23 shows the first and 50th charge and discharge curves of the $NaNi_{1/3}Fe_{1/3}Mn_{1/3}O_2$ cathode at a current density of 11.85 mA g^{-1}. As Fig. 3.23 showed, there are multiple voltage platforms during the charge and discharge process [44]. With the extraction and insertion of Na^+, the phase structure undergoes complex changes. When cycling within the voltage window of 2.0–4.2 V, a reversible specific capacity of approximately 165 mA h g^{-1} can be achieved, accompanied by a phase transition process of O3 $\rightarrow$ O′3 $\rightarrow$ P3 $\rightarrow$ P′3 $\rightarrow$ P3″. As the cycling process progresses, the $NaNi_{1/3}Fe_{1/3}Co_{1/3}O_2$ positive electrode exhibits a reversible specific capacity of around 150 mA h g^{-1} at the 10th cycle, suggesting that the complex phase transition phenomenon during the charge and discharge operations has a substantial influence on its sodium insertion and extraction processes. Hence, how to effectively enhance the structural reversibility of materials and alleviate the irreversible transformation process constitutes an important problem that urgently requires a solution for layered positive electrode materials.

(3) Problems and Development Directions of Layered Oxide Positive Electrodes

Layered transition metal oxide cathodes are regarded as candidate cathode materials for next-generation sodium-ion batteries due to their advantages, such as high capacity, high-rate performance, and low cost. However, the practical application of layered transition metal oxide positive electrodes is still restricted by low energy density and poor cycle performance, and this problem is particularly prominent under extreme temperature conditions.

On one hand, because the ionic radius of Na^+ is much larger than that of Li^+, the layered oxide cathode needs to undergo complex electrochemical behaviors and structural evolutions during the charging and discharging process. The deep de/intercalation procedure of sodium ions is liable to induce irreversible phase transitions and inevitably leads to significant expansion and contraction. This is an issue that must be avoided during the charging and discharging cycle and is one of the crucial factors that affect the cycle performance of layered metal oxides [45]. During the phase transition, the Na^+/vacancy ordering and the interlayer slippage of transition metals unavoidably give rise to volume expansion and contraction during the cycling process, thereby diminishing the structural stability and rate performance. At present, it has been discovered by researchers that optimization approaches, including ion doping, surface modification, and structural design, are capable of retarding the occurrence of complex phase transitions.

On the other hand, layered positive electrode materials that undergo electrochemical cycling in air are prone to deliquescence and have high air sensitivity. Their sample materials will experience volume expansion or even cracking. This is because layered transition metal oxide positive electrode materials are prone to react with water, oxygen, and CO_2 in humid air, and their structures are easily transformed into sodium-deficient phases, accompanied by the dissolution of Na^+ in the form of sodium carbonate and sodium bicarbonate, and the oxidation of transition metals, resulting in poor electrochemical performance. Therefore, many-layered materials require air protection during sample processing, increasing the manufacturing and transportation costs. To date, surface coating is a very effective and common method to improve the air stability of positive electrode materials, but there are only a few systematic studies on the mechanism behind its air instability.

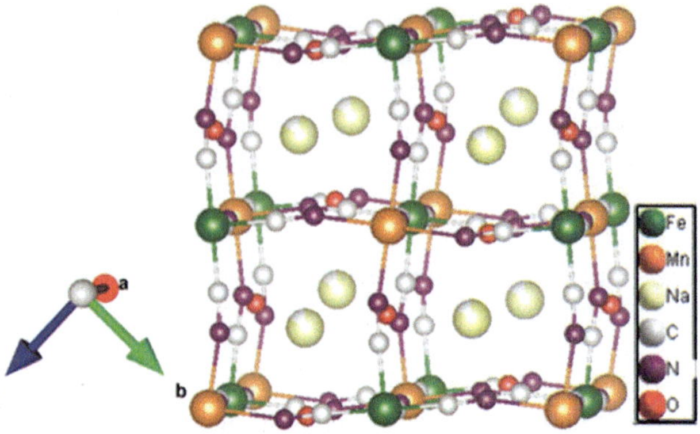

FIGURE 3.24 Schematic diagram of the crystal structure of the $Na_xMnFe(CN)_6$ positive electrode [46].

3.4.4.4 *Prussian blue-based positive electrode materials*

As a kind of sodium-ion battery positive electrode material favored for its low cost, Prussian blue analogs have become the focus of researchers due to their open three-dimensional structural framework and high theoretical capacity, in preparation for their large-scale applications. Nowadays, Prussian blue positive electrode materials are applied in a variety of completely different yet very promising fields, such as sewage treatment, catalysis, biosensors, and rechargeable batteries, and even used as medical nanomaterials for cancer treatment.

Prussian blue and its analogs Prussian Blue Analogs (PBAs) have a general structural formula of $Na_xM_A[M_B(CN)_6]_{1-y} \cdot zH_2O$ $(0 \leq x \leq 2,\ 0 \leq y \leq 1)$, where M is a transition metal element, such as Mn, Fe, Co, Ni, Cu, Zn, Cr, etc., and H_2O represents crystal water, including interstitial water and coordinated water. Hexacyanoferrate compounds possess an open-framework architecture, characterized by an abundance of redox-active sites and robust structural stability. Typically, these materials adopt a face-centered cubic crystal structure with a space group of Fm-3m. Taking the classic $Na_xMnFe(CN)_6$ positive electrode as an example, as shown in Fig. 3.24, the $Na_xMnFe(CN)_6$ positive electrode consists of a three-dimensional network of Fe–C≡N–Mn chains [46]. Mn and Fe are respectively located at the vertices of the face-centered cube, and Na^+ is located in the cubic interstitial sites. This crystal structure is conducive to the insertion and extraction of Na^+ and has a relatively small volume change during the cycling process, thus endowing it with good cycling stability.

At present, within the commercial development of energy-storage batteries, long-term cycling performance persists as a crucial criterion for all intercalation-type positive electrode materials. Even though certain Prussian blue compound materials under development demonstrate relatively remarkable cycling performance, when compared to the outstanding cycling performance of currently commercialized LCO and LFP positive electrodes, their cycling endurance is insufficient. Apart from the structural degradation induced by lattice defects, the dissolution and loss of transition metal ions represent a nonnegligible factor contributing to the long-term cycling instability of Prussian blue compound materials. Currently, a diverse array of surface modification methods is being explored and developed to ensure that Prussian blue compound materials can maintain a stable surface composition and structure throughout the cycling process.

Moreover, a safety concern associated with Prussian blue compound materials pertains to the potential toxicity arising from the release of cyanide. Extensive toxicological investigations have demonstrated that Prussian blue itself poses no harm to the environment and human body. Nevertheless, under high-temperature or strongly acidic conditions, it decomposes to generate free cyanide CN^-, which is highly toxic and polluting. Consequently, utmost care must be taken in the disposal of chemical waste. It is advisable to thoroughly oxidize CN^- into nontoxic products (such as CO_2, N_2, etc.) by employing an excess of oxidants like H_2O_2 and $NaClO_4$ prior to discarding the reaction waste. In contrast to the extensively studied transition metal oxide intercalation – positive electrodes, the research and development of Prussian blue compound materials, specifically Na^+ intercalation – positive electrodes, experienced a resurgence several years ago. As a result, there is a substantial body of fundamental and technical research, covering aspects such as material design, structural systems, and synthesis techniques, aimed at obtaining superior Prussian blue-positive electrodes.

3.4.4.5 Polyanion-type cathode materials

Polyanion-type compounds are compounds with a three-dimensional network structure formed by the connection of polyanion polyhedra and transition metal ion polyhedra through strong covalent bonds, and sodium ions occupy the interstitial positions. Its chemical formula is $Na_xM_y(X_aO_b)_zZ_w$, where M is one or several metal elements such as V, Ti, Mn, Fe, Ni, Al, etc.; X is one of the elements such as Si, S, P, B, etc.; and Z is mainly the F element. At present, the polyanion-type cathode materials that are commonly under investigation encompass phosphate, pyrophosphate, and mixed polyanion material systems.

The $Na_3M_2(PO_4)_3$ material with an open sodium superionic conductor framework structure (NASICON structure) containing variable valence transition metal elements (such as M = V, Fe, Cr, Mn, Ni, Cu, etc.) is a typical phosphate material. Among them, the sodium vanadium phosphate NVP is a typical representative and is also one of the most widely studied polyanion-type materials at present.

The NASICON-structured sodium vanadium phosphate material has high ionic conductivity and excellent chemical and electrochemical stability. Taking the classic $Na_3V_2(PO_4)_3$ cathode as an example, as shown in Fig. 3.25 [47],

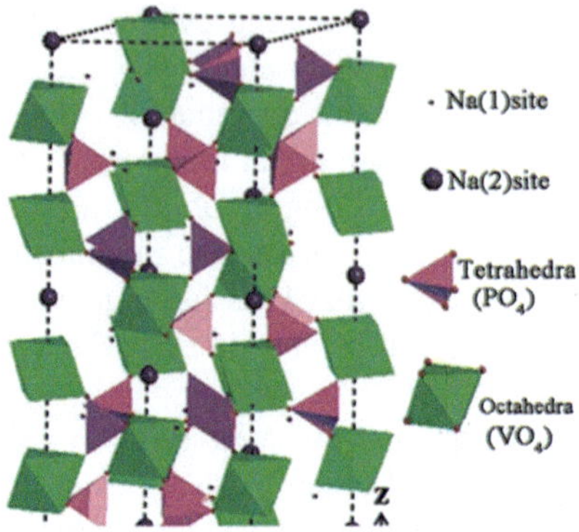

FIGURE 3.25 Schematic diagram of the crystal structure of the polyanion $Na_3V_2(PO_4)_3$ cathode [47].

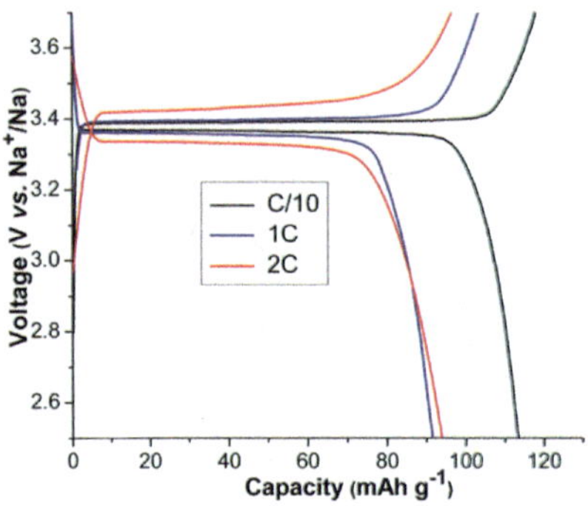

FIGURE 3.26 Charge and discharge curves of sodium vanadium phosphate cathode [48].

its crystal structure is trigonal, with a space group of R-3c. The vertices of VO_6 octahedra and PO_4 tetrahedra are interconnected, establishing a three-dimensional framework with two different Na crystallographic sites. Among them, the Na1 site with six-fold coordination is occupied by one sodium ion, and this sodium ion cannot move freely, while the Na2 site is occupied by two sodium ions, and the sodium ions at this site can be extracted and inserted during the charging and discharging process.

The sodium vanadium phosphate cathode's charge and discharge curves in a half-cell arrangement are displayed in Fig. 3.26 [48]. As illustrated, it has a long voltage plateau at around 3.4 V, indicating that the reaction mechanism of this material is a typical two-phase reaction, corresponding to the transition between V^{3+} and V^{4+}. At this time, one sodium ion located in the VO_6 octahedron will be extracted/inserted, and simultaneously, its variable valence V ion will change its valence state within the material system to maintain the overall charge conservation.

The theoretical specific capacity of the sodium vanadium phosphate cathode is 117 mA h g^{-1}, and it has high cycle stability. During the two-phase transition process of this material, the volume change is relatively minor, and the cycle reversibility is relatively high. The core of the key issues for polyanion-type cathodes mainly resides in their intrinsically low electronic conductivity, which gives rise to poor sodium ion diffusion kinetics. At present, industry predominantly reduces the size of polyanion materials to the nanoscale or coats and composites them with conductive carbon materials and downsizes them to the nanoscale level. This approach shortens the sodium ion diffusion length, improves the electrical contact of materials, and mitigates the phenomenon of particle agglomeration.

Simultaneously, enhancing the voltage of polyanion-type materials to achieve a higher energy density is also a focal point. Currently, the corresponding strategies involve using F atoms to substitute O atoms for the synthesis of sodium vanadium fluorophosphate or forming a mixed polyanion system by blending phosphate and pyrophosphate to strengthen the inductive effect and elevate the voltage plateau. Concurrently, the doping and substitution of metal cations are employed to activate the redox of high-valence transition metals, thereby increasing the reaction voltage. At present, in vanadium-based polyanion materials, the incorporation of vanadium elements leads to relatively high production costs and environmental toxicity during material preparation. Consequently, iron-based and manganese-based polyanion materials, which have the advantage of low cost, hold broad market prospects in large-scale energy storage.

3.4.4.6 *Hard carbon anode materials*

Like the anode materials of lithium-ion batteries, anode materials for sodium-ion batteries mainly include carbon-based, titanium-based, organic, and alloy-based anode materials. Among these materials, hard carbon materials possess the advantages of rich resources, low cost, favorable stability, and high safety. Currently, they are the sole commercialized anode materials utilized in sodium-ion batteries.

Currently, various manufacturers are screening and researching the precursor raw materials of hard carbon materials, mainly focusing on the following aspects: (1) Streamlining the production process and optimizing the cost; (2) Modulating the pore structure of materials to enhance the specific capacity of materials; (3) Augmenting the compacted density to further boost the energy density of battery cells.

Hard carbon is usually obtained by high-temperature pyrolysis of precursor carbon materials and retains the structural characteristics of the precursors to a large extent. Different precursors have unique microstructures (different numbers and structures of pore diameters, pores, and defects) and various heteroatoms (such as nitrogen, phosphorus, sulfur, etc.).

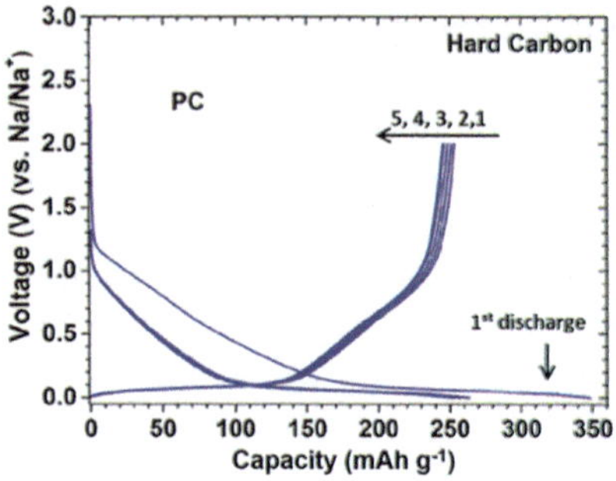

FIGURE 3.27 Charge and discharge curves of the hard carbon anode [49].

The biomass hard carbon anode's charge and discharge curves at 0.01 and 2.0 V are displayed in Fig. 3.27 [49]. The graphic illustrates how the sodium insertion process of hard carbon has characteristics with the lithium insertion process, such as a low potential plateau below 0.1 V and a sloping potential region above 0.1 V. Among them, the first region represents the insertion of sodium ions between carbon layers, while the second step represents the adsorption process of sodium ions in carbon micropores. Generally speaking, the currently researched hard carbon materials have a high reversible specific capacity of up to 300 mA h g^{-1} and a low sodium storage voltage of 0.1 V.

At present, regarding the improvement of the sodium-storage performance of hard carbon materials, the research concepts and modification methods predominantly adopted by researchers can be categorized into three aspects: heteroatom doping, microstructure regulation and modification, and electrolyte regulation. Among these, doping modification represents a frequently utilized modification approach in the field of electrode materials. The doping of nonmetallic heteroatoms (such as N, B, P, and S) can remarkably enhance the performance of anodes. This enhancement encompasses the improvement of the specific charge-discharge capacity and rate performance of anodes. Research findings have indicated that the doping of N and B into the carbon structure can significantly boost the sodium-ion transport and charge-transfer processes. Moreover, the incorporation of electrochemically active elements S and P can engage in reversible reactions with sodium, thereby providing additional sodium ion adsorption sites. Additionally, hard carbon materials possess a rich pore structure suitable for sodium ion filling. Through screening precursors with specific pore structures and devising a rational pyrolysis process, the internal pore structure (including pore size and the ratio of open pores to closed pores) of hard carbon materials can be effectively regulated. This regulation can enhance the first cycle coulombic efficiency and the capacity in the sloping region. Adjusting the formulation of the electrolyte can also minimize side reactions on the material surface, facilitate the formation of a stable SEI film, and improve the coulombic efficiency of the materials.

References

[1] R. Hu, H. Liu, F. Meng, et al., Electrochemical Energy Storage Power Sources: Design and Applications, Chemical Industry Press, 2025 (in Chinese).

[2] S. Koohi-Fayegh, M.A. Rosen, A review of energy storage types, applications and recent developments, Journal of Energy Storage 27 (2020) 101047.

[3] Y. Zhang, C. Zhou, J. Yang, S. Xue, H. Gao, X. Yan, et al., Advances and challenges in improvement of the electrochemical performance for lead-acid batteries: a comprehensive review, Journal of Power Sources 520 (2022) 230800.

[4] V. Arun, R. Kannan, S. Ramesh, M. Vijayakumar, P.S. Raghavendran, M. Siva Ramkumar, et al., Review on Li-ion battery vs nickel metal hydride battery in EV, Advances in Materials Science and Engineering 2022 (2022) 1–7.

[5] T. Kim, W. Song, D. Son, L.K. Ono, Y. Qi, Lithium-ion batteries: outlook on present, future, and hybridized technologies, Journal of Materials Chemistry A 7 (2019) 2942–2964.

[6] A.Z. Weber, M.M. Mench, J.P. Meyers, P.N. Ross, J.T. Gostick, Q. Liu, Redox flow batteries: a review, Journal of Applied Electrochemistry 41 (2011) 1137–1164.

[7] H.D. Yoo, E. Markevich, G. Salitra, D. Sharon, D. Aurbach, On the challenge of developing advanced technologies for electrochemical energy storage and conversion, Materials Today 17 (2014) 110–121.

[8] S.S. Zhang, Status, opportunities, and challenges of electrochemical energy storage, Frontiers in Energy Research 1 (2013).

[9] Scientific Background on the Nobel Prize in Chemistry 2019.

[10] K. Franke, J.F. Garcia, C. Kleinschmitt, F. Sensfuß, Assessing worldwide future potentials of renewable electricity generation: installable capacity, full load hours and costs, Renewable Energy 226 (2024) 120376.

[11] Q. Abbas, M. Mirzaeian, M.R.C. Hunt, P. Hall, R. Raza, Current state and future prospects for electrochemical energy storage and conversion systems, Energies 13 (2020) 5847.

[12] T.S. Mathis, N. Kurra, X. Wang, D. Pinto, P. Simon, Y. Gogotsi, Energy storage data reporting in perspective—guidelines for interpreting the performance of electrochemical energy storage systems, Advanced Energy Materials 9 (2019).

[13] C. Liu, Z.G. Neale, G. Cao, Understanding electrochemical potentials of cathode materials in rechargeable batteries, Materials Today (Kidlington, England) 19 (2016) 109–123.

[14] K. Chen, D. Xue, Materials chemistry toward electrochemical energy storage, Journal of Materials Chemistry A 4 (2016) 7522–7537.

[15] D.A.J. Rand, P.T. Moseley, 3 – Lead–acid battery fundamentals, in: J. Garche, E. Karden, P.T. Moseley, D.A.J. Rand (Eds.), Lead-Acid Batteries for Future Automobiles, Elsevier, Amsterdam, 2017, pp. 97–132.

[16] N. Vangapally, T.R. Penki, Y. Elias, S. Muduli, S. Maddukuri, S. Luski, et al., Lead-acid batteries and lead–carbon hybrid systems: a review, Journal of Power Sources 579 (2023) 233312.

[17] P.P. Lopes, V.R. Stamenkovic, Past, present, and future of lead-acid batteries, Science 369 (2020) 923–924.

[18] X. Zhang, W. Chen, L. Yang, J. Wang, X. Sun, Y. Zhao, et al., PEDOT-coated rice husk-based activated carbon: boosting lead-acid battery performance, Journal of Energy Storage 90 (2024) 111771.

[19] A. Moncada, M.C. Mistretta, S. Randazzo, S. Piazza, C. Sunseri, R. Inguanta, High-performance of PbO2 nanowire electrodes for lead-acid battery, Journal of Power Sources 256 (2014) 72–79.

[20] S.G. Zhou, M.L. Tian, Y. Xu, C. Zhang, Y. Cao, First-principles study of the elastic, thermal and optical properties of α-PbO2 and β-PbO2, Indian Journal of Physics 96 (2022) 3449–3460.

[21] X. Zou, Z. Kang, D. Shu, Y. Liao, Y. Gong, C. He, et al., Effects of carbon additives on the performance of negative electrode of lead-carbon battery, Electrochimica Acta 151 (2015) 89–98.

[22] W. Zhang, H. Lin, H. Lu, D. Liu, J. Yin, Z. Lin, On the electrochemical origin of the enhanced charge acceptance of the lead–carbon electrode, Journal of Materials Chemistry A 3 (2015) 4399–4404.

[23] O.C. Esan, et al., Modeling and simulation of flow batteries, Advanced Energy Materials 10 (31) (2020) 2000758.

[24] Y.K. Zeng, X.L. Zhou, L. An, L. Wei, T.S. Zhao, A high-performance flow-field structured iron-chromium redox flow battery, Journal of Power Sources 324 (2016) 738–744.

[25] Z. Xu, Q. Fan, Y. Li, J. Wang, P.D. Lund, Review of zinc dendrite formation in zinc bromine redox flow battery, Renewable and Sustainable Energy Reviews 127 (2020) 109838.

[26] K. Lourenssen, J. Williams, F. Ahmadpour, R. Clemmer, S. Tasnim, Vanadium redox flow batteries: a comprehensive review, Journal of Energy Storage 25 (2019) 100844.

[27] K. Wang, J. Wan, Y. Xiang, J. Zhu, Q. Leng, M. Wang, et al., Recent advances and historical developments of high voltage lithium cobalt oxide materials for rechargeable Li-ion batteries, Journal of Power Sources 460 (2020) 228062.

[28] J.B. Goodenough, K. Park, The Li-ion rechargeable battery: a perspective, Journal of the American Chemical Society 135 (2013) 1167–1176.

[29] A. Pražanová, Z. Plachý, J. Kočí, M. Fridrich, V. Knap, Direct recycling technology for spent lithium-ion batteries: limitations of current implementation, Batteries 10 (2024) 81.

[30] T. Zhang, D. Li, Z. Tao, J. Chen, Understanding Electrode Materials of Rechargeable Lithium Batteries via DFT Calculations, Progress in Natural Science, vol. 23, Materials International, 2013, pp. 256–272.

[31] G. Zhang, et al., Lithium iron phosphate and layered transition metal oxide cathode for power batteries: attenuation mechanisms and modification strategies, Materials 16 (17) (2023) 5769.

[32] H. Wang, L. Liu, R. Wang, D. Zhang, L. Zhu, S. Qiu, et al., Design and synthesis of high performance $LiFePO_4$/C nanomaterials for lithium ion batteries assisted by a facile H^+/Li^+ ion exchange reaction, Journal of Materials Chemistry A 3 (2015) 8062–8069.

[33] B. Zhang, X. Wang, S. Wang, Y. Li, L. Chen, H. Jiao, et al., High-energy-density lithium manganese iron phosphate for lithium-ion batteries: progresses, challenges, and prospects, Journal of Energy Chemistry 100 (2025) 1–17.

[34] C.M. Julien, A. Mauger, NCA, NCM811, and the route to Ni-richer lithium-ion batteries, Energies 13 (2020) 6363.

[35] A.D. Jara, A. Betemariam, G. Woldetinsae, J.Y. Kim, Purification, application and current market trend of natural graphite: a review, International Journal of Mining Science and Technology 29 (2019) 671–689.

[36] P. Lin, J. Wu, W. Liu, Green and facile synthesis of few-layer graphene via liquid exfoliation process for Lithium-ion batteries, Scientific Reports 8 (1) (2018) 9766.

[37] L. Sun, Y. Liu, R. Shao, J. Wu, R. Jiang, Z. Jin, Recent progress and future perspective on practical silicon anode-based lithium ion batteries, Energy Storage Materials 46 (2022) 482–502.

[38] A. Paravannoor, D. Panoth, P. Pattathil, Silicon anode design for Li ion batteries: synergic effects of Ag nanoparticles and ionic liquid electrolytes, Chemical Engineering Journal Advances 4 (2020) 100037.

[39] T. Wang, D. Su, D. Shanmukaraj, T. Rojo, M. Armand, G. Wang, Electrode materials for sodium-ion batteries: considerations on crystal structures and sodium storage mechanisms, Electrochemical Energy Reviews 1 (2018) 200–237.

[40] C. Peng, X. Xu, F. Li, L. Xi, J. Zeng, X. Song, et al., Recent progress of promising cathode candidates for sodium-ion batteries: current issues, strategy, challenge, and prospects, Small Structures 4 (2023).

[41] X. Wan, Y. Li, S. Chen, W. Duan, W. Lei, Cathode modification of sodium-ion batteries for improved energy density: a review, Advanced Sustainable Systems 8 (2024).

[42] S. Tan, H. Yang, Z. Zhang, X. Xu, Y. Xu, J. Zhou, et al., The progress of hard carbon as an anode material in sodium-ion batteries, Molecules 28 (2023) 3134.

[43] P.F. Wang, Y. You, Y.X. Yin, Y.G. Guo, Layered oxide cathodes for sodium-ion batteries: phase transition, air stability, and performance, Advanced Energy Materials 8 (2018).
[44] P. Vassilaras, A.J. Toumar, G. Ceder, Electrochemical properties of NaNi1/3Co1/3Fe1/3O2 as a cathode material for Na-ion batteries, Electrochemistry Communications 38 (2014) 79–81.
[45] F. Wei, Q. Zhang, P. Zhang, W. Tian, K. Dai, L. Zhang, et al., Review—research progress on layered transition metal oxide cathode materials for sodium ion batteries, Journal of the Electrochemical Society 168 (2021) 50524.
[46] Y. Tang, W. Li, P. Feng, M. Zhou, K. Wang, Y. Wang, et al., High-performance manganese hexacyanoferrate with cubic structure as superior cathode material for sodium-ion batteries, Advanced Functional Materials 30 (2020).
[47] W. Song, X. Ji, Z. Wu, Y. Zhu, Y. Yang, J. Chen, et al., First exploration of Na-ion migration pathways in the NASICON structure Na3V2(PO4)3, Journal of Materials Chemistry A 2 (2014) 5358.
[48] L. Si, Z. Yuan, L. Hu, Y. Zhu, Y. Qian, Uniform and continuous carbon coated sodium vanadium phosphate cathode materials for sodium-ion battery, Journal of Power Sources 272 (2014) 880–885.
[49] J. Wang, L. Xi, C. Peng, X. Song, X. Wan, L. Sun, et al., Recent progress in hard carbon anodes for sodium-ion batteries, Advanced Engineering Materials 26 (2024).

Chapter 4

Energy storage battery systems and recycling

Gui Chen, Zihan Xiong, Yuhao Xia, Haotian Liang, Hui Wang, and Renzong Hu
School of Materials Science and Engineering, South China University of Technology, Guangdong Provincial Key Laboratory of Advanced Energy Storage Materials, Guangzhou, China

Contents

Modern Energy Storage. https://doi.org/10.1016/B978-0-32-395419-8.00010-6

4.1 Battery energy storage system

As the core of an electrochemical energy storage system, batteries are required to meet two major categories of electrochemical performance indicators, namely energy and power. Electrochemical performance indicators are obtained by conducting charge and discharge tests and storage under different conditions, and recording the variation relationships of parameters such as current and voltage over time [1]. Based on the characteristics of electrochemical performance indicators, suitable batteries can be configured for energy-type/power-type energy storage systems for application in different scenarios.

Apart from the performance of the battery module itself, the performance of an energy storage system in use is also closely related to the functions of the battery management system and the PCS it employs. The battery management system of an energy storage power station is the brain of the energy storage battery system. It is mainly used to monitor the energy storage batteries in real time to ensure the safe and reliable operation of the battery system. It is of great significance for preventing battery overcharging/discharging, improving battery utilization, maintaining batteries, and prolonging battery life. The power conversion system is the power interface connecting the energy storage system with the power grid or AC loads. It can control the bidirectional flow of energy between the power grid and the energy storage units to meet the requirements of power control accuracy and the response speed for rapid charge and discharge conversion. The power conversion system determines the power quality and dynamic characteristics of the electrical energy output by the battery system to the outside world and also affects the safety and service life of the batteries to a large extent.

4.1.1 Battery energy storage system

The main internal equipment of a Battery Energy Storage System (BESS) includes batteries, Battery Management Systems (BMS), PCS, local controllers, power distribution units, prefabricated cabins, and other auxiliary equipment such as temperature control and fire protection. Under the unified management of the local controller, it can operate independently or receive instructions from an external Energy Management System (EMS) to complete energy scheduling and power control, thus realizing safe and efficient operation.

4.1.2 Basic characteristics of battery energy storage systems

A battery energy storage system is a relatively complex integrated power unit constructed by integrating relevant equipment such as energy storage batteries, energy storage battery management systems, power conversion devices, local controllers, power distribution systems, temperature, and fire safety systems according to certain application requirements. Its basic characteristics include [2]:

(1) The internal equipment has clear divisions of labor and is also interconnected with each other. Under the premise of safety, high efficiency, and long service life, they jointly achieve energy, power, and voltage control at the grid connection point or output port.

(2) The safety and service life of the energy storage batteries largely determine the safety and service life of the entire system. Moreover, they have strict technical requirements for the working environment, which are key aspects to focus on when designing the internal system.

(3) The power conversion device is a crucial node for the entire energy storage system to conduct power exchanges with the outside world. Its performance directly reflects the working mode, control accuracy, response speed, and grid connection friendliness of the battery energy storage system, and also affects the most intuitive usage experience of customers with the energy storage system in a short period.

(4) The energy storage batteries, power conversion devices, and equipment such as air conditioners and fire protection are each equipped with independent controllers to achieve self-operation, alarm, or protection. The realization of system functions, the linkage and coordination between equipment, start-stop, and fault protection operations, external communication, and effective information transmission are completed by the local controller so that the energy storage system can participate in grid scheduling as a whole or achieve the application goals of the project.

(5) As a power execution unit that is unified externally and autonomous internally, the energy storage system accepts scheduling and executes power or mode control instructions from the upper energy management system. Therefore it should have abundant external communication interfaces and flexible and diversified working modes. By means of functions like energy transfer in accordance with demand, rapid power ascension, and stable voltage regulation, it is capable of enhancing the overall operational efficacy of application scenarios such as power generation, power grid, and load, and thus manifesting its value.

(6) Control and management are the keys to the value realization of the energy storage system, which largely depends on the understanding of the original system, control, or development direction of the application field by the system integrator.

4.1.3 Key equipment of battery energy storage systems

The key equipment included in a battery energy storage system is battery cells, battery management systems, and power conversion systems.

4.1.3.1 Battery cells

The batteries of a battery energy storage system consist of several parallel or series-connected battery units, and each battery unit is a battery cell. Among

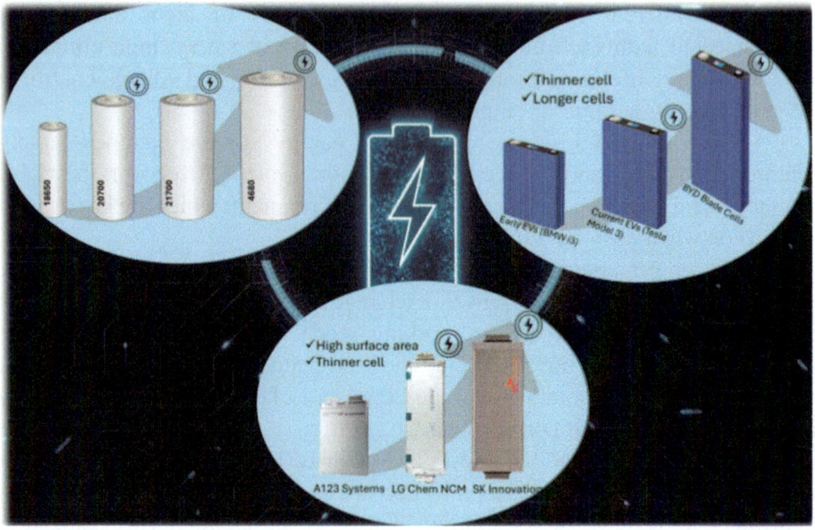

FIGURE 4.1 Appearance of energy storage Li-ion battery cells [3].

them, the energy storage lithium-ion battery cells can be mainly divided into three forms in appearance: square hard shell, cylindrical hard shell, and soft package. Because the capacity of each battery is limited, the high-capacity requirements for applications are met by combining multiple batteries to form an energy storage system. Taking the square hard-shell Li-ion battery as an example, several battery cells are first connected in series and parallel to form a battery module (also called a battery pack), and then multiple battery modules are connected in series to form a battery cluster for use. The schematic schematics of the various energy storage Li-ion battery cells are displayed, respectively, as seen in Fig. 4.1 [3].

Generally speaking, the specification information of batteries is marked on the product's appearance or nameplate in the form of easily recognizable and readable codes or texts. Fig. 4.2 takes lithium-ion batteries for electric energy storage as an example, and the identification of specifications complies with the following rules:

Example 1. For Li-ion batteries used in electric energy storage, a hard-shell square battery cell with lithium iron phosphate as the cathode material, graphite as the anode material, and a liquid electrolyte. The nominal voltage is 3.2 V, the rated charging power is 80 W, the rated discharging power is 160 W, the rated charging energy is 320 Wh, the rated discharging energy is 300 Wh, and the model number is A1B2C3. The code is EES-LIB-LFP/C-L-HS-Cell_3.2 V-80 W-160 W-320 Wh-300 Wh-A1B2C3.

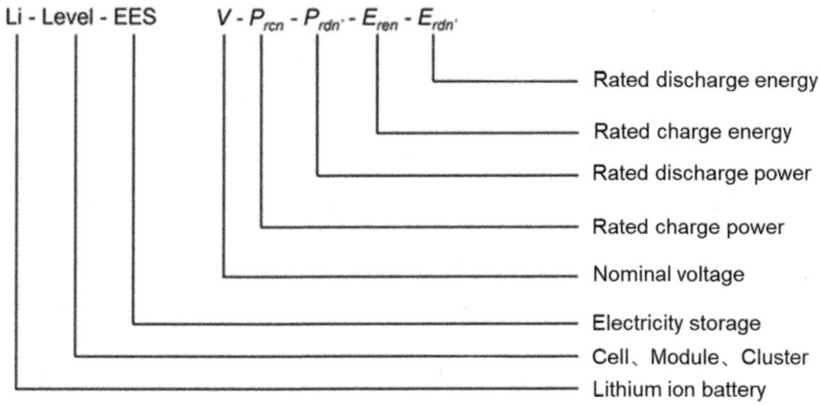

FIGURE 4.2 Specification information for Li-ion batteries for power storage.

Example 2. For Li-ion batteries used in electric energy storage, an air-cooled battery module composed of hard-shell cylindrical battery cells with Li iron phosphate as the cathode material, graphite as the anode material, and a solid electrolyte. The nominal voltage is 48 V, the rated charging power is 1.5 kW, the rated discharging power is 3 kW, the rated charging energy is 6 kWh, and the rated discharging energy is 5.8 kWh.

Example 3. For Li-ion batteries used in electric energy storage, a liquid-cooled battery cluster composed of soft package battery cells with Li iron phosphate as the cathode material, lithium titanate as the anode material, and a solid-liquid hybrid electrolyte. The nominal voltage is 650 V, the rated charging power is 250 kW, the rated discharging power is 500 kW, the rated charging energy is 1000 kWh, and the rated discharging energy is 950 kWh.

4.1.3.2 Battery management system

To meet the application requirements of large-scale energy storage in practical scenarios, it is necessary to connect multiple stacks or cells in parallel and in series to boost the voltage and output power. Beyond the parameter characteristics of battery modules, the performance of energy-storage batteries is intimately associated with the functions of the applied battery management system. The battery management system plays a crucial role in safeguarding battery safety, enhancing battery utilization, and prolonging battery lifespan. As depicted in Fig. 4.3, which presents a schematic diagram of the composition of an energy-storage system, the battery management system of the batteries within an energy-storage power station functions as the "control center" of the energy-storage battery system [4]. Its primary functions include real-time monitoring, fault diagnosis, state-of-charge estimation, short-circuit protection, leakage detection, and display alarm for energy-storage batteries. These functions ensure

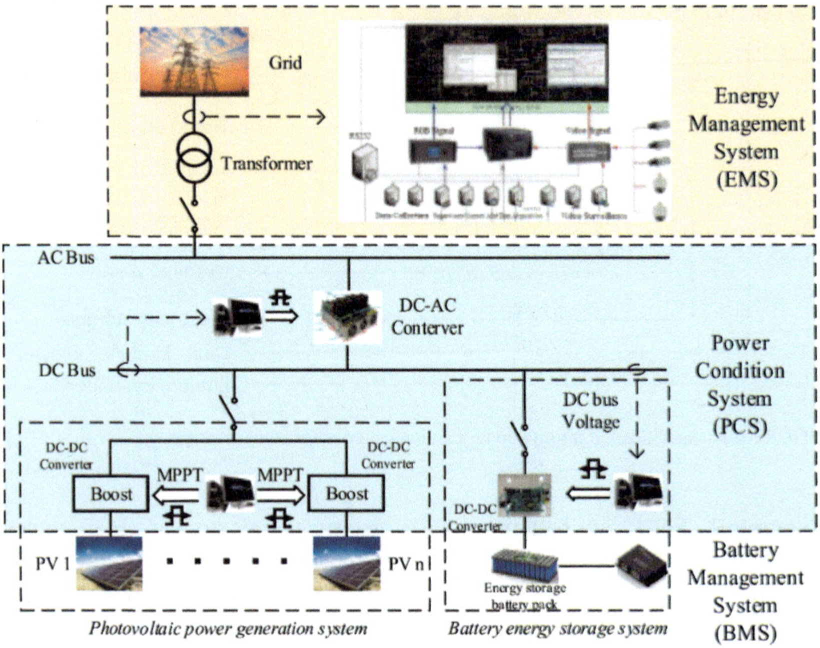

FIGURE 4.3 Schematic diagram of the energy storage system [4].

the safe and reliable operation of the battery system and make it an essential component of the entire energy-storage system.

The battery management system is capable of conducting real-time monitoring and collection of the state parameters of battery modules. It undertakes essential calculations and processing of the relevant state parameters to derive additional system parameters, and then effectively controls the battery system in accordance with specific control strategies. Simultaneously, the battery management system can communicate with other external devices (such as converters, energy management units, and fire-protection systems) via its communication interfaces, analog/digital input and output interfaces. This interaction enables the formation of a coordinated operation within the entire energy storage system. By integrating all system components and establishing reliable physical and logical connections, the battery management system can efficiently and reliably execute the monitoring function of the entire energy storage system.

(1) Introduction to the Functions of the Battery Management System

The main functions of the energy storage battery management system include state monitoring and evaluation, cell balancing, control and protection, communication, and log recording. Its structure and main functions are shown in Fig. 4.4 [4].

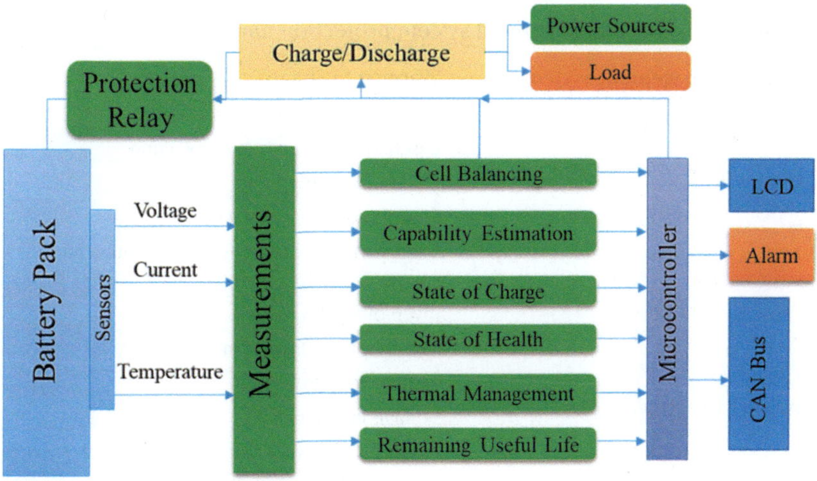

FIGURE 4.4 Schematic diagram of the main functions of the battery management system [4].

State monitoring and evaluation is the method of evaluating battery-level related parameters by analyzing and testing data such as the voltage and current of each battery cluster, cell voltage, total system current, the temperature of battery packs or cells, and ambient temperature. It mainly includes state of charge, state of health, battery internal resistance, and capacity. Among them, state monitoring is the most fundamental function of the battery management system and also the basis for subsequent balancing, protection, and communication with external information. Parameter evaluation, on the other hand, is a relatively complex function possessed by the battery cluster management system. Table 4.1 is an introduction to the protection functions of the battery cluster management system and the system management system.

Control and protection refer to the control functions of the battery cluster management system and the system management system. It is principally demonstrated by the normal connection and disconnection of battery packs by operating the contactors within the switch box. The protection function chiefly aims to prevent the voltage, current, and temperature of cells from exceeding the safety thresholds by actively halting, diminishing the battery current, or providing feedback to halt and reduce the battery current.

The battery system realizes external communication through the system management system. The information it transmits externally not only includes the aforementioned state monitoring or estimation information, but also can include relevant statistical information or security information, such as the voltage, current, or state of charge of the battery system.

Log records can use built-in storage devices in the system management system, such as simple SD cards, to store the necessary key data of battery operation, including voltage, current, state of charge/state of health, maximum

TABLE 4.1 Battery management system protection function.

Battery Management System	Warning Message
Battery Cluster Management System	Charge Overcharge Protection
	Discharge Undercurrent Protection
	Overtemperature Protection
	Low-Temperature Protection
	Battery Management System Communication Failure
System Management System	Overvoltage Protection
	Undervoltage Protection
	Overcurrent Protection
	Voltage Imbalance
	Uneven Cell Temperature
	System DC Contactor Fault Protection
	System DC Sensor Failure
	System Management System Communication Failures

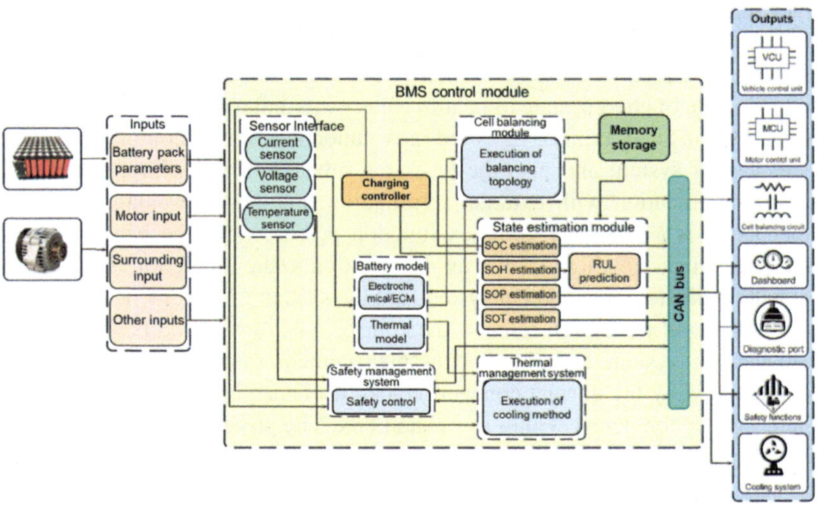

FIGURE 4.5 Typical architecture diagram of the battery management system [5].

and minimum single cell voltages, lowest and highest temperatures, as well as alarm and error information.

(2) Architecture of the Battery Management System

The battery management system's architecture is schematically depicted in Fig. 4.5 [5]. As shown in the figure, a complete battery management system of an energy storage system consists of three parts: the Battery Management Unit (BMU), the Battery Cluster Management Unit (BCU), and the Battery Array Management Unit (BAU). The BMU is responsible for the management of individual battery cells and battery modules, integrating functions such as the collection of information, such as the voltage and temperature of individual battery cells, cell balancing, information uploading, and thermal management. The BCU is responsible for managing all the BMUs in a battery cluster. Meanwhile, it has functions such as a current collection of the battery cluster, total voltage collection, insulation resistance detection, and state of charge estimation. When the state of the battery pack is abnormal, it drives to disconnect the high-voltage direct current contactor to make the battery cluster exit from operation and ensure the safety of battery use. The BAU performs numerical calculations, performance analysis, and data storage on the data uploaded by the BMC and the BCU, and also conducts information interaction with the power conversion system for energy storage and the monitoring background.

Battery Management Unit: The battery management unit serves as the bottom-level unit of the energy storage battery management system and plays a decisive role in the safe use and lifespan extension of battery modules. It can conduct real-time monitoring of the voltage, balancing current, and temperature of the batteries under its jurisdiction and report them, ensuring the healthy, safe, and stable operation of the battery cluster. The battery management unit is responsible for the management of battery modules and has functions such as collecting information on the voltage and temperature of individual batteries within the group, balancing batteries within the group, uploading information, and thermal management. When a fault is detected, the battery management unit can issue alarms for abnormal phenomena such as excessively high single-cell voltage, excessively low single-cell voltage, differential pressure of single-cell voltage, excessively high temperature, excessively low temperature, excessive temperature difference, charging current, and discharging current. Moreover, the alarm levels can all be set openly.

Battery Cluster Management Unit: The main control unit is an intermediate component of the energy storage battery management system. On the one hand, it collects information such as the voltage, temperature, and balancing status of individual batteries uploaded by all battery management units. On the other hand, it collects information such as the charging and discharging current, terminal voltage, and insulation resistance of the entire battery cluster, completes the estimation of the state of charge and state of health, and uploads the integrated information from the battery management units to the battery array management unit. Based on this, it realizes the charging and discharging management, thermal management, individual cell balancing management, and fault alarm for the battery pack [6].

The main functions of the battery cluster management unit include: 1. Collection of the terminal voltage of the battery cluster; 2. Detection of the charging and discharging current of the battery cluster; 3. Detection of the insulation resistance; 4. Charging and discharging management of the battery cluster; 5. Monitoring the temperature of individual cells during the system's charging and discharging processes, issuing alarms when conditions such as excessively high individual cell temperature, excessively low individual cell temperature, or excessive temperature difference among individual cells. When a second-level alarm occurs, it actively reports the alarm information. When a first-level alarm occurs, the system automatically disconnects the contactor 6. Real-time and dynamic estimation of the state of charge and state of health; 7. Fault diagnosis and alarm for the battery cluster; 8. Security protection for various abnormal and faulty situations; 9. A powerful system self-check function to ensure the normal operation of the system itself [7].

Battery Array Management Unit: The battery array management unit is the top-level control unit of the energy storage battery management system. As a core component, the battery array management unit receives all the battery state information of the entire battery array and uploads it to the monitoring background. The battery array management unit is connected to the battery cluster management unit, communicates with the power conversion system for energy storage and the energy management system, and, according to the control instructions of the system, completes the communication with the power conversion system for energy storage to realize the charging and discharging processes of each battery cluster. The battery array management unit has the functions of clock and data storage. It stores key battery information and fault information according to requirements. Meanwhile, it is equipped with a touch screen. On the one hand, it can display detailed battery data locally. On the other hand, it can locally realize functions such as upgrading the control program for charging and discharging contactors.

The battery array management unit is the assembly control module of the battery management system. Its main functions include: 1. Charging and discharging management of the battery pack; 2. Self-check and fault diagnosis alarm of the battery management system; 3. Fault diagnosis and alarm of the battery pack; 4. Security protection for various abnormal and faulty situations; 5. Communication with other equipment, such as the power conversion system for energy storage and the energy management system; 6. Data storage, transmission, and processing. It can store the latest alarm information, reset information, and sample abnormal information of the system and export the stored information as needed; 7. Big data storage and processing. It stores all the collected information, alarm information, reset information, and various abnormal information of the system (the storage capacity can also be optionally configured). The system is equipped with a robust self-checking mechanism that serves to guarantee the proper functioning of the system itself.

Working Principle of the Energy Storage Battery Management System

The operating principle diagram for the energy storage battery management system is shown in Fig. 4.6. The battery array management unit is taken as

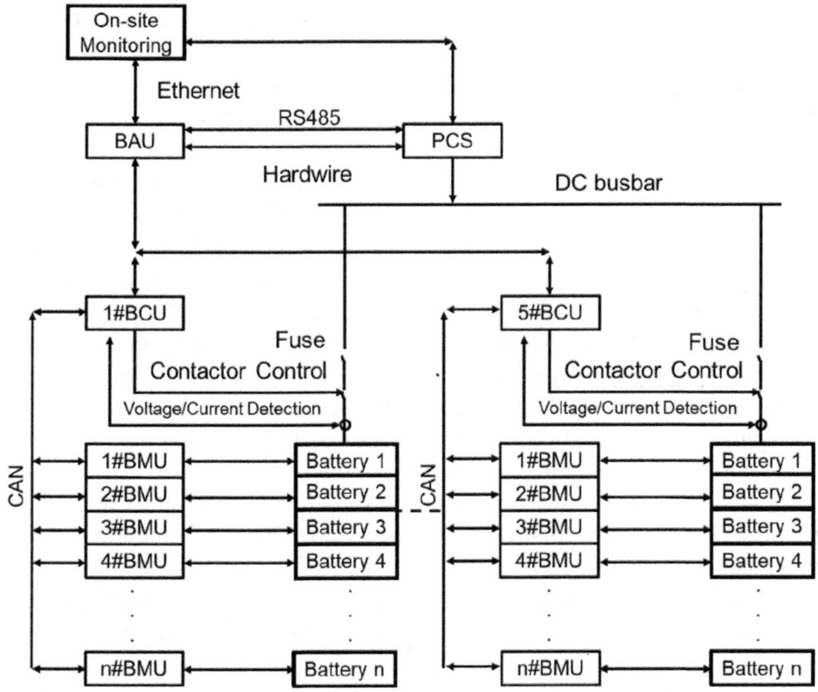

FIGURE 4.6 Schematic diagram of the energy storage battery management system.

a unit by the control layer of the battery management system. One battery array management unit controls several battery clusters connected in parallel (battery cluster management units). Each battery cluster management unit obtains information such as battery voltage and temperature through the battery management unit. The battery management unit undertakes the tasks of gathering battery voltage and temperature data, as well as performing balancing control operations. The battery cluster management unit is charged with the management of all battery management units within the battery pack. It acquires the individual cell voltage and temperature details of all battery management units via the Controller Area Network (CAN) bus. Additionally, it has the capabilities of collecting the current of the battery cluster, measuring the total voltage, and conducting electric leakage detection and alarm determination. If the state of the battery pack deviates from normal, the battery cluster management unit will disconnect the high-voltage power contactor, causing the battery cluster to cease operation and ensuring the safe utilization of the battery. The battery array management unit is tasked with the management of all battery clusters. When a severe fault occurs in a battery cluster, the battery array management unit will actively direct the battery cluster management unit to cut off the relay.

The battery management system is classified into two modes: the automatic operation mode and the maintenance mode. Within the automatic operation mode, the battery array management unit performs automatic control over connection and disconnection operations in line with the statuses of the battery clusters within its jurisdiction.

Automatic Operation Mode Control Strategies: 1. Detection of the Number of Battery Cluster Management Units at Power-on. When the battery array management unit is powered on, it detects the number of battery cluster management units in place. When all N groups of battery cluster management units are in place, the battery array management unit allows full-power charging and discharging. When the number of battery cluster management units in place is less than N groups, the battery array management unit operates with limited power according to the specific number of units in place (the battery management system sends the maximum charging and discharging current to the power conversion system for energy storage/the energy management system). When the number is less than the minimum supported number of groups (which can be set on the upper computer), the battery array management unit is not in place, and charging and discharging cannot be carried out.

2. Detection of the Total Voltage Difference at Power-on. After the battery array management unit passes the detection of being in place, it judges the total voltage difference. When the voltage difference between the maximum total voltage and the minimum total voltage of the battery packs is less than the "maximum total voltage difference allowed for the battery packs to close", the battery management system judges that the voltage difference of all the battery packs in place is small and meets the conditions for the relay to close. Then, it closes the main positive and negative relays of all battery cluster management units and enters the precharge and balancing process. When the battery array management unit detects that the current total voltage difference in place exceeds the allowable value, at this time, the battery array management unit reports a fault of excessive total voltage difference, and manual intervention is required. The faulty battery pack needs to be shut down, or the maintenance mode should be enabled, and the battery pack should be balanced manually.

3. Pre-charge Control at Power-on. Before each relay is closed, the capacitors in the high-voltage system connected to the battery cluster must be precharged. Only after it is determined that the precharging process is completed can the relay be closed. Otherwise, the relay contacts are prone to adhesion and damage due to the heat generated by excessive current. When the battery array management unit conducts precharge control, it first controls all battery cluster management units to close the precharge relays. When the battery cluster management units detect that the precharge current and the voltage difference before and after precharging are less than certain values and the precharge time is greater than a certain value, the battery cluster management units report that the precharge is complete. At this time, after the battery array management unit detects that all precharges are complete, it controls to close the main positive

relay, disconnects the precharge circuit, and enters the normal grid-connected operation state.

4. Balancing Control. During system operation, the balancing control evenly charges the batteries based on battery voltage, improving the consistency of the battery pack and alleviating performance degradation in the battery system caused by the “short-board effect”.

5. Charging and Discharging Management. Throughout the system’s operation, the voltage of each cell and the temperature of the battery pack are continuously monitored in real-time. Based on the state evaluation of the battery system, the upper-limit voltage value for charging, the lower-limit voltage value for discharging, the maximum chargeable current, and the maximum dischargeable current are computed. These values are then transmitted to the energy-storage power conversion system via messages. Subsequently, the energy-storage power conversion system performs charging and discharging operations, with the strict guarantee that the charging and discharging currents do not surpass the maximum values specified by the battery management system.

In the charging mode, when the single-cell voltage reaches the “single-cell voltage for current reduction during charging”, the battery management system sends a current reduction request according to the current charging current of the power conversion system for energy storage. After reaching the “single-cell voltage for current reduction during charging” multiple times, the current will reach the “minimum limited charging current”. At this point, the battery management system will no longer control the current reduction and will maintain the charging of the power conversion system for energy storage until the charging reaches the “single-cell voltage for charging stop”. Then, the battery management system sets the full-charge flag and limits the charging current to 0. The power conversion system for energy storage stops charging. Only when the “first-level charging alarm disappears” will the battery management system allow charging again [8].

In the discharging mode, when the single-cell voltage drops to the “single-cell voltage for current reduction during discharging”, the battery management system makes a current reduction request according to the current discharging current of the PCS. After reaching the “single-cell voltage for current reduction during discharging” multiple times, the current will reach the “minimum limited discharging current”. At this time, the battery management system will no longer control the current reduction and will maintain the discharging of the power conversion system for energy storage until the discharging reaches the “single-cell voltage for discharging stop”. Then, the battery management system sets the empty-discharge flag and limits the discharging current to 0, and the power conversion system for energy storage stops discharging. Only when the “first-level discharging alarm disappears” will the battery management system allow discharging again.

When a third-level severe fault occurs in the battery system, the battery management system will implement a time delay and then forcibly disconnect the relay to safeguard the battery. If the single-cell voltage is below or above the limit voltage, the battery management system will forcibly cut off the relay for battery protection purposes.

Maintenance Mode Control Strategies: When a large single-cell voltage difference, a large total voltage, or a third-level alarm occurs in a battery cluster and maintenance is required, through the upper computer of the battery array management unit, the faulty cluster's battery pack can be manually controlled to conduct individual charging and discharging for maintenance with a small manual current. When the battery pack platforms are consistent, the maintenance mode can be stopped. After reapplying low voltage to the battery management system, the battery management system will automatically recognize and enter automatic control mode.

(3) Protection Strategies of the Battery Management System

The safety protection of batteries is the most important function of the battery management system. Through monitoring and analyzing the state of batteries, the battery management system can protect against various abnormal states during the battery operation process, and it can also send alarm signals or trip instructions to implement on-site fault isolation. The battery management system includes functions such as current protection, voltage protection, temperature protection, and state-of-charge protection [9].

(1) Current Protection: Current protection, which is also referred to as overcurrent protection, implies that in the course of charging and discharging procedures, if the operational current surpasses the safe level, steps ought to be taken to restrain the growth of the current. Current protection includes overcurrent protection during charging of battery clusters and overcurrent protection during the discharging of battery clusters.

(2) Voltage Protection: Voltage protection refers to taking measures to limit the voltage from exceeding the set value when the voltage exceeds the set value during the charging and discharging processes. Voltage protection includes high/low battery cluster voltage, high/low single-cell voltage, and single-cell voltage difference.

(3) Temperature Protection: The processes of battery charging and discharging possess particular requisites regarding the ambient temperature range. Temperature protection pertains to the implementation of protective measures for batteries when the temperature exceeds a specific threshold value. Temperature protection comprises single-cell overtemperature protection, single-cell undertemperature protection, protection against excessive single-cell temperature variation, and pole overtemperature protection.

(4) State of Charge Protection: To prevent batteries from overdischarging, when the state of charge is lower than the set value, protective measures should be taken to limit the discharging.

The protection action strategies of the battery management system are shown in Fig. 4.7. Its current protection, voltage protection, and temperature

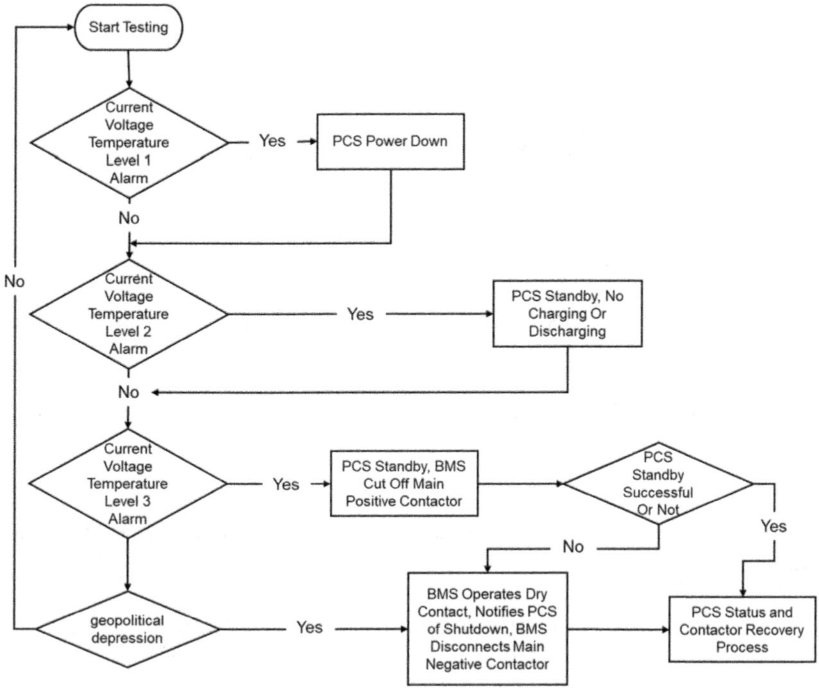

FIGURE 4.7 Protection strategy action flow.

protection all use a three-level protection mechanism. Upon the occurrence of a first-level alarm, the battery management system issues a notification to the energy-storage power conversion system, instructing it to operate at a reduced power level. In the event of a second-level alarm, the battery management system signals the energy-storage power conversion system to halt the charging or discharging processes. When a third-level alarm is triggered, the battery management system commands the energy-storage power conversion system to shut down completely. After a predefined time delay, the battery management system will autonomously disconnect the relay.

4.1.4 Power conversion system for energy storage

PCS serves as a crucial power interface device that links the energy storage system to the power grid or AC load. It is endowed with the capacity to regulate the bidirectional flow of energy between the power grid and the energy storage unit. Through this function, it can satisfy the demands for power control precision and the response speed during rapid charging-discharging transitions. The PCS plays a decisive role in defining the power quality and dynamic characteristics of the external output of the battery energy storage system.

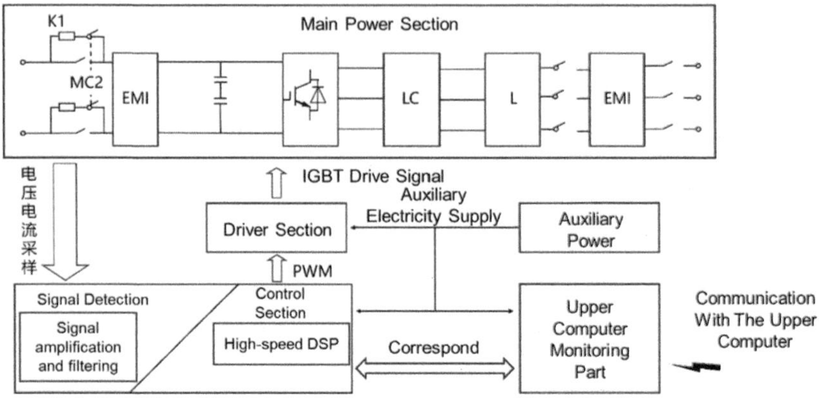

FIGURE 4.8 Structural block diagram of 100 kW bidirectional power conversion system for energy storage.

Moreover, it exerts a significant influence on the safety and service life of the battery.

4.1.4.1 Introduction to the principle and composition of the power conversion system for energy storage

The operational principle of the energy storage power conversion system involves an AC/DC side controllable four-quadrant operation converter device, which facilitates bidirectional AC/DC conversion of electrical energy. This principle entails performing constant-power or constant-current control in accordance with microgrid monitoring instructions to charge or discharge the battery. Simultaneously, it aims to smooth the output of fluctuating power sources such as wind and solar energy. The energy storage power conversion system ought to possess multiple control modes, and the appropriate control mode can be selected based on requirements. Among them, the "virtual generator control" mode enables the energy storage power conversion system to function like that of a traditional generator. Specifically, when the energy storage system absorbs electrical energy from the power grid, the energy storage power conversion system operates in the rectification mode. Conversely, when the energy storage system transmits electrical energy to the power grid, the energy storage power conversion system operates in the inversion mode [10].

A 100 kW bidirectional power conversion system for energy storage consists of the main power part, signal detection part, control part, drive part, upper computer monitoring part, auxiliary power supply part, etc. Its structural block diagram is shown in Fig. 4.8.

(1) Main Power Part: The main power part is mainly composed of a precharge circuit, bus capacitor, Insulated Gate Bipolar Transistor (IGBT) power switch module, LCL filter, AC contactor, etc. The main power part is the main body of the bidirectional converter, and also the path of energy flow. The transformation of energy forms and the bidirectional flow of energy are realized through the conduction and turn-off of the insulated-gate bipolar transistor.

(2) Signal Detection Part: The principal signal detection component primarily fulfills the high-precision sampling and signal processing functions of voltage and current signals, as well as the detection function of fault signals.

(3) Control Part: The control segment constitutes the core portion of the bidirectional converter. Diverse high-speed chips are adopted as the central processor (DSP) and the supplementary controller (FPGA). The primary functions actualized by the control part encompass signal sampling and computation, converter control, fault assessment, and safeguarding of the converter, as well as communication with the upper computer monitoring interface.

(4) Insulated Gate Bipolar Transistor Drive Part: The converter system drive adopts a special drive for insulated gate bipolar transistors, making the insulated gate bipolar transistors work in the optimal switching state and effectively improving their working reliability. At the same time, the drive itself detects abnormal states such as overcurrent and overtemperature in the insulated-gate bipolar transistor power devices. When an abnormal state occurs, it turns off the power devices to realize the function of protecting the devices.

(5) Upper Computer Monitoring Part: The upper computer monitoring interface adopts a high-definition liquid crystal touch screen as the input and output interface. Based on the Microsoft operating system, it provides a friendly human-computer interaction interface and also provides multiple communication interfaces to realize the function of the converter local controller.

(6) Auxiliary Power Supply Part: T The auxiliary power supply furnishes power to the drive component, control component, and upper computer monitoring component, thereby guaranteeing the uninterrupted and proper operation of the equipment.

4.1.4.2 *Main circuit topology structure of the power conversion system for energy storage*

The grouping method and connection topology of battery packs should match the topology structure of the power conversion system. The common topology structures of the power conversion system can be divided into single-stage conversion topology type, two-stage conversion topology type, and H-bridge cascaded topology type [11].

(1) Single-stage Conversion Topology Type

A schematic diagram of a typical single-stage conversion topology construction may be found in Fig. 4.9. It is a DC/AC link-only single-stage power conversion system. After being connected in parallel and series, the batteries are connected straight to the AC/DC's DC side. This power conversion system has a simple topology structure and relatively low energy consumption. How-

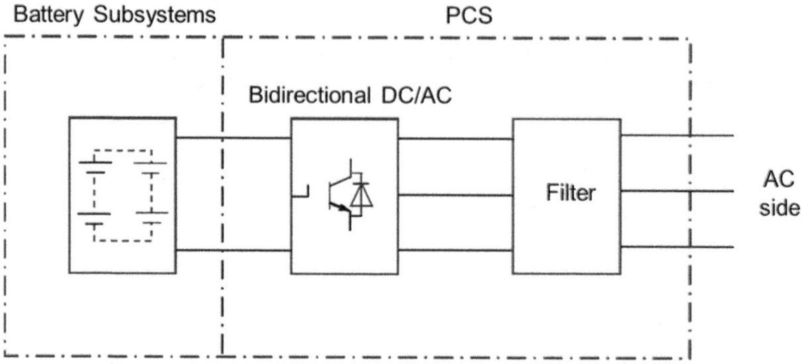

FIGURE 4.9 Topology structure of the power conversion system for energy storage with only DC/AC link.

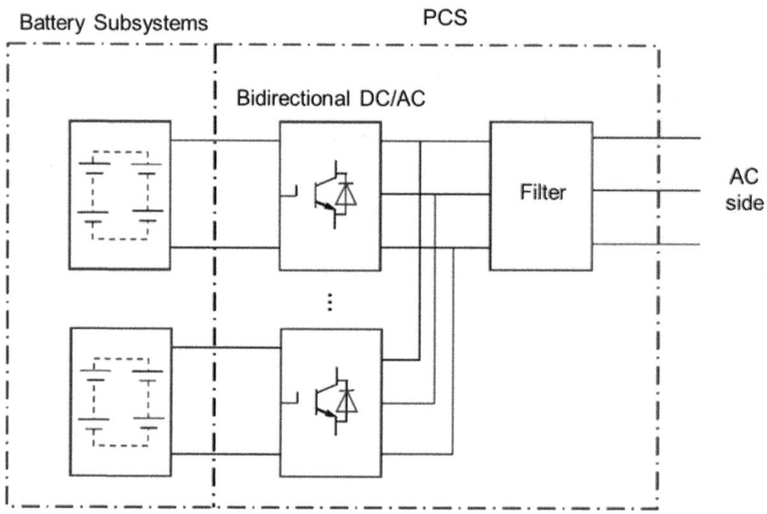

FIGURE 4.10 Topology structure of the power conversion system with only DC/AC link sharing the AC side.

ever, the capacity selection of the energy storage unit lacks flexibility, and it is suitable for the grid connection of independent distributed energy storage [12].

For the convenience of capacity expansion, the power conversion system with only the DC/AC link can be extended to a topology structure sharing the AC side with only the DC/AC link. Its structural diagram is shown in Fig. 4.10. The topology structure sharing the AC side with the DC/AC link adopts a modular connection and has a more flexible composition and configuration. When individual battery packs or DC/AC links malfunction, the energy storage system can still operate. However, this will lead to an increase in the power of electronic devices and complicate the design of the control system.

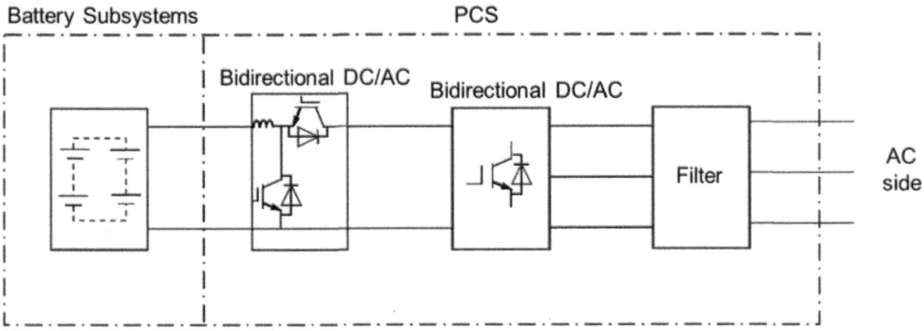

FIGURE 4.11 Topology structure of the bipolar power conversion system containing DC/AC and DC/DC links.

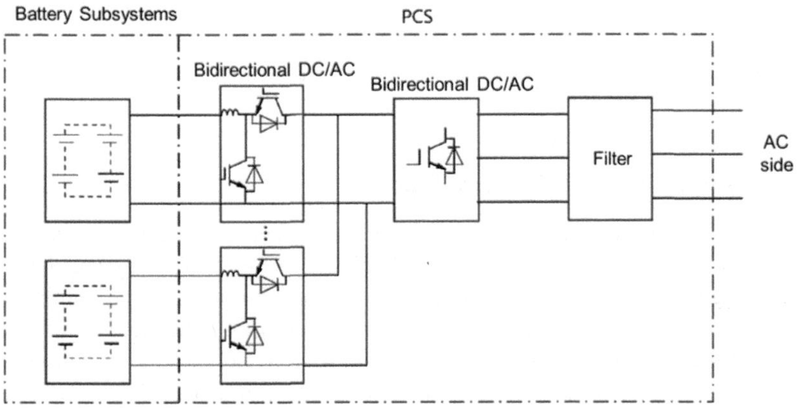

FIGURE 4.12 Topology structure of the bipolar power conversion system with DC/AC and DC/DC links sharing the DC side.

(2) Two-stage Conversion Topology Type

The two-stage conversion topology type is a bipolar power conversion system that contains DC/AC and DC/DC links (Fig. 4.11). The bidirectional DC/DC link is mainly used for step-up and step-down conversions to provide a stable direct current voltage. This kind of power conversion system with such a topology has strong adaptability. Since the DC/DC link realizes the step-up and step-down of the direct current voltage, the capacity allocation becomes more flexible. It is suitable for being connected with distributed power sources with strong intermittency and volatility, and can suppress the voltage fluctuations that may be brought about by their direct grid connection. However, due to the existence of the DC/DC link, the efficiency of the power conversion system is reduced [13].

For the convenience of capacity expansion, the bipolar power conversion system can be extended to a topology structure sharing the DC side or the AC side that contains DC/AC and DC/DC links. Fig. 4.12 and Fig. 4.13 exhibit their structures, respectively.

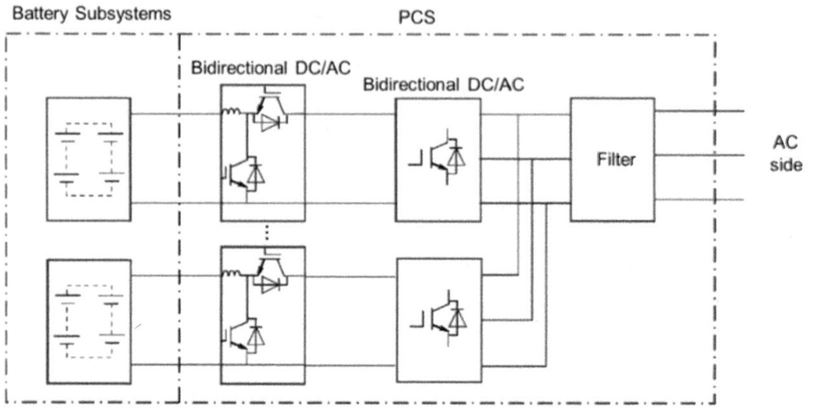

FIGURE 4.13 Topology structure of the bipolar power conversion system with DC/AC and DC/DC links sharing the AC side.

(3) H-bridge Cascaded Topology Type

The H-bridge cascaded power conversion system adopts the method of connecting multiple power modules in series to achieve high-voltage output. When high voltage is required, it is only necessary to increase the number of power modules, avoiding excessive series connection of batteries. Each power module has the same structure and is easy to design and package in a modular way [14]. Each power module is a separate DC power source, independent of the others. The control of one unit will not affect other units.

Power conversion systems with single-stage conversion topology and two-stage conversion topology are generally applicable to occasions where the capacity of the energy storage unit is not greater than 1 MW. When the energy of the energy storage unit is relatively high, to avoid the parallel connection of multiple battery packs, the two-stage conversion topology structure can be adopted.

The power conversion system with the H-bridge cascaded topology structure is generally applicable to applications where the capacity of the energy storage unit is greater than 1 MW. For voltage levels of 35 kV or lower and without considering the regulation of three-phase imbalance, the H-bridge cascaded topology structure can adopt the Y-connection method. For higher or lower voltage levels where the regulation of three-phase imbalance needs to be considered, the H-bridge cascaded topology structure can adopt the delta-connection method [15].

In specific engineering design, the topology of the power conversion system can be comprehensively considered based on the actual situation of the project, the capacity, energy, battery type of the energy storage unit, and the manufacturing level, as well as the performance requirements of the power conversion system.

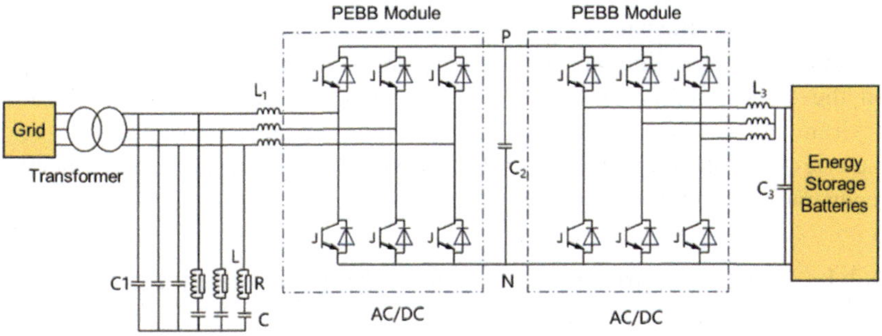

FIGURE 4.14 Main circuit topology structure of the bidirectional converter system based on power electronic component units.

As shown in Fig. 4.14, it illustrates the main circuit topology structure of the 100 kW bidirectional converter system. In the topology, two power electronic component unit modules are used as the converter power devices. Among them, the AC/DC module on the AC side is a three-phase voltage source rectifier, which realizes the bidirectional active rectification function [16]. The DC/DC module on the DC side is a DC/DC converter, which realizes the step-up and step-down functions. When the grid load is small and the energy storage battery has a low charge, the grid charges the battery through the converter. The AC/DC module functions in the rectification mode, with a DC output voltage of 700 V. Meanwhile, the DC/DC module operates in the step-down mode, and its output voltage corresponds to the charging voltage demanded by the energy storage battery. When the grid load is substantial and the energy storage battery is fully charged, the energy storage battery discharges power back to the grid via the converter. At this time, the DC/DC module operates in step-up mode, raising the voltage of the energy storage battery to 700 V, and the AC/DC module operates in inverter mode, outputting 380 V AC power at the line voltage. The inclusion of the DC/DC module within the system is intended to meet the charging and discharging requirements of diverse battery types, as set by the converter. Nevertheless, the introduction of the DC/DC module also increases the complexity of system control to some extent. Additionally, it will decrease the overall conversion efficiency of the system. Consequently, for energy storage batteries whose charging and discharging voltages fall within the output voltage range of the AC/DC in the system, the DC/DC module can be excluded.

4.2 Safety and recycling of electrochemical energy storage batteries

Electrochemical energy storage batteries play an increasingly vital role in modern industry and society, with applications expanding from early portable communication devices to power tools, electric vehicles, and now to large-scale grid energy storage systems. However, with the rapid increase in market deployment over the past few years, issues such as battery safety, environmental pollution, and recycling have become critical concerns. This section will focus on the safety challenges and preventive strategies of lithium-ion batteries, environmental pollution during the production of lithium-ion and lead-acid batteries, and their control measures, as well as the recycling and material regeneration of lithium-ion batteries.

4.2.1 Safety and protection of electrochemical energy storage systems

The market share of electrochemical energy storage systems within emerging energy storage technologies continues to expand, highlighting increasingly complex safety challenges. Therefore a comprehensive understanding of the safety issues associated with electrochemical energy storage systems is essential. On this basis, it is imperative to establish robust safety protection measures and response strategies to effectively prevent various safety incidents.

4.2.1.1 Safety issues in electrochemical energy storage systems

During the operation of electrochemical energy storage systems, safety concerns primarily due to internal damage within the battery cells. This internal damage is particularly pronounced under conditions such as overcharging, overdischarging, and significant temperature fluctuations. The continued accumulation of such damage can ultimately lead to battery failure and even internal short circuits, which pose serious safety risks. Typically, electrochemical energy storage systems contain thousands of small cells, and the failure of any single cell can result in severe safety incidents, such as fires or explosions [17].

In recent years, dozens of safety incidents involving electrochemical energy storage stations have occurred both domestically and internationally. On April 16, 2021, a fire occurred at an energy storage station in Fengtai District, Beijing, leading to the tragic deaths of two firefighters. After on-site inspection, forensic analysis, experimental studies, simulation modeling, and expert evaluations, it was determined that the direct cause of the fire was an internal short circuit fault in the lithium iron phosphate batteries from second-life use, which triggered thermal runaway and combustion. Experimental analysis of the first faulty battery revealed that during thermal runaway, it emitted jetting gases, primarily ethylene carbonate vapor along with hydrogen, methane, carbon monoxide, and carbon dioxide [18].

On July 30, 2023, a Tesla Megapack energy storage system in Victoria, Australia, experienced a catastrophic fire. The battery units burned intensely, and it took firefighters four days of continuous effort to bring the fire under control. This incident once again raised significant concerns about the safety of energy storage systems. The project used Tesla's Megapack storage units, which contain ternary lithium-ion batteries, with each Megapack providing an energy storage capacity of approximately 3 MWh. Reports indicated that the affected Megapack storage unit lacked proper firebreaks in its vicinity, and the inadequate layout hindered the firefighters' ability to address the blaze efficiently.

Thermal runaway, overcharging/overdischarging, short circuits, and other issues are among the common causes of safety hazards in electrochemical energy storage devices, as explained below:

1. Battery thermal runaway occurs when the battery cycles between charging and discharging. Thermal runaway can result in fire or explosion if there is inadequate heat dissipation or if a fault happens. Problems with battery design, incorrect charging and discharging, and extremely high outside temperatures are some of the main culprits.

2. Overcharging and overdischarging: These conditions can impair the functionality of electrode materials or possibly cause structural damage, which could result in an explosion or fire. Frequently, this problem is brought on by malfunctions in the charging apparatus or battery management system.

3. Short Circuits: Internal or external short circuits can cause the battery to overheat, catch fire, or explode. Short circuits may originate from internal structural damage or external short-circuiting.

4. Environmental Factors: Energy storage stations are subject to environmental influences such as temperature, humidity, and dust, which can accelerate equipment aging or lead to malfunctions, increasing safety risks.

5. Electrical Faults: Electrical issues within or outside the battery pack, such as short circuits or poor electrical connections, can trigger fires or other safety incidents.

6. Human Error: Operational errors are also a significant source of safety concerns in energy storage stations, including improper handling and inadequate maintenance.

4.2.1.2 Safety regulations for electrochemical energy storage systems in China

Regions such as Europe and North America, where electrochemical energy storage technologies have been applied earlier, have already established mandatory safety standards for energy storage systems. For example, the UL9540A certification in North America mandates thermal runaway risk assessments for energy storage stations before grid integration. Although China started later in this regard, recent years have seen a growing focus on the safety of energy storage systems, with a series of safety regulations and technical standards being issued for battery systems in energy storage stations. Some industry standards also address safety distances for energy storage containers and fire protection requirements.

In 2022 the General Office of the National Energy Administration issued the Notice on Strengthening the Safety Management of Electrochemical Energy Storage Stations. By the end of 2023, the National Standardization Administration of China published the 2023 No. 20 China National Standards

Announcement, which includes 13 new standards related to energy storage batteries, all of which will be officially implemented from July 1, 2024. Furthermore, by 2025, six new standards, including the Commissioning Code for Electrochemical Energy Storage Stations, Technical Guidelines for Black Start of Electrochemical Energy Storage, Guidelines for Supervising the Manufacturing of Lithium-ion Batteries for Energy Storage, Technical Regulations for the Integration of Electrochemical Energy Storage Systems into Distribution Networks on the User Side, Technical Requirements for Battery Management Communication in Electrochemical Energy Storage, and Technical Requirements for the Retirement of Lithium-ion Batteries for Energy Storage will fill gaps in the industry. The remaining seven standards will replace existing ones. The current standards will be replaced by the remaining seven. A large number of standards will be created and released over the next two years to support the superior growth of China's energy storage sector.

4.2.1.3 Thermal runaway and protection in lithium-ion batteries

Thermal runaway is a critical safety issue in electrochemical energy storage systems and represents a system failure with significant potential hazards. During regular operation, various factors may trigger a series of chain reactions within energy storage batteries, leading to a rapid accumulation of heat inside the battery [19]. This results in a continuous temperature rise, causing the electrolyte to decompose and generate large amounts of gas. These gases increase internal pressure, potentially leading to battery swelling or even rupture.

Currently, lithium-ion batteries are being widely used in both power batteries and energy storage batteries. The thermal runaway of lithium-ion batteries is highly contagious, meaning that thermal runaway in one cell or module can easily spread to neighboring units, causing a more extensive failure. The occurrence of thermal runaway in battery packs can lead to severe damage to energy storage equipment, such as battery module fires and equipment destruction, and may even result in catastrophic outcomes like building fires or explosions, posing a significant threat to the surrounding environment and human safety. This section will focus on the mechanisms and causes of thermal runaway in lithium-ion batteries, along with common monitoring and fire suppression systems used to mitigate these risks.

(1) Mechanism of Thermal Runaway

Thermal runaway and the resulting chain reactions in lithium-ion batteries typically follow a sequential mechanism. Taking the NCM cathode/graphite anode battery system with a PE separator as an example, thermal runaway begins with the decomposition of the SEI on the anode surface. This is followed by the reaction between the graphite anode and the electrolyte, the melting of the PE separator, the decomposition of the NCM cathode, electrolyte decomposition, short-circuiting, and finally, the combustion of the electrolyte

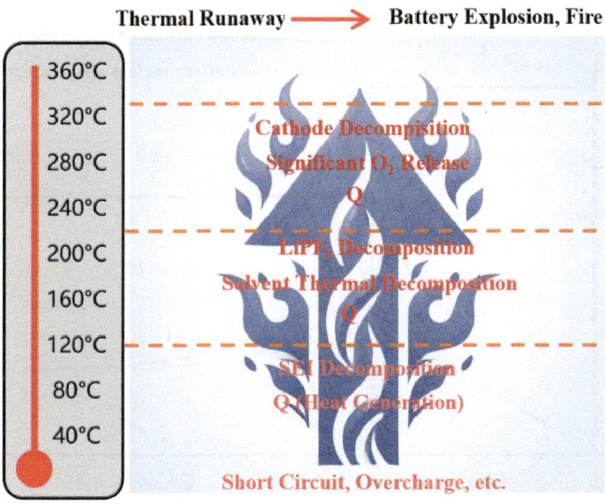

FIGURE 4.15 Chain reactions during thermal runaway.

[20]. As the reactions progress, the temperature continuously rises, eventually leading to thermal runaway. Therefore the heat-temperature-reaction (HTR) cycle is the fundamental cause of this chain reaction.

Specifically, the core of thermal runaway lies in the positive feedback loop of heat within the battery. The chain processes during thermal runaway are depicted in Fig. 4.15. It has been demonstrated that abusive situations such as short circuits and overcharging result in aberrant heat creation inside the battery, raising its temperature and causing adverse effects. These side reactions release even more heat, creating a cumulative effect that forms a "heat-temperature-reaction" loop. This loop drives the battery temperature to extreme levels, ultimately resulting in thermal runaway. It is important to note that these side reactions do not occur at a single, specific temperature. Instead, various side reactions have overlapping temperature ranges. For instance, the decomposition and reconstruction of the SEI layer spans a wide temperature range, during which thermal reactions between the electrolyte, anode, and cathode components occur.

The internal and external temperature progression curves of a battery during thermal runaway, as measured and simulated by international researchers, are shown in Fig. 4.16. The self-heating stage (50–150 °C), the thermal runaway stage (150–800 °C), and the conclusion of thermal runaway (800 °C to room temperature) are the three stages of thermal runaway [21]. The self-heating stage, also known as the thermal accumulation stage, begins with the dissolution of the SEI layer. When the internal battery temperature reaches 90 °C, the SEI layer starts to dissolve, exposing the lithiated carbon anode directly to the electrolyte. The lithiated carbon anode continues to react exothermically

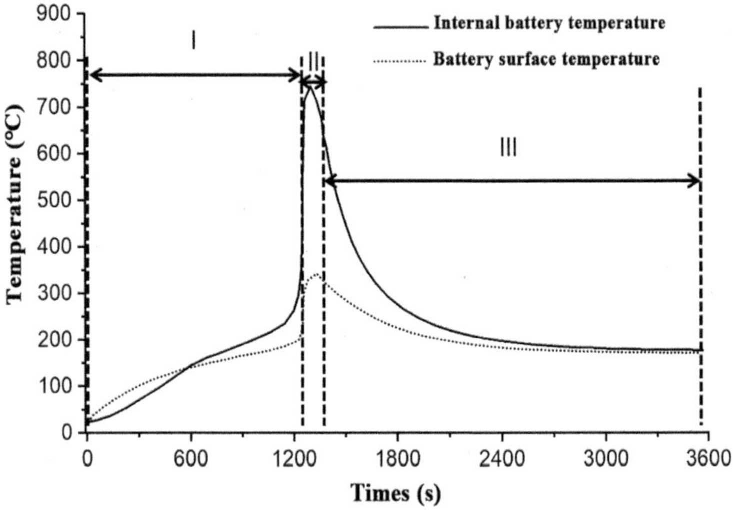

FIGURE 4.16 Temperature-time curve of thermal runaway.

with the electrolyte, further increasing the temperature and accelerating the decomposition of the SEI layer. Without external cooling measures, this process leads to a sustained positive feedback loop until the SEI layer is completely dissolved.

Once the temperature exceeds the critical point where the separator begins to melt significantly, complex electrochemical reactions in both the anode and cathode materials generate additional heat, entering the thermal runaway stage. At this point, the temperature rise becomes uncontrollable, as the amount of reactants increases significantly, leading to a rapid spike in temperature. Externally, this manifests as a sharp drop in battery voltage. By this time, thermal runaway can no longer be mitigated by external cooling measures. In a short period, violent reactions produce large amounts of gas and heat, causing the expanded gases to rupture the battery casing, ejecting materials that carry away some of the heat. This is the most severe stage of thermal runaway, during which the battery reaches its peak temperature. Additionally, heat can be transported directly between neighboring cells when their enlarged casings come into contact, or it can be passed through associated conductive components.

Once the battery enters thermal runaway, the process can only end when all reactive materials inside the battery are fully consumed. Reports from firefighting departments indicate that for high-energy devices like lithium-ion batteries enclosed in a sealed casing, conventional firefighting methods cannot effectively terminate ongoing thermal runaway. This is because extinguishing agents cannot penetrate and reach the reacting materials inside the battery. Thus, the only available measures are isolating the area and waiting for the reactive

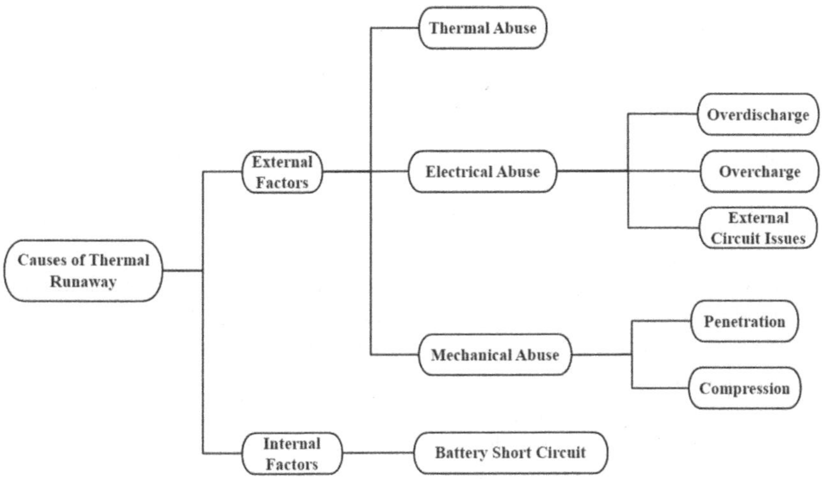

FIGURE 4.17 Classification of thermal runaway triggers: internal and external factors.

materials in the thermal runaway cell to be fully depleted before the thermal runaway process can naturally end.

(2) Causes of Thermal Runaway

The factors inducing thermal runaway in lithium-ion batteries can generally be divided into internal short circuits and external triggers, as shown in Fig. 4.17. External triggers include thermal abuse, electrical abuse, and mechanical abuse.

Internal Short Circuits:

When the positive and negative electrodes of a battery come into direct contact, the subsequent reactions vary greatly depending on the extent of contact. Typically, large-scale internal short circuits caused by mechanical or thermal abuse directly lead to thermal runaway. In contrast, internal short circuits that develop spontaneously are relatively mild and generate little heat, not immediately leading to thermal runaway. Such internally driven conditions often result from manufacturing defects, battery aging (such as increased internal resistance), and long-term mild improper use, leading to lithium metal deposition. Over time, the risk of internal short circuits caused by these internal factors gradually increases.

Mechanical Abuse:

Mechanical abuse refers to the deformation of individual cells or battery packs due to external forces, resulting in the relative displacement of different parts. For battery cells, the main forms of mechanical abuse include collisions, compression, and punctures. For example, when a vehicle strikes a foreign object while driving at high speed, it can rupture the separator within the battery, causing an internal short circuit and, within a short time, leading to spontaneous combustion.

Electrical Abuse:

Electrical abuse in lithium batteries generally includes external short circuits, overcharging, and overdischarging, with overcharging being the most hazardous form of electrical abuse. Heat and gas generation are common characteristics of overcharging. The heat arises from Ohmic heating and side reactions. First, excessive lithium intercalation leads to the growth of lithium dendrites on the anode surface, with the onset of dendrite formation determined by the stoichiometric ratio of the cathode to the anode. Second, excessive lithium deintercalation causes the cathode structure to collapse due to heating and oxygen release (e.g., oxygen release from NCA cathodes). The release of oxygen accelerates electrolyte decomposition, generating large amounts of gas. As internal pressure increases, the safety valve opens, and the battery begins to vent. Once the active materials inside the cell come into contact with air, they react violently, releasing significant heat.

Thermal Abuse:

Localized overheating is a typical scenario of thermal abuse within battery packs. Thermal abuse rarely occurs in isolation and is often a result of mechanical or electrical abuse. It is a key factor that directly triggers thermal runaway. In addition to overheating caused by mechanical or electrical abuse, overheating can also result from loose electrical connections. Thermal abuse is currently the most commonly simulated condition, where controlled heating of the battery is used to observe its reactions during the heating process.

Since 2020, China's national standards for traction batteries have added a requirement for thermal propagation tests. These tests stipulate that after a single cell undergoes thermal runaway, the battery system must not catch fire or explode within five minutes, allowing time for safe evacuation. The comprehensive safety measures currently adopted by automakers for traction batteries can also provide insights into the design, manufacturing, and operation of energy storage battery systems. These measures include:

(1) Reducing the heat generated by exothermic side reactions within the cell, for example, using lithium iron phosphate cathodes.

(2) Increasing the ignition temperature inside the cell, such as by adding flame retardants to the electrolyte or using ceramic separators.

(3) Adding flame-retardant materials between cells to prevent thermal runaway in one cell from affecting others.

(4) Implementing other measures, such as oxygen-locking technology, to moderate oxygen release at high temperatures.

(5) Continuously monitor cell data around the clock, and improve the heat dissipation and cooling capabilities at the cell, module, and pack levels to prevent heat accumulation.

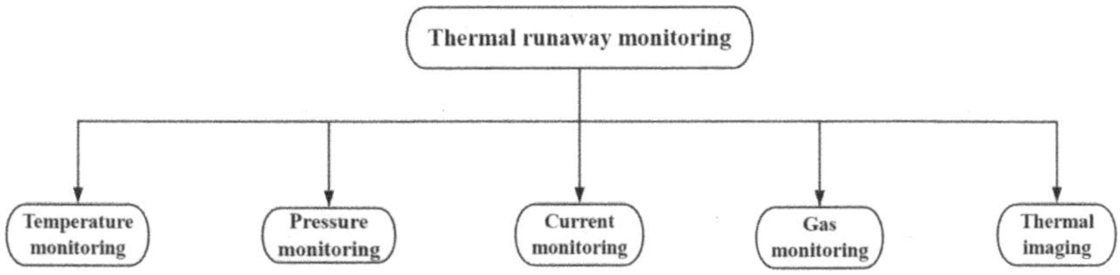

FIGURE 4.18 Main monitoring methods for thermal runaway.

(3) Thermal Runaway Monitoring and Cooling System

Thermal runaway monitoring is one of the key measures to ensure the safe operation of batteries. By monitoring parameters such as temperature, pressure, and current in real-time, early signs of thermal runaway can be detected, allowing appropriate interventions to be taken to prevent safety incidents and protect personnel and property [22].

Temperature, pressure, current, and gas monitoring, and prediction using thermal imaging technology and data analysis are the primary techniques for thermal runaway monitoring [23], as shown in Fig. 4.18. Temperature monitoring involves real-time measurement of the temperature changes on the surface or inside the battery to detect signs of thermal runaway. Pressure monitoring focuses on changes in internal battery pressure, triggering alarms when abnormal pressure rises are detected. Current monitoring tracks the charging and discharging currents of the battery, preventing thermal runaway caused by overcharging, overdischarging, or short circuits. Gas monitoring detects specific gases released by the battery to assess the risk of thermal runaway. Additionally, infrared thermal imaging technology can be used to obtain surface temperature distribution images of the battery, further analyzing the possibility of thermal runaway. Lastly, by combining various monitoring data and utilizing big data analysis and machine learning techniques, it is possible to predict and assess the risk of thermal runaway, achieving real-time monitoring and early warning to effectively prevent safety incidents.

The cooling system of a battery is a key component for ensuring battery performance, safety, and lifespan. An efficient cooling system can effectively manage the heat generated during battery operation, keeping the battery temperature within an optimal range [24]. The cooling system mainly consists of heat sinks, thermal conductors, air or liquid cooling systems, and temperature monitoring and control units. The designs of liquid and air cooling systems for battery cooling configurations are shown in Fig. 4.19. As demonstrated, the air cooling system speeds up heat dissipation by increasing the airflow velocity near the heat sink through forced convection. On the other hand, the liquid cooling system uses circulating coolant to transfer the heat generated by the battery to the external environment via heat exchangers.

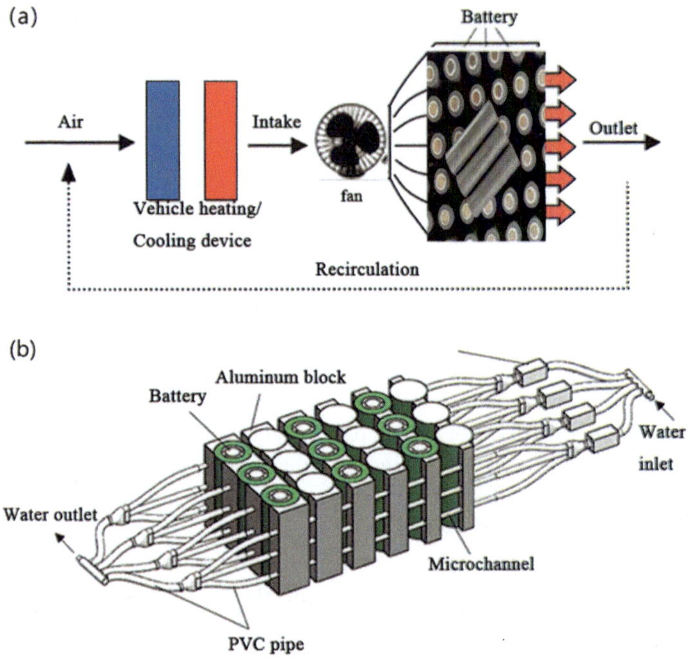

FIGURE 4.19 Schematic diagrams of battery cooling systems: (a) Air cooling system and (b) Liquid cooling system.

Air cooling systems are characterized by low cost, ease of maintenance, lightweight, long lifespan, and absence of leakage issues. However, air cooling systems face challenges such as uneven air distribution and low heat capacity, resulting in poor thermal performance. Recent improvements in air cooling systems focus on battery pack design, inlet and outlet configurations, and airflow channel optimization. Compared to air cooling, liquid cooling systems have higher thermal conductivity, making them a more competitive choice for cooling media. Under high thermal loads, liquid cooling systems offer lower energy consumption and better cooling efficiency. Currently, liquid-cooling systems for batteries can be classified into direct and indirect systems, depending on whether the coolant contacts the battery. Indirect cooling systems typically employ cold plates, fins, and microchannels to exchange heat between the battery pack and the coolant, preventing leakage and short circuits. Direct cooling systems, in contrast, offer superior cooling performance, better temperature uniformity, and greater compactness. To avoid short circuits at the battery-coolant contact, the liquid coolant used in direct liquid cooling must be electrically insulated, nontoxic, and chemically stable.

Apart from common cooling technologies, other methods warrant attention. Thermoelectric cooling, an advanced battery-cooling technology, uses the Peltier effect in semiconductor materials to achieve energy conversion for cooling or heating [25]. This technology features noiseless operation, small size, no moving parts, ease of operation and maintenance, and adjustable cooling capacity, making it suitable for applications requiring high cooling performance. In battery cooling, thermoelectric cooling can be combined with other technologies, such as composite phase change materials, to achieve more efficient cooling. As technology advances, thermoelectric cooling is expected to play a greater role in battery thermal management systems.

The combination of thermal runaway monitoring and cooling systems can form an intelligent battery safety management system. In practical applications, when the thermal runaway monitoring system detects abnormal temperature rises or potential thermal runaway risks in the battery, the cooling system is immediately activated. The cooling system, through liquid circulation or air cooling, rapidly lowers the battery's temperature, preventing thermal runaway. Additionally, the system automatically adjusts cooling operations based on temperature changes in the battery, such as increasing the fan speed or coolant flow rate, to optimize the cooling effect. This combined approach has the advantage of timely detection and mitigation of battery safety hazards, effectively preventing thermal runaway and other safety incidents. Furthermore, through intelligent management, the system enables real-time monitoring and prediction of battery status, enhancing safety and stability.

(4) Automatic Fire Suppression System

The fire hazards of lithium-ion battery energy storage systems can be broadly categorized into two types: fire risks associated with the main battery system and those related to the auxiliary electrical system [26]. The risk factors of the energy storage electrical system, in particular, require heightened attention. The high degree of integration of electrical equipment and the numerous communication lines contribute to the risk of high-voltage and high-current surges (from lightning or electrical faults), which can damage or even destroy energy storage components, potentially triggering electrical system fires. Thus, electrical fire suppression technology plays a critical role in enhancing the safety of lithium-ion battery energy storage systems.

The automatic fire suppression system in a lithium-ion battery energy storage system is an essential safety feature. This system is typically equipped with intelligent monitoring and response mechanisms, capable of detecting abnormalities such as temperature spikes or smoke in real-time. Upon detecting a potential hazard, the system swiftly activates, releasing specific extinguishing agents, such as dry powder or heptafluoropropane, to quickly extinguish any potential fire, prevent its spread, and minimize damage and danger. The system also integrates with other safety systems to form a comprehensive safety protection network, ensuring the safety and stability of the energy storage system under various conditions.

A novel fire suppression technology involves the use of perfluorohexanone, which operates by rapidly deploying extinguishing agents once a fire source

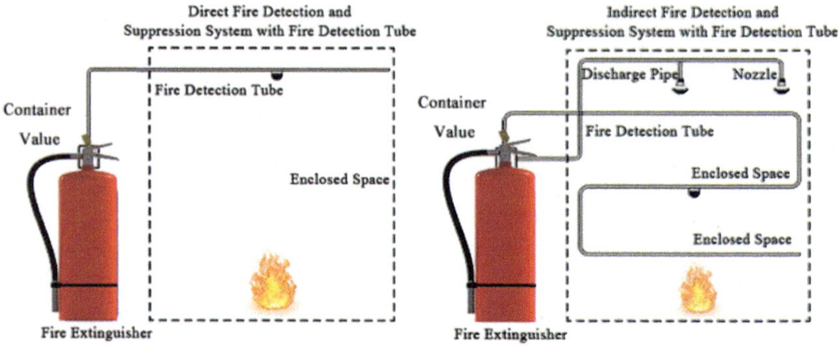

FIGURE 4.20 Schematic diagram of the working principles of direct and indirect fire detection tube suppression systems.

is detected by automatic sensors [27]. Upon contact with the fire source, the perfluorohexanone significantly reduces the surrounding temperature and suppresses the flames, achieving quick fire suppression. In recent years, fire detection tube systems have greatly improved the speed of extinguishing fires in lithium-ion battery systems. The basic operation of both direct and indirect fire detection tube automatic suppression systems is shown in Fig. 4.20. These systems can rapidly extinguish fires inside battery enclosures using the detecting tube, which automatically releases extinguishing agents when a fire starts.

Compared to traditional fire suppression systems, fire detection tubes respond quickly, identifying fires in their early stages and deploying extinguishing agents within seconds. These systems are highly targeted, effectively suppressing lithium-ion battery fires without damaging other equipment. Additionally, fire detection tubes can be installed inside energy storage containers without requiring significant modifications to the battery enclosures. However, the fire detection tube system is considered a supplementary fire suppression method and should be used in conjunction with conventional fire safety measures, such as fire alarms and evacuation routes, to ensure comprehensive safety.

4.2.2 Recycling of lithium-ion batteries

Electrochemical energy storage batteries, after usage, can become potential sources of pollution and require proper recycling and treatment. In particular, some metal elements within the batteries are scarce resources, while certain heavy metals and other hazardous substances pose threats to the environment and human health. Therefore the recycling and regeneration of batteries are not only crucial for environmental protection but also for the efficient utilization of resources and the sustainable development of the economy.

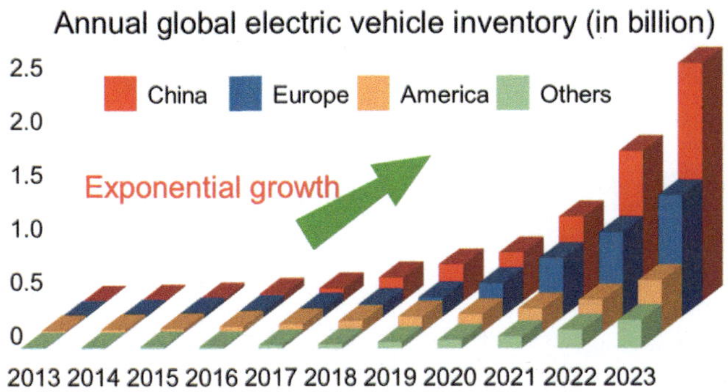

FIGURE 4.21 Changes in global electric vehicle stock across regions from 2013 to 2023.

4.2.2.1 Recycling of energy storage batteries

With the rapid development of China's new energy vehicle and advanced energy storage industries, the demand for power battery assemblies has reached record highs. This surge has significantly increased demand for metals such as lithium, cobalt, and nickel, further increasing China's dependency on foreign resources. At the same time, the growing need for energy storage batteries to support clean energy generation has exacerbated carbon emissions and carbon footprint associated with new battery production [28]. As a result, China places great emphasis on the recycling and reuse of batteries, as well as the establishment of a second-life battery energy storage system. Governments at all levels are promoting the concept of green, low-carbon, and circular development, while gradually shifting their focus to the downstream of the new energy vehicle industry chain.

The global inventory of Electric Vehicles (EVs) has exhibited an exponential growth trend over time, as shown in Fig. 4.21. As early generations of new energy vehicles progressively reach the end of their service life, the volume of retired power batteries is increasing sharply, thereby raising significant safety and environmental concerns. By the end of 2022, the cumulative volume of retired power batteries had reached 20.7 GWh, and it is projected to exceed 380.3 GWh by 2030. On this basis, the power battery recycling industry is expected to enter a period of rapid growth after 2025. Market forecasts indicate that the power battery recycling market will reach approximately RMB 100 billion by 2028 and exceed RMB 300 billion by 2033.

From a global perspective, the European Union's proposed "New Battery Law" requires automotive manufacturers and battery producers to further increase the recycling rates of end-of-life batteries. Specifically, the annual recycling rate is expected to reach 73% by 2030, while the minimum recovery rate for critical metallic materials in waste batteries must increase to 95% by 2031. Consequently, future recycling of retired EV batteries must comply with these regulatory and emission requirements to maintain access to the

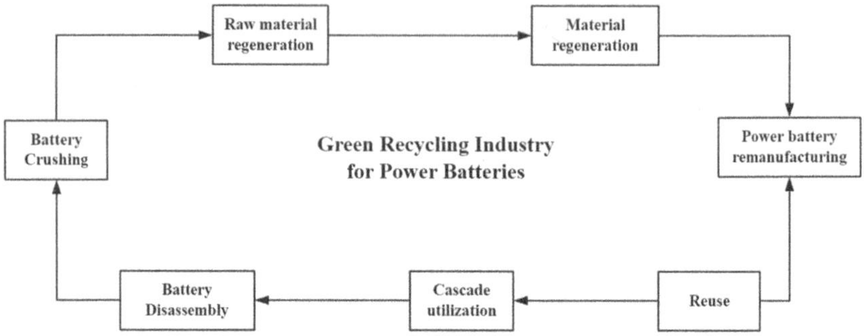

FIGURE 4.22 Recycling and regeneration process of energy storage batteries.

European market. These regulations are expected to have a significant impact on the industrial strategies of battery manufacturers and electric vehicle original equipment manufacturers (OEMs).

Battery material recycling is currently emerging as a key element of the power and energy storage battery supply chain of the future [29]. One of the most important aspects of the battery industry is already the recycling of spent lithium-ion batteries. The energy storage battery recycling and regeneration industrial chain is depicted in Fig. 4.22. Among the significant steps and production processes in the long and complex industrial chain that is involved in recycling and regeneration are battery system recovery, system disassembly, second-life utilization, battery crushing, raw material remanufacturing, material remanufacturing, and battery remanufacturing.

With the continuous increase in the production of energy storage battery systems, the number of retired energy storage lithium-ion batteries is also rising, making recycling an urgent issue. Currently, traditional recycling methods include mechanical, pyrometallurgical, and hydrometallurgical processes. Although these methods have achieved substantial industrialization, the purity of the products recovered is relatively low, and recycling costs are high. There remains significant room for optimization in the energy storage lithium-ion battery recycling industry, making the development of a rational, effective, and low-cost recycling process increasingly important.

4.2.2.2 Recycling methods

The recycling process of electrochemical energy storage batteries primarily includes two stages: pretreatment and valuable metal recovery. Valuable metal recovery is further divided into mechanical recycling, pyrometallurgical recycling, and hydrometallurgical recycling methods.

(1) Pretreatment

Pretreatment is crucial for ensuring the efficiency and safety of battery recycling. Energy storage batteries contain various recyclable materials [30],

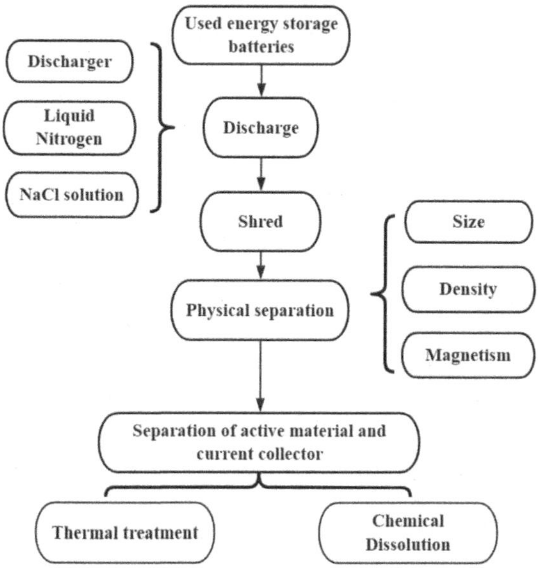

FIGURE 4.23 Pre-treatment process.

such as liquid electrolytes, high-value active materials, and high-purity metals in the casing and current collectors. Different recycling strategies are required to address the distinct properties of these materials. By separating materials with different characteristics during pretreatment, the recycling efficiency can be significantly improved.

The process flow for pretreating retired batteries, which consists of crushing, sorting, discharging, and disassembly, is shown in Fig. 4.23. In general, retired batteries may still have between 0% and 40% of their original charge. Batteries in this state pose risks of electric shock, corrosion, and explosion during storage and recycling. Therefore to ensure safety and improve recycling efficiency, most recycling methods require that waste energy storage batteries undergo a discharging process before further operations to deplete the residual energy and prevent hazards such as short circuits or spontaneous combustion in subsequent recycling steps [31].

Common methods for discharging waste energy storage batteries include physical short-circuiting, low-temperature treatment, and saltwater soaking [32]. Compared to the risks associated with physical short-circuiting and the high costs of low-temperature methods, saltwater soaking offers safety, cost-effectiveness, and simplicity, making it the most commonly used discharging method in the industry. Disassembly involves breaking down the discharged energy storage batteries into components such as casings, positive electrodes, negative electrodes, separators, and electrolytes. Crushing involves breaking down hard parts like electrodes or casings to facilitate subsequent recycling.

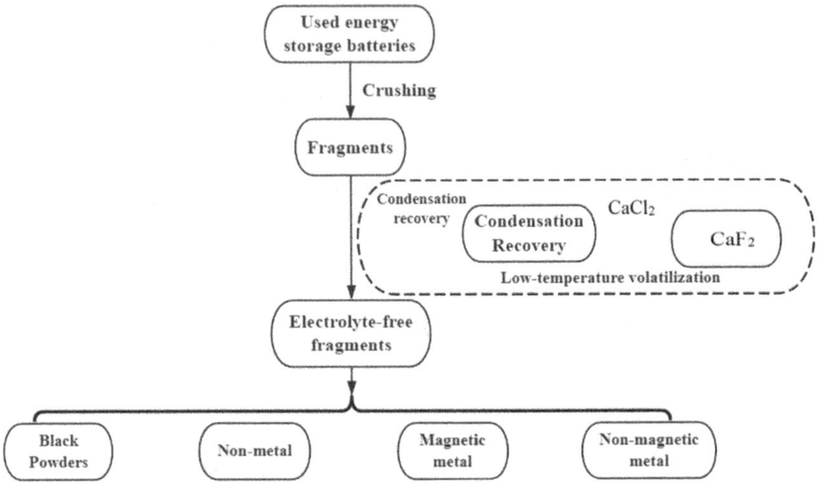

FIGURE 4.24 Mechanical recycling process.

Finally, depending on the subsequent recycling methods for the fragments, physical, chemical, or combined separation methods are used to isolate these materials from one another.

(2) Mechanical Recycling Method

The mechanical recycling method includes steps such as crushing, screening, magnetic separation, fine crushing, and sorting. In industrial practice, mechanical methods are widely used due to their simplicity and efficiency, typically serving as pretreatment before pyrometallurgical processes [33]. The primary goal is to isolate the energy storage batteries' valuable interior electrode components from their external casing. The typical process flow for mechanical recycling techniques is shown in Fig. 4.24.

Mechanical recycling methods effectively separate the casing from the positive and negative electrode sheets. However, since the active materials and current collectors are tightly bound, mechanical methods can only crudely crush both the active materials and current collectors into powder. Therefore the separation of active materials and current collectors requires subsequent separation methods, including acid/alkali leaching, organic solvent dissolution, and high-temperature pyrolysis.

Both acidic and alkaline solutions can dissolve the surface of the current collectors, thereby achieving the separation of active materials from current collectors. However, this method results in a high concentration of aluminum ions in the acidic/alkaline solution and may also lead to the dissolution of some active materials. While the separation process is rapid, the recovery rates for both current collectors and active materials are low.

The organic solvent method involves using organic solvents to dissolve the organic binders, leveraging the mutual solubility of organic solvents to separate the active materials from the electrodes while preserving their integrity. However, the high cost, volatility, and toxicity of organic solvents pose significant challenges to the industrial application of this method.

The high-temperature pyrolysis method utilizes the property of organic binders to decompose at high temperatures to achieve the separation of active materials from current collectors. At temperatures above 400 °C, the binders melt or decompose, allowing the active materials and current collectors to be easily separated under mechanical force. Although this high-temperature pyrolysis method is straightforward and achieves good separation results, it involves significant energy consumption due to the high temperatures required.

(3) Pyrometallurgical Recycling Method

Large amounts of waste batteries can be processed quickly using pyrometallurgy, and the metal products produced are of excellent quality [34]. However, this process uses a lot of energy, is unable to recover electrolytes, and is prone to losing some metals to slag. Furthermore, there are serious environmental dangers associated with calcination since it can release toxic gases, including sulfur dioxide and carbon dioxide.

The electrode materials obtained after pretreatment undergo high-temperature processes to remove volatile components and other impurities. The carbothermal reduction method is frequently employed in pyrometallurgy to recycle cathode materials. In this approach, the cathode materials are cocalcined with reducing agents such as coke. During calcination, the high-valence metal elements in the transition metal oxides are reduced to lower valence states by the carbon materials, as shown in Fig. 4.25. The crystal structure of the metal oxides also undergoes certain changes, facilitating subsequent metal refining. Additionally, carbon materials, as excellent microwave absorbers, enhance heat absorption and reduce thermal losses. However, the carbothermal reduction method cannot process lithium-containing energy storage batteries due to the formation of carbonates, such as lithium carbonate, which has low solubility and consequently lowers the lithium recovery rate in the subsequent refining steps.

Salt-assisted calcination is another effective calcination method that utilizes noncarbonate salts characterized by low melting points, high solubility, and high volatility. Salt-assisted calcination can be classified into processes such as chlorination, sulfidation, and nitration. The added salts convert the insoluble transition metal oxides into soluble metal salts, enabling both to dissolve in the subsequent refining stages. Comparatively, salt-assisted calcination requires lower calcination temperatures, making it more energy-efficient and environmentally friendly.

Large amounts of wasted energy storage batteries may be processed quickly by pyrometallurgy, and the metal products produced are of excellent quality. However, this process uses a lot of energy, is unable to recover electrolytes,

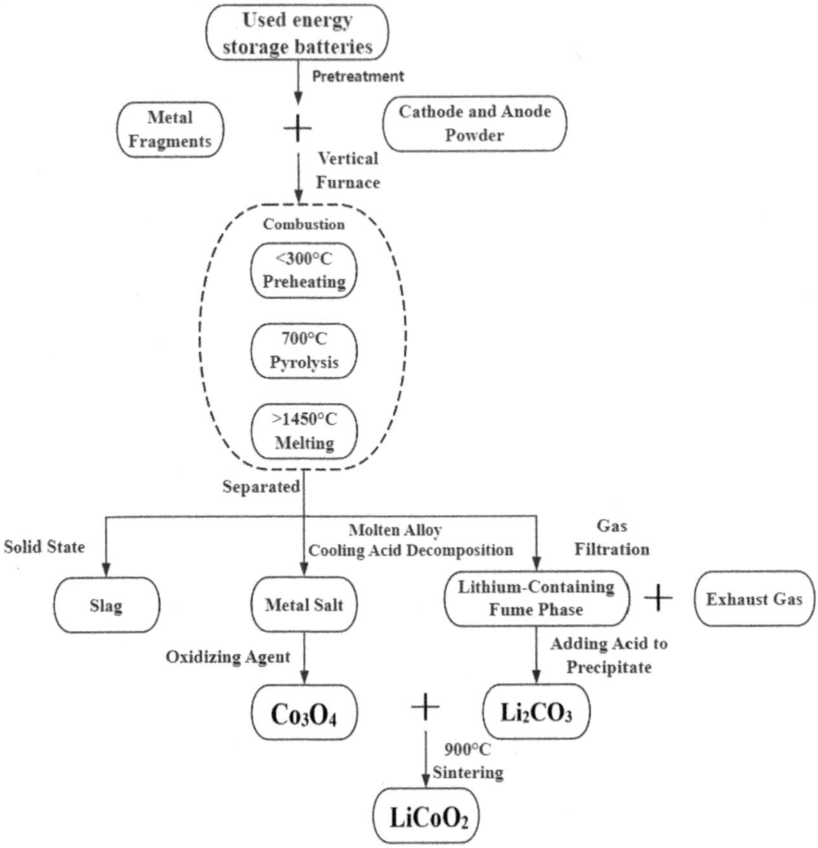

FIGURE 4.25 Fire recycling process.

and is prone to losing some metals to slag. Furthermore, there are serious environmental dangers associated with calcination since it can release toxic gases, including sulfur dioxide and carbon dioxide.

(4) Hydrometallurgical Recycling Method

The hydrometallurgical recycling method is also applicable to the recycling of spent energy storage batteries [35], with its process flow illustrated in Fig. 4.26. This process includes leaching, extraction, precipitation, and electrolysis. The core of hydrometallurgical recycling is to separate metals from the leachate through solvent extraction or precipitation, and then recover the purified metals via electrolysis. Hydrometallurgical processes can be carried out at room temperature, offering advantages such as low energy consumption, minimal greenhouse gas emissions, high metal selectivity, and superior metal purity [36]. Among industrial practices, acid leaching is the most widely used

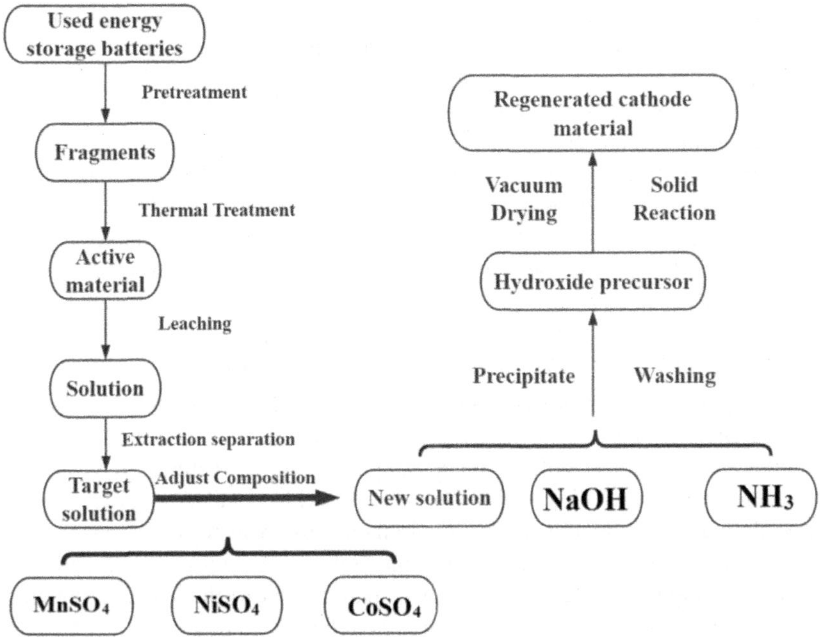

FIGURE 4.26 Hydrometallurgical recovery process.

and is considered the most effective method. Acid leaching can be further divided into inorganic acid leaching and organic acid leaching. The subsequent extraction and precipitation steps separate different metals from the leachate, followed by electrolysis to obtain purified metals.

Inorganic acids can leach nearly all metals from spent energy storage batteries. Commonly used inorganic acids in leaching processes include hydrochloric acid, sulfuric acid, nitric acid, and phosphoric acid. However, inorganic acid leaching suffers from low selectivity, often resulting in leachates containing multiple metal elements, complicating the subsequent separation of these elements. Furthermore, strong inorganic acids can corrode equipment during the leaching process, leading to equipment degradation and the production of harmful gases and acidic wastewater, posing environmental hazards.

Organic acids, on the other hand, are more environmentally friendly and serve as effective leaching agents. The functional groups in organic acids form stable chelates or complexes with metal ions. Additionally, some organic acids, such as ascorbic acid, have reducing properties, allowing them to act without additional reducing agents during the leaching process. However, the generally low acidity of organic acids results in limited efficiency and selectivity for metal leaching.

Aside from acid leaching, ammonia leaching is another effective method for separating and recovering materials from retired batteries. Ammonia leaching exhibits high selectivity in separating nickel, cobalt, and manganese, as manganese has low solubility in ammonia solutions, while nickel and cobalt readily form soluble complexes with ammonia reagents. Therefore ammonia leaching can effectively separate nickel, cobalt, and manganese. Its advantages include low toxicity to both the environment and human health. However, the downside is that ammonia leaching is only applicable to a limited range of elements and types of spent energy storage batteries, restricting its broader applicability.

4.2.2.3 Recycling standards for lithium-ion batteries

The recycling standards follow the current "Recycling and Reuse of Vehicle Power Batteries Part 2: Material Recycling Requirements" and the soon-to-be-implemented "Retirement Technical Requirements for Lithium-Ion Batteries for Energy Storage". These two standards provide detailed specifications for calculating recycling efficiency, requirements for battery retirement, general processing requirements, and pollution control measures. Specifically, the standards stipulate that, from the power battery cell to the stage of obtaining purified metal solution, the comprehensive recovery rate of nickel, cobalt, and manganese in lithium-ion power battery materials should not be less than 98%, and the recovery rate of lithium should not be less than 85%. For nickel-metal hydride power battery materials, the nickel recovery rate should be no less than 98%, while the recovery rates for rare earth elements and other elements should preferably be no less than 95%.

The calculation of the comprehensive recovery rate of copper, iron, and aluminum, as well as the calculation of the recovery rate of nickel, cobalt, manganese, lithium, and rare earth elements, is as follows [37].

The formula for the comprehensive recovery rate of copper, iron, aluminum, and the recovery rate of nickel, cobalt, and manganese:

$$R_j = \frac{\sum_{t=1,2,3} m_{jt}}{\sum_{t=1,2,3} M_{jt}} \times 100\% \tag{4.1}$$

where m_{jt} represents the mass (kg) of element jt obtained from the recycling of a unit mass of a power battery; m_{jt} represents the mass (kg) of element jt in the power battery before recycling. When j = a, a1, a2, and a3 correspond to copper, iron, and aluminum, respectively. When j = b, b1, b2, and b3 correspond to nickel, cobalt, and manganese, respectively.

$$R_k = \frac{\rho_k \times V_k}{M_k} \times 100\% \tag{4.2}$$

where ρ_k represents the mass concentration (kg/m^3) of metal element k in the pure solution obtained from the recovery of a unit mass of target power battery;

V_k represents the volume (m^3) of the pure solution obtained from the recovery of a unit mass of target power battery; MK represents the mass (kg) of metal element k in a unit mass of target power battery, where k denotes lithium or rare earth elements.

The cost, efficiency, speed, and recovery rates of different recycling methods vary significantly. Among them, the mechanical method has the highest recovery rate, achieving a comprehensive recovery rate of over 98% for components such as casings, separators, current collectors, and active materials. Pyrometallurgy can only recover metals from spent energy storage batteries, with recovery rates for different metals ranging from 90% to 98%. Hydrometallurgy, a relatively mature industrial technology, achieves recovery rates of over 99.5% for nickel, cobalt, and manganese from spent energy storage batteries, while advanced techniques for recovering metals like iron and lithium can yield recovery rates of over 92% [38].

The current "Recycling and Reuse of Vehicle Power Batteries-Part 2: Material Recycling Requirements" stipulates that the comprehensive recovery rate for nickel, cobalt, and manganese should not be less than 98%, the recovery rate for lithium should not be less than 85%, and the recovery rate for rare earth elements and other metals should preferably exceed 95%. Under these requirements, most recycling companies tend to combine mechanical pretreatment with hydrometallurgical recycling methods. Additionally, the "Notice on Promoting Coordinated and Stable Development of the Lithium-Ion Battery Industry Supply Chain", issued by the Ministry of Industry and Information Technology (MIIT) in November 2022, emphasizes the need for upstream metal resource companies (e.g., lithium, nickel, cobalt), midstream energy storage battery manufacturers, downstream battery recycling companies, and energy storage battery end-users to work together, strengthen communication, and establish a systematic approach. It also highlights the importance of deepening collaboration with channel distribution and logistics management companies to build a healthy and well-developed energy storage battery supply chain.

4.2.2.4 Second-life utilization of energy storage batteries

Energy storage batteries used in various application fields have strict performance requirements. In the case of vehicle energy storage batteries, they are decommissioned when their capacity drops to 70%–80% of their initial capacity, and their use in vehicles is prohibited thereafter [39]. However, first-round retired energy storage batteries often still have decent performance. By collecting, diagnosing, classifying, and even reassembling these batteries into new battery packs, they can be reused in applications with lower performance requirements. This process of reusing retired energy storage batteries is known as second-life utilization.

Second-life utilization maximizes the economic and environmental value of energy storage batteries by extending their operational life through downgraded applications [40]. This significantly reduces costs in industries related to energy storage batteries and promotes the sustainable use and healthy development of the energy storage battery industry. It also alleviates the pressure of large-scale battery recycling, reduces raw material consumption, and lessens the need for mining lithium, cobalt, and other metal resources, thereby minimizing environmental damage. Moreover, second-life utilization decreases the direct disposal or landfilling of large numbers of retired batteries, reducing the risk of soil and water contamination from waste batteries.

As the energy storage battery industry is currently experiencing rapid growth, second-life utilization has become an important research focus, particularly in areas such as feasibility, economic and environmental benefits, and scalable, systematic management [41]. Although the scale of second-life utilization is still small, with applications mainly limited to demonstration projects or small-scale energy storage and power supply, it is undeniable that most decommissioned batteries have significant potential for second-life use. This process offers considerable economic and environmental value. Therefore it is essential to conduct comprehensive analyses of second-life utilization from technical, economic, environmental, and industrial perspectives to fully realize the benefits of energy storage batteries.

4.2.2.5 Regeneration and restoration of energy storage battery materials

At present, the regeneration and restoration of energy storage battery materials primarily focus on the treatment of cathode active materials [42–44]. Regeneration and restoration refer to the direct processing of extracted cathode materials from spent batteries to restore their original comprehensive performance. Compared to traditional methods where recycled raw materials are reprocessed to prepare precursors, regeneration and restoration can significantly reduce energy consumption and production costs, thereby maximizing the efficient reuse of retired batteries. Currently, the regeneration and restoration of cathode materials remain at the laboratory stage, with reported methods including high-temperature solid-state methods, hydrothermal methods, molten salt methods, chemical methods, and electrochemical methods.

(1) High-Temperature Solid-State Method

The high-temperature solid-state method is the most widely used approach for repairing cathode materials in energy storage batteries. Under high temperatures, elements that deviate from the crystallographic lattice diffuse back to their original positions through thermal driving forces, thereby restoring the complete structure of the cathode material. Irregular cathode material particles also re-fuse at high temperatures, forming a new, coherent whole. This method can also reduce high-valence metal elements to their lower oxidation states. When the content of a particular element in the spent cathode material is insufficient, precise amounts of additional raw materials of the corresponding element must be introduced to restore it into a new cathode material.

The high-temperature solid-state method is often combined with the hydrothermal method or molten salt method to repair spent cathode materials. Compared to other repair methods, the solid-state method improves the grain size of the spent cathode material and reduces material defects through fusion. However, the high energy consumption associated with high temperatures presents a challenge for its industrialization.

(2) Hydrothermal Method

The hydrothermal method is an effective technique for repairing cathode materials. In this method, lithium-rich solutions are used as the medium for mass and heat transfer. Under sealed, heated conditions, the reaction vessel creates a high-pressure environment, allowing high concentrations of lithium ions in the solution to enter lithium vacancies within the spent cathode material, thereby repairing the material's structure. Since the lithium-rich solution contains excess lithium, precise quantification of the lithium source is unnecessary. Common lithium-rich solutions include LiOH, Li_2SO_4, and LiCl. The hydrothermal method typically involves heating to temperatures below 250 °C. After hydrothermal treatment, short-term high-temperature annealing is usually required to improve the material's crystallinity. This repair technique is commonly used for moisture-insensitive cathode materials such as LFP, NCM111, and NCM523. However, high-nickel layered cathode materials are not suitable for hydrothermal repair due to their susceptibility to proton damage in aqueous solutions.

(3) Molten Salt Method

The molten salt method is a homogeneous system that facilitates element dissolution and uniform diffusion, making it particularly suitable for the elemental replenishment of spent cathode materials. The eutectic point of molten salts is lower than the melting point of any single component, providing a highly efficient mass and heat transfer environment at atmospheric pressure, making it more energy-efficient than the high-temperature solid-state method. The molten salt method applies to most cathode materials, and by adjusting the salt ratio and sintering process, single-crystal cathodes can also be obtained. However, the excess lithium salts used to create the molten environment require water washing for removal and are difficult to recycle. Moreover, the strong alkalinity of molten salts leads to severe corrosion of containers, making the overall cost relatively high.

(4) Electrochemical Method

The electrochemical method involves assembling the spent cathode material with a counter electrode in a lithium-containing electrolyte. By applying an external electric field, ion migration and structural reorganization of the cathode material are promoted, thereby replenishing active substances and repairing lattice defects. In addition, this method can dissolve by-products generated during the cathode's operation or introduce new dopants to improve material performance. This method is highly efficient and environmentally friendly, but requires precise control of external electric field parameters. It is also less suitable for nonlithium-ion battery cathode materials.

(5) Chemical Method

The chemical method involves reacting chemical reagents with spent cathode materials to facilitate recovery. It can quickly dissolve by-products formed during the cycling of the spent cathode materials, adjust the oxidation state of elements through redox reactions, and allow for the easy introduction of other elements to enhance performance. The principles and processes of the chemical method are straightforward, offering high processability and scalability, making it one of the most promising methods for large-scale industrial applications. However, the chemical method cannot directly restore the crystal structure of the spent cathode materials, necessitating additional thermal treatment steps. Moreover, the use of chemical reagents may result in waste liquids that pose environmental hazards. In summary, traditional pretreatment and material regeneration routes for retired lithium-ion batteries involve long processes, high waste emissions, and elevated energy consumption and costs, which will not be sustainable in coping with the upcoming peak of battery retirement. On October 22, 2023, the China Association for Science and Technology released the "2023 Major Scientific, Engineering, and Industrial Technology Problems," identifying "How to achieve clean and high-value recycling of new energy waste" as one of the nine major engineering technology challenges. Thus, further development of high-efficiency, high-safety, and low-cost recycling processes for retired electrochemical energy storage batteries could alleviate the shortage of metal resources to some extent. It could also encourage battery manufacturers and upstream raw material companies to focus more on battery recycling, ensuring the market competitiveness and accessibility of battery products in China. The regeneration and reuse of spent power batteries will be a vital strategy to reduce China's dependence on external metal resources and bridge the resource supply gap, making it of significant strategic importance.

References

[1] R. Hu, H. Liu, F. Meng, et al., Electrochemical Energy Storage Power Sources: Design and Applications, Chemical Industry Press, 2025 (in Chinese).

[2] C. Zhang, Y. Wei, P. Cao, M. Lin, Energy storage system: current studies on batteries and power condition system, Renewable and Sustainable Energy Reviews 82 (2018) 3091–3106.

[3] F. Yeganehdoust, A.K. Madikere Raghunatha Reddy, H. Zaghib, Cell architecture design for fast-charging lithium-ion batteries in electric vehicles, Batteries 11 (2025) 20.

[4] A.K.M.A. Habib, M.K. Hasan, G.F. Issa, D. Singh, S. Islam, T.M. Ghazal, Lithium-ion battery management system for electric vehicles: constraints, challenges, and recommendations, Batteries 9 (2023) 152.

[5] B. Ashok, C. Kannan, B. Mason, S.D. Ashok, V. Indragandhi, D. Patel, et al., Towards safer and smarter design for lithium-ion-battery-powered electric vehicles: a comprehensive review on control strategy architecture of battery management system, Energies 15 (2022) 4227.

[6] S. Park, J. Ahn, T. Kang, S. Park, Y. Kim, I. Cho, et al., Review of state-of-the-art battery state estimation technologies for battery management systems of stationary energy storage systems, Journal of Power Electronics 20 (2020) 1526–1540.

[7] L. Lu, X. Han, J. Li, J. Hua, M. Ouyang, A review on the key issues for lithium-ion battery management in electric vehicles, Journal of Power Sources 226 (2013) 272–288.

[8] M. Shen, Q. Gao, A review on battery management system from the modeling efforts to its multiapplication and integration, International Journal of Energy Research 43 (2019) 5042–5075.

[9] S. Nazaralizadeh, P. Banerjee, A.K. Srivastava, P. Famouri, Battery energy storage systems: a review of energy management systems and health metrics, Energies 17 (2024) 1250.

[10] S. Hwangbo, B. Kim, J. Kim, Application of economic operation strategy on battery energy storage system at Jeju, in: IEEE, 2013, pp. 1–8.

[11] R. Kaiser, Optimized battery-management system to improve storage lifetime in renewable energy systems, Journal of Power Sources 168 (2007) 58–65.

[12] P. Siewniak, B. Grzesik, A brief review of models of DC-DC power electronic converters for analysis of their stability, International Journal of Electronics 101 (2014) 1405–1426.

[13] V. Fernão Pires, E. Romero-Cadaval, D. Vinnikov, I. Roasto, J.F. Martins, Power converter interfaces for electrochemical energy storage systems – a review, Energy Conversion and Management 86 (2014) 453–475.

[14] Z. Rehman, I. Al-Bahadly, S. Mukhopadhyay, Multiinput DC–DC converters in renewable energy applications – an overview, Renewable and Sustainable Energy Reviews 41 (2015) 521–539.

[15] N. Zhang, D. Sutanto, K.M. Muttaqi, A review of topologies of three-port DC–DC converters for the integration of renewable energy and energy storage system, Renewable and Sustainable Energy Reviews 56 (2016) 388–401.

[16] Ş. Demirbaş, H. Fidanboy, E. Kurt, Exploration of the chaotic behaviour in a buck–boost converter depending on the converter and load elements, Journal of Electronic Materials 45 (2016) 3889–3899.

[17] D. Mishra, R. Tummala, T. Brigmon, A. Jain, Simulations-based investigation of the effectiveness of fire suppression techniques for safe, large-scale storage of Li-ion batteries, Journal of Energy Storage 84 (2024) 110870.

[18] N.S. Spinner, C.R. Field, M.H. Hammond, B.A. Williams, K.M. Myers, A.L. Lubrano, et al., Physical and chemical analysis of lithium-ion battery cell-to-cell failure events inside custom fire chamber, Journal of Power Sources 279 (2015) 713–721.

[19] P. Huang, P. Ping, K. Li, H. Chen, Q. Wang, J. Wen, et al., Experimental and modeling analysis of thermal runaway propagation over the large format energy storage battery module with Li4Ti5O12 anode, Applied Energy 183 (2016) 659–673.

[20] M. Börner, A. Friesen, M. Grützke, Y.P. Stenzel, G. Brunklaus, J. Haetge, et al., Correlation of aging and thermal stability of commercial 18650-type lithium ion batteries, Journal of Power Sources 342 (2017) 382–392.

[21] D.P. Finegan, E. Darcy, M. Keyser, B. Tjaden, T.M.M. Heenan, R. Jervis, et al., Identifying the cause of rupture of Li-ion batteries during thermal runaway, Advanced Science 5 (2018).

[22] S. Mallick, D. Gayen, Thermal behaviour and thermal runaway propagation in lithium-ion battery systems – a critical review, Journal of Energy Storage 62 (2023) 106894.

[23] Z. Zeng, X. An, C. Peng, X. Ruan, Z. Song, C. Dang, et al., Study on the thermal runaway behavior and mechanism of 18650 lithium-ion battery induced by external short circuit, Applied Thermal Engineering 258 (2025) 124569.

[24] A.H. Mohammed, R. Esmaeeli, H. Aliniagerdroudbari, M. Alhadri, S.R. Hashemi, G. Nadkarni, et al., Dual-purpose cooling plate for thermal management of prismatic lithium-ion batteries during normal operation and thermal runaway, Applied Thermal Engineering 160 (2019) 114106.

[25] S. Sirikasemsuk, S. Wiriyasart, P. Naphon, N. Naphon, Thermal cooling characteristics of Li-ion battery pack with thermoelectric ferrofluid cooling module, International Journal of Energy Research 45 (2021) 8824–8836.

[26] I. Cho, S. Park, J. Kim, A fire risk assessment method for high-capacity battery packs using interquartile range filter, Journal of Energy Storage 50 (2022) 104663.

[27] Y. Guo, X. Wang, J. Gao, Z. He, S. Yao, X. Zhou, et al., In situ extinguishing mechanism and performance of self-portable microcapsule fire extinguishing agent for lithium-ion batteries, Journal of Energy Storage 93 (2024) 112393.

[28] Q. Tan, J. Li, L. Yang, G. Xu, Cascade use potential of retired traction batteries for renewable energy storage in China under carbon peak vision, Journal of Cleaner Production 412 (2023) 137379.

[29] J. Lin, X. Zhang, E. Fan, R. Chen, F. Wu, L. Li, Carbon neutrality strategies for sustainable batteries: from structure, recycling, and properties to applications, Energy & Environmental Science 16 (2023) 745–791.

[30] S. Nazari, A.B. Vakylabad, K. Asgari, J. Li, H. Khoshdast, Y. He, et al., Bubbles to batteries: a review of froth flotation for sustainably recycling spent lithium-ion batteries, Journal of Energy Storage 84 (2024) 110702.

[31] T. Fan, W. Liang, W. Guo, T. Feng, W. Li, Life cycle assessment of electric vehicles' lithium-ion batteries reused for energy storage, Journal of Energy Storage 71 (2023) 108126.

[32] C. Fleuriault, X. Guan, J. Grogan, Extraction and recycling of battery materials, JOM 71 (2019) 4445–4446.

[33] M. Liu, X. Pan, X. Sun, H.C. Kim, W. Shen, D. De Castro Gomez, et al., Lithium-ion battery recycling: bridging regulation implementation and technological innovations for better battery sustainability, Environmental Science & Technology 58 (2024) 21385–21388.

[34] C. Liu, F. Xiao, Z. Yang, L. Jiang, D. Gao, W. Zhang, Electrochemical synergistic recycling: coupling lithium extraction with regeneration of spent $LiFePO_4$ cathode materials, Industrial & Engineering Chemistry Research 63 (2024) 15291–15300.

[35] D. Saju, J. Ebenezer, N. Chandran, N. Chandrasekaran, Recycling of lithium iron phosphate cathode materials from spent lithium-ion batteries: a mini-review, Industrial & Engineering Chemistry Research 62 (2023) 11768–11783.

[36] P. Gao, P. Yuan, T. Yue, X. Zhao, B. Shen, Recycling metal resources from various spent batteries to prepare electrode materials for energy storage: a critical review, Journal of Energy Storage 68 (2023) 107652.

[37] S. Windisch-Kern, A. Holzer, L. Wiszniewski, H. Raupenstrauch, Investigation of potential recovery rates of nickel, manganese, cobalt, and particularly lithium from NMC-type cathode materials (LiNixMnyCozO2) by carbo-thermal reduction in an inductively heated carbon bed reactor, Metals 11 (2021) 1844.

[38] Y. Wang, H. Yin, L. An, An upcoming global challenge: efficient recycling for end-of-life lithium-ion batteries, Global Challenges 6 (2022).

[39] X. Gu, H. Bai, X. Cui, J. Zhu, W. Zhuang, Z. Li, et al., Challenges and opportunities for second-life batteries: key technologies and economy, Renewable and Sustainable Energy Reviews 192 (2024) 114191.

[40] Y. Yang, W. Zhu, C. Xie, Y. Shi, F. Liu, W. Li, et al., A layered bidirectional active equalization method for retired power lithium-ion batteries for energy storage applications, Energies 13 (2020) 832.

[41] T. Zhao, Y. Choi, C. Wu, Z. Zhang, C. Wang, D. Liu, et al., A review on direct regeneration of spent lithium iron phosphate: from waste to wealth, Science of the Total Environment 957 (2024) 177748.

[42] J. Wu, L. Xiao, P. Liu, Y. Zhu, J. Li, Direct regeneration and upcycling of cathode material from spent lithium ion batteries: recent advances and perspectives, Separation and Purification Technology 355 (2025) 129574.
[43] Y. Shin, S. Kim, S. Park, J. Lee, J. Bae, D. Kim, et al., A comprehensive review on the recovery of cathode active materials via direct recycling from spent Li-ion batteries, Renewable and Sustainable Energy Reviews 187 (2023) 113693.
[44] Y. Yin, C. Li, X. Hu, D. Zuo, L. Yang, L. Zhou, et al., Rapid, direct regeneration of spent $LiCoO_2$ cathodes for Li-ion batteries, ACS Energy Letters 8 (2023) 3005–3012.

Chapter 5

Hydrogen energy storage and utilization

Huai-Jun Lin
Institute of Advanced Wear & Corrosion Resistance and Functional Materials, Jinan University, Guangzhou, China

Contents

5.1 Introduction

Hydrogen energy is a versatile renewable energy carrier that may be extracted from water and stored for extended periods. It can be incorporated into the convergence point of electricity, heat, and gas networks via the technical integration of hydrogen energy fuel cells [1]. The combustion heat of hydrogen is significantly higher than that of many other fuels. The heat of combustion of hydrogen is 142 MJ/kg, approximately three times greater than that of gasoline. It is globally regarded as the optimal method for interconnecting power and gas grids, while simultaneously addressing the energy resource and environmental crises. Since the 1970s, hydrogen has been presented as a viable energy carrier to partially or fully replace fossil fuels, and it is considered the most promising energy source for replacing carbon-based energy in pursuit of a totally sustainable society and economy. Despite the challenges in advancing

Modern Energy Storage. https://doi.org/10.1016/B978-0-32-395419-8.00011-8

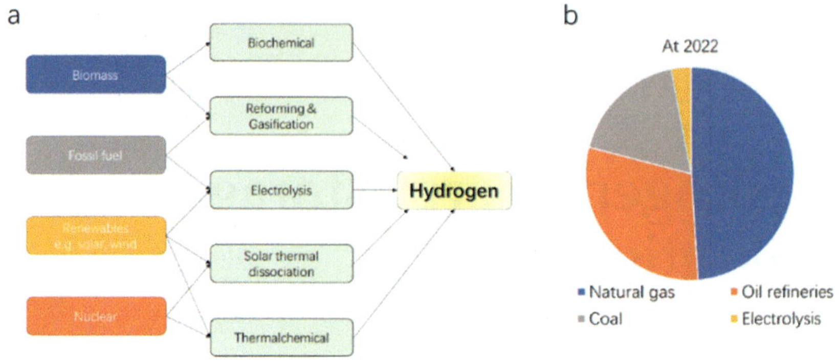

FIGURE 5.1 a) Different routes of hydrogen production, b) Sources of hydrogen production in 2022 [6].

hydrogen energy technology over the past 50 years, there has been a growing global interest in these technologies due to the pressing demands for carbon peaking and carbon-neutrality plans in recent decades [2]. Hydrogen energy technology is crucial for most countries globally to mitigate carbon emissions and attain carbon peaking and neutrality [3–5].

Although hydrogen is the most plentiful element in the universe, hydrogen gas is nearly absent in the Earth's atmosphere (less than 1 ppm). Consequently, hydrogen can be acquired alone through artificial processing. Hydrogen gas can be generated from several feedstocks, including water, coal, natural gas, biomass, and hydrides, via thermal, electrolytic, or photolytic methods. **Fig. 5.1a** shows that hydrogen production can be classified into four primary categories: biomass, fossil fuels, renewables, and nuclear energy [6]. Hydrogen can be generated from several sources using diverse ways, each possessing distinct advantages and downsides. At present, it is challenging to identify a single technology that may simultaneously satisfy the diverse requirements of various working environments. Large-scale hydrogen production can be achieved through fossil fuel upgrading; however, this process relies on nonrenewable fossil fuels and emits pollutants, imposing a significant environmental burden. Techniques like water electrolysis or biomass conversion are ecologically sustainable and can be directly or indirectly derived from solar energy. Nonetheless, these approaches continue to encounter significant technological limitations, and the expense of hydrogen production remains excessively elevated. **Fig. 5.1b** illustrates the current sources of hydrogen generation, indicating that hydrogen is predominantly derived from fossil fuels, including coal, natural gas, and oil. Electrolysis accounts for about 3%–4% of overall hydrogen generation. Researchers and engineers around are actively advancing renewable energy hydrogen generation technologies and implementing it extensively. By 2050, it is anticipated that hydrogen generation from renew-

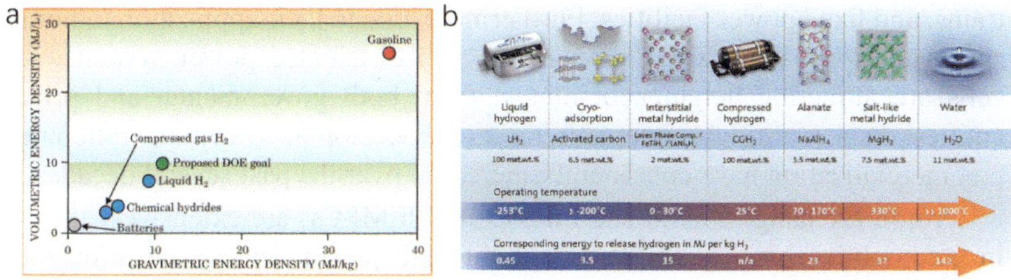

FIGURE 5.2 Diagram illustrates the energy densities of hydrogen stored in different phases and materials, including the weight of the storage containers and the equipment necessary for filling and dispensing the fuel [7]. Typical operating regimes for various hydrogen storage technologies [9].

able energy sources will comprise around 50-70% of total hydrogen production. [https://www.iea.org/fuels-and-technologies/hydrogen]

The generated hydrogen is subsequently refined, stored, and transported in gaseous, liquid, or solid states, and ultimately combusted to produce electricity or energy. The sole output of this cycle process is purified water. Nonetheless, in terms of energy density, both the volumetric and gravimetric energy densities of hydrogen are significantly lower than those of oil or gasoline. As illustrated in **Fig. 5.2a**, gasoline considerably surpasses lithium-ion batteries and hydrogen in gaseous, liquid, or compound states [7]. The low energy density of hydrogen results from its minimal physical density of about 0.09 g L^{-1}. Consequently, a significant issue of hydrogen energy storage is achieving high-density hydrogen storage under appropriate conditions. **Fig. 5.2b** illustrates the standard operating regimes for several hydrogen storage methods [8]. The temperature for hydrogen storage rises as the energy required to liberate hydrogen escalates. Two primary approaches that function under ambient settings are compressed gaseous hydrogen and interstitial metal hydride. The requirements for hydrogen energy storage vary under different operating situations. On-board hydrogen energy storage necessitates high volumetric and gravimetric hydrogen density, rapid hydrogenation kinetics, adjustable hydrogen desorption, excellent reversibility, low cost, and additional criteria. Currently, prototype fuel-cell vehicles utilize high-pressure gaseous hydrogen storage methods; however, the associated space and weight constraints render this alternative less appealing. Hydrogen storage for stationary applications necessitates stringent safety standards. Moreover, there are rigorous specifications for materials, including hydrogen absorption/desorption temperature and pressure, kinetic performance, cost, cycle life, and environmental dependence, among others.

In the realm of hydrogen usage, hydrogen fuel cells are paramount. Hydrogen fuel cells convert hydrogen or hydrogen-derived fuels into electricity and heat through an electrochemical reaction with oxygen. These cells have demonstrated their benefits, including high energy conversion efficiency, minimum noise, and zero or low carbon emissions. They are adaptable, appropriate for many transportation modalities, including automobiles, aircraft, trains, and fixed power facilities. Furthermore, due to their application in human space missions, underwater submarines, and distributed power systems, fuel cells are attracting heightened interest from both governmental and private entities. As a result, investments in research, development, demonstration, and commercialization have consistently increased over the past several decades.

Proton Exchange Membrane Fuel Cells (PEMFCs) are extensively utilized in automobile vehicles. PEMFCs exhibit high power density and efficiency; however, even minimal impurities in the fuel (e.g., CO, CO_2, H_2S, and NH_3), air streams (e.g., NO_x, SO_x, CO, and CO_2), or components of the fuel cell system (e.g., Fe^{3+} and Cu^{2+} from corrosion) can significantly poison the anode, membrane, and cathode [10]. Furthermore, the expense of PEMFCs has not adequately aligned with the long-term cost objectives established by funding organizations, including the U.S. Department of Energy (DOE). The cost of the Toyota Mirai PEMFC system is approximately $233 per kW with a manufacturing volume of 1000 stacks annually, whereas the Department of Energy has established a target system cost of $30 per kW by 2025. Conversely, solid oxide fuel cells (SOFCs) have exceptional temperature tolerances and improved efficiency, rendering them optimal for power generation and hybrid power applications. These fuel cells are the subject of intensive study and development due to their high efficiency and capability to utilize various hydrocarbon fuels, as well as their versatile electrolyte materials [11].

Hydrogen energy can serve as a conduit for energy interconnection, facilitating synergistic optimization among energy networks, including electricity, thermal energy, wind energy, solar energy, and gas. In recent years, hydrogen energy storage has demonstrated exceptional attributes in various domains, including transportation, energy structure optimization, and power grid enhancements globally. In Japan, hydrogen energy vehicles Mirai and Clarity have been effectively commercialized for several years. Hydrogen-powered houses have been showcased in many locations, demonstrating their remarkable potential as hotels, office buildings, and residential structures. California, United States, now has approximately 40 hydrogen refueling stations, and the U.S. government aims to establish 200 hydrogen stations in the state by 2025. Germany and France serve as international centers for hydrogen and fuel cell demonstration, research, and development in Europe. The German Energiewende seeks to nearly entirely substitute fossil fuels with renewable energy sources by 2050, with hydrogen and fuel cell technologies providing significant support. The Chinese government published the National Development and Reform Commission's medium- and long-term strategy for the hydrogen energy industry's development (2021–2035). A hydrogen energy supply system is proposed, utilizing industrial by-product hydrogen and nearby renewable hydrogen production. Approximately 50,000 fuel cell vehicles will be deployed, along with several hydrogen refueling stations, aiming for a renewable hydrogen production capacity of 100,000-200,000 tons per year. Hydrogen energy storage is increasingly garnering global attention. This chapter will provide a detailed review of hydrogen

TABLE 5.1 Summary lists of some typical hydrogen energy storage demonstration projects in the past decade.

Year	Project name	Hydrogen production source	Storage technology
2010	Wind-Hydrogen-Diesel on Ramea, Canada	Wind	Gaseous
2011	ENERTRAG Hybrid-Power-Plant, Germany	Wind	Gaseous
2021	Hydrogen energy installation on Bozcaada, Turkey	Wind	Gaseous
2012	RH2-WKA, Germany	Wind	Gaseous
2014	Hydrogen Energy Storage Demonstration, China	Solar	Solid
2016	Hydrogen Office Project, UK	Wind	Gaseous
2021	Lu'an MW Hydrogen Comprehensive Utilization Demonstration, China	Wind and Solar	Gaseous

energy storage systems, summarizing the latest research and applications in the field. Finally, the remaining obstacles and future directions in hydrogen energy storage will be addressed.

5.2 Hydrogen energy storage

5.2.1 Hydrogen energy storage system

Renewable energy sources are recognized as pivotal in the transition to a less carbon-intensive and more sustainable energy system. According to the International Energy Agency (IEA), renewable energy sources encompass sun, wind, hydro, biofuels, and more. Nonetheless, the unpredictability and variability of wind and solar energy can result in issues such as voltage fluctuations, frequency deviations, and power harmonics when integrated into large-scale grids. Hydrogen energy storage is an optimal solution for large-scale comprehensive development, consumption, and storage of energy.

The hydrogen economy is an appealing notion that has evolved since its proposal in the 1970s, influenced by advancements in social, economic, and technological domains. This concept entails a novel system that use hydrogen for energy transport, encompassing the integration of hydrogen generation, storage, transit, distribution, and utilization. Table 5.1 encapsulates several representative hydrogen energy storage demonstration projects over the past decade. Hydrogen serves as an energy carrier that facilitates the recycling of carbon-free energy and efficiently harnesses variable renewable energy through the electricity-hydrogen-electricity (or chemical raw material) process, which is both environmentally sustainable and effective.

To achieve hydrogen energy storage, three primary components must be thoroughly considered: hydrogen generation, hydrogen storage, and hydrogen consumption. Hydrogen storage is integral to the hydrogen energy storage process. The subsequent section will address hydrogen storage technologies and materials.

5.2.2 Hydrogen storage technologies

Hydrogen storage methods are categorized into three forms based on the state of hydrogen: 1) high-pressure gaseous hydrogen storage, 2) liquid hydrogen storage, and 3) solid-state hydrogen storage materials. This section provides a concise overview of three hydrogen storage systems.

The method for gaseous high-pressure hydrogen storage offers a straightforward equipment design, minimal energy expenditure for hydrogen compression, and rapid hydrogen charging and discharging rates. It has been extensively utilized in large-scale high-pressure hydrogen storage vessels for transportation, hydrogen storage units for refueling stations, and high-pressure hydrogen storage containers for fuel cell vehicles, among others. There are three primary categories of commonly utilized high-pressure hydrogen storage vessels [8]: stationary, vehicle, and bulk transportation. Stationary storage is typically utilized in hydrogen refueling stations. For vehicle applications, it is advantageous for hydrogen storage to possess high volumetric and gravimetric density, lightweight characteristics, and appropriate cost-effectiveness. For bulk transportation, which conveys hydrogen from producers to end users, it is optimal for storage containers to possess substantial hydrogen storage capacity and lightweight characteristics to save costs. Nonetheless, the hydrogen storage density of the high-pressure gaseous technique is inadequate. The mass hydrogen storage density of a 15 MPa high-pressure steel cylinder is around 0.5%–1%, which inadequately fulfills the hydrogen source needs for mobile devices.

Hydrogen was first liquefied by Sir James Dewar, who compressed gaseous hydrogen to 180 bar and precooled it to $-250\,^{\circ}C$. The volumetric density of liquid hydrogen reaches 71 kg/m^3, and the mass hydrogen storage efficiency is substantial, with the mass of hydrogen storage constituting almost 40% of the entire mass of the container. Nonetheless, hydrogen liquefaction is an energy-intensive procedure characterized by low equipment efficiency. The boiling point of liquid hydrogen is around 20.3 K, and its latent heat is relatively low, making liquid hydrogen highly susceptible to evaporation [12]. Despite significant advancements in hydrogen liquefaction technologies, they are unsuitable for long-term storage, resulting in limited possibilities in the civilian sector. The primary application domain of liquid hydrogen is in rocket and aerospace engineering.

A further approach to storing hydrogen in a liquid form utilizes Liquid Organic Hydrogen Carriers (LOHCs) [13], including cyclohexane, methylcyclohexane, decalin, and perhydro-N-ethylcarbazole, among others. The LOHC systems comprise pairs of hydrogen-deficient and hydrogen-saturated organic molecules capable of storing hydrogen via reversible hydrogenation and dehydrogenation reactions, aided by catalysts. Typically, precious metals like platinum or palladium serve as catalysts to expedite hydrogenation and dehydrogenation reactions. Recently, nonprecious metal catalysts have been extensively investigated as effective catalysts for LOHCs, with metal hydrides demonstrating particularly noteworthy performance attributed to the hydrogen spillover effect [14]. A primary advantage of utilizing LOHCs is their compatibility with the current gasoline infrastructure. Recent demonstrations have shown the viability of LOHC technology in commercial stationary energy storage, hydrogen distribution to refueling stations, and direct LOHC fuel cell applications [15].

The storage of hydrogen in solid-state materials is regarded as one of the most promising and secure ways for the global implementation of hydrogen energy. There are two fundamental bonding methods for material-based hydrogen storage: physical adsorption and chemical absorption. Physical sorption materials typically consist of porous substances characterized by a significantly large specific surface area, including porous carbon materials, mesoporous silica, Metal-Organic Frameworks (MOFs), and Covalent-Organic Frameworks (COFs) [16]. The storage of hydrogen in these porous materials is reversible, and the kinetics of adsorption and desorption can be easily regulated by pressure. Nonetheless, their hydrogen absorption and desorption temperatures typically reach as low as liquid nitrogen temperature (77 K), which is far from practical application.

The chemical hydrogen absorption materials mostly consist of complicated chemical hydrides and metal hydrides. Despite the capacity of complex chemical hydrides to store significant quantities of hydrogen, such as 18.4 wt.% for LiBH4 and 10.5 wt.% for LiAlH4, their hydrogen storage reversibility is inadequate, and the material costs are prohibitively high for practical uses. Conversely, metal hydrides are considered advantageous materials for practical hydrogen storage due to several factors: 1) a volumetric capacity exceeding 110 kg/m^3, 2) a gravimetric capacity ranging from 2 to 7 wt% of hydrogen, 3) preparation methods typically involving metallurgical processes, making them suitable for large-scale applications, and 4) excellent hydrogen storage reversibility, exceeding thousands of cycles, along with an extended operational lifespan. The subsequent part will address materials-based hydrogen storage, concentrating on metal hydrides, which are considered the most promising hydrogen storage materials to date.

5.2.3 Metal hydrides for hydrogen storage

Hydrogen storage in alloys is regarded as a promising method for practical applications [17]. Palladium and its alloys were discovered over a century ago to absorb hydrogen under ambient circumstances in nature. Nevertheless, noble

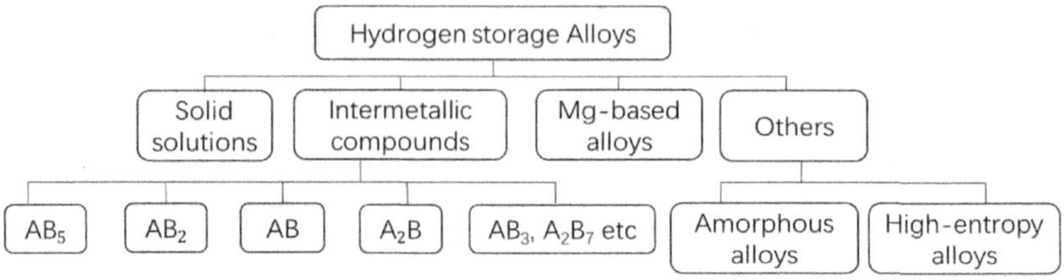

FIGURE 5.3 A family tree of hydrogen storage alloys.

metals have not been regarded for extensive hydrogen storage applications. Until the late 1960s, the initial hydrogen storage alloy SmCo5 was intermittently developed at the Philips Laboratory in the Netherlands [18]. Over the subsequent fifty years, numerous hydrogen storage alloys have been engineered and created. The family of hydrogen storage alloys can now be categorized into the following kinds, as seen in **Fig. 5.3**. Hydrogen storage alloys are typically designated as various A_xB_y kinds, with the A component generally comprising elements that can efficiently absorb hydrogen, such as Li, Mg, Ca, Ti, V, Pd, and Rare Earth (RE) metals. The B side elements are those that naturally exclude hydrogen, such as Al, Cr, Mn, Fe, Co, Ni, Cu, Zn, Sn, and Si [19].

AB_5-type alloys represent the inaugural generation of hydrogen storage alloys. The $SmCo_5$ alloy, initially investigated as a magnetic material, was inadvertently discovered to be effective for hydrogen absorption. Subsequently, $LaNi_5$ was developed as a highly promising reversible hydrogen storage alloy under ambient conditions [18]. Researchers have discovered that replacing La with Ce or Mm (mischmetal) can modify thermodynamic properties and significantly lower costs [20]. Additionally, substituting Ni with Al can regulate hydrogen desorption pressure across extensive ranges of pressure and temperature while maintaining superior hydrogen storage kinetics and capacity [21]. It has been conclusively shown that partial substitution of Al or Sn on the B-side of AB_5 alloys substantially mitigates the disproportionation problem in $LaNi_5$/$MmNi_5$-based alloys. Over the past fifty years, multiple iterations of Mm(Ni, Co, Mn, Al) have constituted the foundation of the majority of contemporary commercial Ni-MH battery anodes [19].

AB_2-type alloys represent a diverse and versatile class of hydrogen storage materials, distinguished by thermodynamic features that render them appropriate for functioning under standard climatic circumstances. Prominent instances within this category encompass $TiCr_{1.8}$, $TiMn_{1.5}$, and $ZrMn_2$ alloys. The AB_2 alloys are predominantly organized into two Laves phase structures: the hexagonal (prototype $MgZn_2$, C_{14}, space group P63/mmc) and the cubic (prototype $MgCu_2$, C_{15}, space group Fd-3m). These materials can cyclically absorb and desorb hydrogen within temperature ranges of 0 to 100 °C and pressure ranges of 1 to 10 atm. In situations necessitating wider temperature and pressure ranges, AB_2 alloys typically demonstrate superior storage capabilities relative to AB_5 alloys. The AB_2 alloys, especially those with titanium as the principal A-element, offer significant economic advantages compared to the AB5 alloys [19].

The initial functional AB hydrides were exhibited using TiFe by Reilly and Wiswall in the 1970s [22]. TiFe and its substitutional variations currently represent the most superior of the AB alloys. Activating TiFe-based AB-type alloys is challenging due to the inherent oxidation layer of Ti-alloys. Edalati et al. indicated that substituting Fe with Mn and employing High-Pressure Torsion (HPT) significantly improves activation performance and air resistance for hydrogenation [23]. The enhancement of hydrogen storage characteristics is attributed to the creation of lattice defects and amorphous regions within the alloy matrix, which facilitate numerous pathways for hydrogen diffusion and hydride production.

A variety of crystal structures can exist within the A_2B-type hydrogen storage alloys family. In one subfamily, the A side components may include Ti, Zr, or Hf, whereas the B side element is commonly Ni. Another member of the A_2B alloys family is derived from the Mg_2Ni alloy, as developed by Reilly and Wiswall [24]. Regrettably, A_2B-type alloys exhibit minimal applicability within the operational range of 0–100 °C and 1–10 atm due to the excessive stability of the hydrides, which inhibits hydrogen desorption. Consequently, they are deemed unsuitable for practical on-board applications.

AB_3-type alloys have a stratified configuration of AB_5 and AB_2 (A_2B_4) unit cells along the c-axis. Kadir and colleagues [25] characterized the RMg_2Ni_9 compounds (where R is La, Ce, Pr, Nd, Sm, or Gd) that display a $PuNi_3$ structure. Over the past decade, research on AB_3-type hydrogen storage alloys has increased, notably for usage in Nickel-Metal Hydride (Ni-MH) batteries. AB_3-type alloys provide superior hydrogen storage capacity within the same volume relative to the more often used AB_5-type alloys. Nonetheless, their longevity is frequently undermined by fragmentation and deterioration resulting from oxidation or corrosion in electrolytic environments [26]. Apart from AB_3-type alloys, A_2B_7- and A_5B_{19}-type alloys have been recognized as viable materials for hydrogen storage, broadening the selection in this domain [27].

Magnesium-based materials are acknowledged as viable candidates for hydrogen storage, with a capacity of up to 7.6 wt.% in MgH_2 [28]. Nonetheless, the considerable durability of these materials requires temperatures ranging from 300 to 350 °C for hydrogen storage. Extensive research has been undertaken over the years to improve the hydrogen storage capabilities of magnesium-based materials. Methods, including nanostructuring, catalytic enhancement, alloying, and composite production, have been proposed as highly effective for improving performance. Recent studies indicate that ambient-temperature reversibility for MgH_2 can be effectively attained by producing nonconfined ultrafine nanoparticles [29], resulting from both thermodynamic instability and kinetic enhancements. **Table 5.2** delineates various methodologies associated with these tactics and their effects on hydrogen storage kinetics and thermodynam-

TABLE 5.2 Strategies to tune the hydrogen storage properties of Mg-based materials.

Strategies	Preparation methods	Typical materials	Advances in hydrogen storage	Refs.
Nanostructuring	High-energy ball milling	Mg, MgNi alloys	Fast kinetics	[30,31]
	Rapid solidification	Mg-RE-Ni alloys	Fast kinetics,	[32–34]
	Chemical Vapor Deposition	Mg-Ni alloys	Fast kinetics, Air-stability, good cycling ability	[35]
	Magnetron sputtering	Mg-Ni thin film alloys, Mg-Ni/MmM_5	Fast kinetics, destabilized thermodynamics	[36,37]
Alloying	Induction melting	$LaMg_2Ni$ alloy	Fast kinetics, Altered thermodynamics	[38]
	High energy ball milling	Mg-In, Mg-Sn alloys	Fast kinetics, Altered thermodynamics	[39,40]
	Hydrogen combustion synthesis	Mg-Ni-Cu alloys	Fast kinetics	[41]
	hydrogen plasma-metal reaction	Mg-Ni, Mg-Cu alloys	Fast kinetics	[42,43]
Catalyzing	High energy ball milling	Mg-Nb_2O_5	Fast kinetics	[44]
	In situ reaction formation	Mg-Ni-$CeH_{2.73}$, Mg-CeO_2/$CeH_{2.73}$	Fast kinetics	[45,46]
	Solution synthesis	Mg-Tm	Fast kinetics	[47]
Formation of composites	High-energy ball milling	MgH_2-$LiBH_4$ composites	Altered reaction path, destabilized thermodynamics	[48]

ics. High-energy ball milling is the most often utilized and efficient technique for the preparation of magnesium-based hydrogen storage materials.

Currently, hydrogen storage alloys have broadened their applications beyond mere hydrogen storage and Ni-MH batteries. They are currently engaged in several fields, including hydrogen purification, hydrogen compression, thermal engines, and pumps. Moreover, these alloys are utilized in specialized areas such as liquid hydrogen management, boil-off capture, cryogenic cooling, chemical catalysis, ammonia synthesis, methane production, diamond fabrication, permanent magnet manufacturing, and various other applications [19]. The following section will address typical successful demonstrations and uses of hydrogen energy storage.

5.3 Hydrogen energy storage utilizations

5.3.1 Nickel metal hydride battery

The Ni-MH batteries represent one of the most successful and widely utilized applications of hydrogen energy storage. The Ni-MH battery is a commercially viable rechargeable battery technology utilized in both consumer and industrial sectors, owing to its great design flexibility, superior energy and power efficiency, nontoxic environmental impact, and reasonable cost [49]. Ni-MH batteries have several significant advantages, including extended cycle life, superior performance across a wide range of operating temperatures, elevated charge rates, and high energy density.

$LaNi_5$ alloys, possessing a theoretical specific capacity of 372 mA·h/g, have been extensively utilized as the negative electrode material in Ni-MH batteries. The drawback of $LaNi_5$ material is the excessive volume growth of the unit cell during hydrogenation, approximately 23.5%. The frequent expansion and contraction of alloy lattices during hydrogen absorption and desorption lead to capacity degradation. To address this issue, various multielement alloy materials have been engineered by substituting La and Ni in LaNi5, resulting in reduced volume expansion of the alloy's unit cell upon hydrogen absorption, thus enhancing the cyclic stability of the hydrogen storage alloy electrode material. The commercially produced Ni-MH battery since 1991 has had significant advancements, enhancing its specific energy from around 50 Wh kg^{-1} to 100 Wh kg^{-1}, increasing energy density, and delivering exceptional performance across a broad temperature range of −40 to 80 °C.

Over the past few decades, various hydrogen storage alloys, such as AB_5-type alloys, AB_2-type alloys, Mg-based alloys, Ti-V-based alloys, and RE-Mg-Ni alloys, have been utilized as anode materials for Ni-MH battery applications [50]. Despite the progressive replacement of Ni-MH batteries by Lithium-Ion Batteries (LiBs) owing to their superior volumetric and gravimetric energy density, LiBs have come to dominate the battery market, particularly in mobile devices, laptops, and various other applications in recent years. Ni-MH batteries continue to constitute a significant share of portable electronics, including camcorders, as well as Hybrid Vehicles (HVs) and Electric Vehicles (EVs).

5.3.2 Hydrogen fuel cell vehicles

Hydrogen Fuel Cell Vehicles (HFCVs) utilize fuel-cell stacks where pure hydrogen traverses a membrane to react with oxygen, generating electricity that propels the vehicle and emits water vapor [51]. Since 2015, three HFCVs have been available for purchase from three manufacturers: the Toyota Mirai, the Honda Clarity Fuel Cell, and the Hyundai Nexo SUV. To date, an increasing number of HFCVs have entered the market.

Toyota is the preeminent maker of hydrogen fuel cell vehicles, with its fuel cell development originating in the early 1990s. The company is not only the world's first manufacturer of hybrid electric vehicles and hydrogen fuel cell vehicles, but also possesses a complete array of power solution assets, along with extensive technological reserves and a robust patent portfolio for Hybrid Electric Vehicles (HEV), Plug-in Hybrid Electric Vehicles (PHEV), Battery Electric Vehicles (BEV), and Fuel Cell Electric Vehicles (FCEV). Originating in 1996, the FCEV-1 was the inaugural model in a series of first-generation Toyota fuel cell vehicles based on the RAV4, marking the debut of Toyota's fuel cell concept car. The concept, originally outfitted with a storage unit containing the BCC-type hydrogen storage alloy, was further improved in 1997 with the introduction of the FCEV-2, which featured a reformer that extracted hydrogen from methanol. In 2015, Toyota unveiled the inaugural mass-produced commercial hydrogen fuel cell vehicle, the Mirai [52]. The initial Mirai is equipped with two hydrogen storage tanks that hold 4.6 kg of hydrogen at a pressure of up to 70 MPa. In 2021, the updated Mirai features three hydrogen storage tanks. The enhancement in fuel economy and the augmented total volume of the three fuel tanks, now approximately 5.6 kg, guarantee the new Mirai's cruising range of up to 650 km.

Over the past decade, China has achieved significant advancements in the HFCV industry, influenced by governmental regulatory initiatives and corporate manufacturing efforts [53]. In October 2021, Guangzhou Automobile Group Co., Ltd. (GAC) introduced car-hailing vehicles powered by hydrogen fuel cell systems. In September 2022, Shanghai Automotive Industry Corporation (SAIC) MAXUS MIFA officially introduced the Xiangdao travel platform, with an initial deployment of 80 vehicles mostly utilized in the Shanghai region.

The global market for fuel cell electric vehicles (FCEVs) attained a valuation of \$1.5 billion in 2024, with a projected annual growth rate of 27.2% over the subsequent ten years. The expansion is mostly attributable to the rise in hydrogen-refueling stations and governmental investments in hydrogen infrastructure development. In 2024, the Asia Pacific region is projected to comprise over 70% of the global market for fuel cell electric vehicles (FCEVs). Substantial investments in hydrogen refueling infrastructure and production facilities in nations including Japan, South Korea, China, and Australia are propelling market expansion in the area.

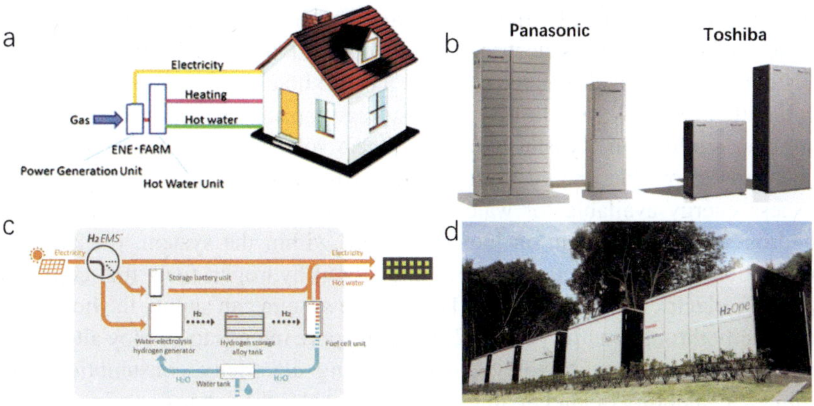

FIGURE 5.4 a) Schematic illustration and b) Images of ENE-FARM. c) System structure and d) Image of the Toshiba H_2One system.

5.3.3 Hydrogen powered houses

Hydrogen energy storage can be utilized in hydrogen-powered residences. Japan occupies a prominent position in hydrogen-powered homes, influenced by governmental laws as well as circumstances such as its remote regions and frequent seismic activity. In 2009, the initial commercialization of home fuel cell systems utilizing Polymer Electrolyte Fuel Cells (PEFCs) occurred. The combined heat and power (CHP) systems utilizing fuel cells for residential usage in Japan are referred to as "ENE-FARM," derived from the term energy farm [54]. The apparatus uses natural gas reforming to generate hydrogen. The power generation efficiency can attain approximately 40%, while the heat utilization efficiency is over 60%, and the operational lifespan may exceed 100,000 hours. The ENE-FARM system utilizes natural gas to on-site manufacture hydrogen, which subsequently enters the fuel cell to generate electricity. The heat produced during this process is then employed for heating and hot water delivery (**Fig. 5.4a-b**). The ENE-FARM system exhibits exceptional energy-saving efficiency (~95%), decreases utility expenses, and promotes an environmentally sustainable ecosystem. Furthermore, ENE-FARM is capable of generating electricity during power outages to meet residential energy demands.

The Toshiba H2One system is among the most effective independent hydrogen energy systems for residential usage. Numerous cities in Japan have implemented H2One systems in parks to supply power for lighting, air conditioning, and various other devices. H2One utilizes solely solar and water as fuel, enabling it to autonomously supply energy and hot water during emergencies. The H2One system comprises a hydrogen generator, a hydrogen storage tank, hydrogen fuel cells, and several batteries (**Fig. 5.4c-d**). The H2One system optimizes solar energy utilization by converting and storing variable solar power, influenced by time of day and weather conditions, into hydrogen, which is then supplied as electric power on demand. In 2016, the Toshiba H2One system commenced electricity supply to a resort hotel in Kyushu. The peak fuel cell output is 54 kW, and the electricity storage capacity is 1.8 MWh. The ample summer sunlight in Kyushu allows H2One's solar power system to generate adequate renewable energy to completely sustain the hotel's 12 rooms, with excess energy available for water electrolysis and hydrogen generation. The hydrogen is contained in an integrated tank within the system, prepared for utilization when required. In winter, the stored hydrogen fuels fuel cells that produce electricity and hot water. The H2One system can supply the hotel with electricity throughout the year. In 2019, Singapore intends to employ alternative energy sources to sustainably meet its rising electricity consumption. The limited availability of land in Singapore complicates the expansion of equipment capacity or the installation of new systems to equilibrate power supply and demand. The H2One's tiny 20-foot container design, featuring a small metal hydride hydrogen storage, is well-suited for constrained environments and can efficiently address these difficulties.

5.3.4 Hydrogen energy supplemented grid system

Hydrogen gas functions as a supplementary energy carrier, similar to electricity. Hydrogen energy storage technique leverages the interconversion of electricity and hydrogen energy, converting intermittently variable and surplus electricity into hydrogen energy for storage. When electricity output is inadequate, hydrogen can be utilized to produce electricity via fuel cells or other power generation technologies and reintegrated into the grid system. Despite Power-to-Gas (P2G) systems exhibiting inferior short-term round-trip efficiency compared to most battery storage alternatives, they provide the distinct benefit of independently scaling power and energy capacities, hence enabling extensive and prolonged storage. Colbertaldo et al. [55] conducted a study analyzing the California power system to ascertain the requisite capacities of electrolyzers and fuel cells for a scenario reliant entirely on renewable energy, especially in light of substantial augmentations in wind and solar generation. The study revealed that this shift would require a significant increase in generation and storage, including 94 GW of solar PV, 40 GW of wind, and 77 GW of electrolysis capacity. The analysis indicated that a varied combination of generation technologies could diminish the total necessary capacity in comparison to scenarios predominantly reliant on either wind or solar energy alone.

González et al. [56] conducted an energy assessment of a solar hydrogen storage facility situated in the R&D facilities of INTA (Spanish National Institute for Aerospace Technology) in Huelva, Spain. **Fig. 5.5** illustrates the general configuration of INTA's hydrogen storage system. Three distinct storage methods have been selected for the storage of produced hydrogen: low-pressure gaseous, metal hydride, and high-pressure gaseous. The results indicated that the

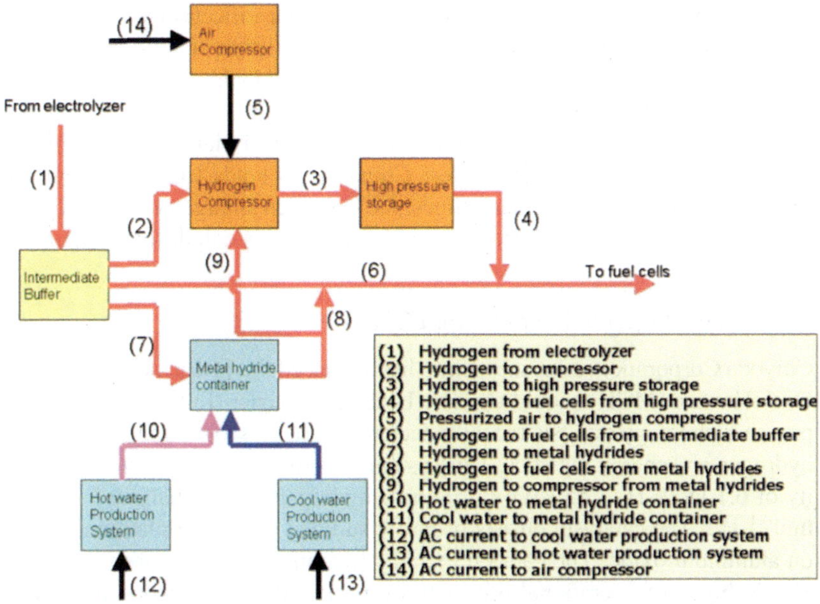

FIGURE 5.5 General layout for INTA's hydrogen storage system [56].

electric energy density storage per mass of the entire hydrogen storage system is comparable to that of lithium-ion batteries and exceeds the normal values of lead-acid batteries. The volumetric electric energy density of the entire hydrogen storage system is comparable to that of conventional wet lead-acid batteries, but lower than that of lithium-ion batteries.

A collaboration between the National Grid and the Town of Hempstead has been undertaken in the United States to construct one of the foremost and most extensive clean hydrogen projects. The HyGrid Project, located on Long Island, seeks to diminish carbon emissions by incorporating green hydrogen into the existing distribution infrastructure, potentially supplying heat to approximately 800 households and powering 10 municipal vehicles.

The Chinese government demonstrates significant interest in integrating the grid with hydrogen energy storage. The National Energy Administration of China has recently recommended enhancing the development of a new power grid system, concentrating on renewable hydrogen production, electricity-hydrogen integration, and fuel cell technologies. In October 2022, a 100 MW/200 MWh energy storage demonstration project in Lu'an City, Anhui Province, began operations. This initiative is a collaborative investment by Shanghai Electric Group, State Grid Comprehensive Energy Company, and China Energy Construction Anhui Electric Power Design Institute. This signifies China's inaugural integration of the complete hydrogen production, storage, and electricity generation process at a megawatt scale. In March 2023, two hydrogen power stations operated by China Southern Electricity Grid were integrated into the grid, signifying the inaugural utilization of solid hydrogen for electricity generation in China. Seven advanced solid hydrogen storage facilities are located in the Nansha Smart Hydrogen Station in Guangzhou, Guangdong province, southern China. The second hydrogen power plant is located in Kunming, Yunnan province, southwestern China, and it holds 165 kg of previously stored hydrogen energy. At peak times, the hydrogen storage system can generate 2300 kilowatt-hours of electricity and maintain this output for 23 hours.

5.3.5 Hydrogen delivery using LOHCs

Chiyoda Corporation, a Japanese firm, has advanced the TOLuene (TOL)/MethylCycloHexane (MCH) Liquid Organic Hydrogen Carrier (LOHC) system into a proven technology and has recently declared its large-scale implementation for hydrogen and energy logistics. This system possesses a hydrogen storage capacity of 6.1 wt.%-H (1.55 kWh L^{-1}) and a hydrogenation enthalpy of −68.3 kJ mol^{-1} H2. The catalyst consists of partly sulfided nanoscale platinum clusters on aluminum oxide. The catalyst exhibits exceptional performance, exceeding 10,000 hours of operational life, with over 95% MCH conversion and greater than 99% toluene selectivity in the MCH dehydrogenation reaction. Chiyoda said that the Advanced Hydrogen Energy Chain Association for Technology Development (AHEAD) has achieved a significant milestone by transporting hydrogen, transformed into MethylCycloHexane (MCH), across international waters with a specialized chemical tanker. This procedure entails the synthesis of MCH from toluene and hydrogen in Brunei, and is thereafter transported to an ENEOS oil refinery in Japan. AHEAD established the requisite infrastructure in Brunei as a component of its worldwide hydrogen supply chain demonstration initiative. ENEOS employed the MCH for an ongoing demonstration project, with assistance from the Consortium for Resilient Oil Supply System (CROS). The first chemical tanker arrived at the refinery's receiving terminal in February 2022, following several months of temporary storage in outdoor tanks at a transit port in Singapore.

5.3.6 Others

There are some other applications of hydrogen energy storage, such as hydrogen fuel cell backup power [57]. In August 2023, a novel emergency power vehicle utilizing solid-state hydrogen storage, created by China Southern Power Grid Co., Ltd., commenced trial operations in Guangzhou, China. This vehicle incorporates a solid-state hydrogen storage system capable of holding 100 kg of hydrogen. Unlike high-pressure gaseous hydrogen storage, this system operates at a much lower pressure of 3 MPa, which is just one-tenth of the pressure in traditional high-pressure gaseous systems, significantly enhancing its safety. The vehicle is fitted with four proton exchange membrane fuel cells, each with a rated power of 50 kW, enabling it to provide continuous power for six hours at a total output of 200 kW. A hydrogen energy heat compressor is another application example of hydrogen energy storage [58,59]. The thermal sorption compression utilizing metal hydrides is a robust and effective technique for transforming thermal energy into hydrogen. Besides the storage characteristics of metal hydrides, such as weight- and volume-based hydrogen storage capacity, the kinetics of hydrogen absorption and desorption, and heat transfer efficiency, the thermodynamic properties of the metal-hydrogen interaction are essential. These properties lead to a pressure for absorption and desorption that varies with temperature. Hydrogen energy storage is also widely used in other industrial fields, including oil refining, chemical industry, steel production, and ammonia manufacturing [60,61]

5.4 Hydrogen safety

Hydrogen is extremely combustible, devoid of odor, and ignites with a practically imperceptible, faint blue flame in daylight. The hydrogen safety issue is paramount in all facets of hydrogen energy consumption, including hydrogen production, storage, transit, and usage [62]. The primary risk is in its leakage, which may result in a fire or explosion, especially during instances involving high-pressure hydrogen. Hydrogen energy is perceived by many as inadequate or insufficiently safe, despite the increasing presence of HFCVs on the roads. HFCVs are as safe as conventional vehicles, as their high-pressure tanks are engineered to withstand high-speed collisions without spilling or breaching. The hydrogen tanks and their components would probably endure even if the remainder of the vehicle were obliterated in a collision. Fortunately, there have been no known injuries or fatalities related to the hydrogen components in the sold HFCVs. Recent years have witnessed hydrogen refueling station explosions in many countries, including Norway, the United States, and South Korea. Concerns with hydrogen safety are increasingly significant in hydrogen production, storage, transportation, refueling stations, and consumption. Researchers have indicated that human errors are a substantial contributor to hydrogen-related incidents. Even minor mistakes can result in significant repercussions. In addition to improving the safety design of hydrogen systems and using appropriate materials, it is essential to elevate the understanding of hydrogen safety among managers and operators across the hydrogen industry [63].

5.4.1 Hydrogen embrittlement

Hydrogen embrittlement is a well-documented phenomenon that causes subcritical crack propagation in materials, fracture initiation, and catastrophic failure, leading to a deterioration in mechanical qualities such as ductility, toughness, and strength. Multiple methods can generate hydrogen embrittlement in materials, including hydride production, hydrogen-enhanced depolymerization, hydrogen-enhanced local plasticity, and adsorption-induced dislocation emission [64]. The precise process responsible for the material's embrittlement remains unidentified, and the study will continue to ascertain the underlying source of this phenomenon. A range of hydrogen permeation protection technologies has been established to mitigate hydrogen embrittlement, including the application of coatings on material surfaces to obstruct hydrogen atom infiltration into the substrate; the removal of detrimental elements in steel to decrease hydrogen atom traps; and the development of high-purity, hydrogen-resistant steel.

5.4.2 Hydrogen sensors

Reliable and economical hydrogen sensors should be employed to detect the occurrence of a leak. The operational principles of hydrogen sensors are fundamentally grounded in the distinct interactions between hydrogen and materials like platinum and palladium. The detection and quantification of hydrogen gas concentration can be achieved through either the reaction itself (reaction heat, swapped charge carriers) or alterations in the sensing materials' properties (resistance and volume expansion) [65]. Hydrogen gas detection systems can be categorized into eight primary types: 1) catalytic; 2) thermal conductivity; 3) electrochemical; 4) resistance-based; 5) work function-based; 6) mechanical; 7) optical; and 8) acoustic. Several of these have been commercially utilized for years, manufactured by numerous producers with diverse performance capabilities and price ranges. Other methods are not as commercially developed; nonetheless, current research suggests that certain upcoming hydrogen detection applications necessitate additional performance criteria, including expedited response, reduced power consumption, and multipoint detection, among others.

5.5 Challenges and future prospective

5.5.1 Challenges

Hydrogen energy storage technology represents a novel form of large-scale energy storage with significant potential, appropriate for both extremely short-term and extremely long-term energy storage applications. Hydrogen energy storage technology, akin to chemical battery energy storage, exhibits reduced dependence on external conditions, facilitates site selection for project development, and has a diminished environmental impact. Nonetheless, significant unresolved technical impediments persist in comparison to existing energy storage methods.

The energy conversion efficiency of hydrogen energy storage is comparatively poor. The present technological efficiency of hydrogen production via water electrolysis is approximately 60%, with a 5% loss in hydrogen storage, transportation, and retail processes, and a fuel cell gas-to-electricity conversion efficiency of around 70%. The overall energy conversion efficiency is below 40%, far lower than the around 80% efficiency of power-to-power systems utilizing battery energy storage.

Second, the expense associated with electrochemical water electrolysis is prohibitively high for practical applications. The current energy conversion efficiency of hydrogen generation via water electrolysis is 20%–30% lower than that of fossil-energy hydrogen production methods, including reforming and industrial by-product purification.

Third, high-capacity hydrogen storage technology has not yet achieved widespread commercialization. High-pressure gaseous hydrogen storage technology is well developed; nonetheless, its storage density remains poor. Hydrogen storage with metal hydrides is an optimal compact storage solution; yet, the research and development of efficient and cost-effective reversible hydrogen storage materials remains unestablished. All existing hydrogen storage systems are inadequate for optimal hydrogen energy storage across diverse operational situations.

Fourthly, the transportation of hydrogen across great distances presents significant challenges. Hydrogen transportation typically employs vehicles or pipes. Truck transportation is exceptionally versatile and appropriate for short-distance or small-scale deliveries. Long-distance pipeline transportation necessitates stringent technical specifications, elevated construction expenses, and substantial maintenance costs.

Furthermore, the economic model for hydrogen energy storage remains ambiguous. Concerns persist regarding the ecological impact of hydrogen energy storage.

5.5.2 Future prospective

Hydrogen energy serves as an optimal new energy carrier, and the hydrogen energy storage system, when integrated with electric energy, exhibits significant advantages, particularly in the realm of renewable energy generation [66]. Currently, technical and material constraints result in issues of low efficiency and elevated costs. **Fig. 5.6** illustrates a pathway delineating the phases of hydrogen development across diverse applications [67]. The comprehensive advantages of hydrogen and fuel cell technologies are realized when implemented at scale and across many applications.

The electrochemical electrolysis of seawater is an attractive and environmentally sustainable method for large-scale hydrogen production. The practical application of this approach is impeded by electrode side reactions and corrosion, which are intensified by the intricate composition of seawater. Xie et al. [68] presented an innovative direct seawater electrolysis method that effectively addresses these challenges. They exhibited a system that maintained steady functioning at a current density of 250 mA cm^{-2} for more than 3200 hours. This novel method facilitates efficient, adaptive, and scalable direct seawater

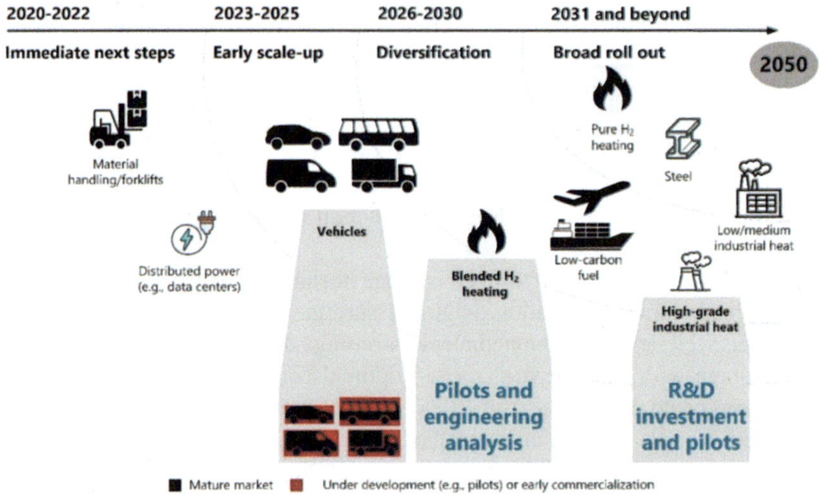

FIGURE 5.6 Hydrogen application road Map of the United States [67].

electrolysis, equivalent to freshwater splitting, without substantially raising operational expenses, rendering it highly promising for practical applications.

Hydrogen storage materials and technologies are essential for the practical and large-scale application of hydrogen energy. Research on hydrogen storage materials and technologies has been conducted for fifty years; yet, it still fails to fully satisfy practical needs. Novel materials possessing elevated hydrogen storage capacity impose significant technical prerequisites for the extensive implementation of hydrogen energy. The research and development of hydrogen storage materials and systems characterized by high hydrogen storage density, near-room temperature functionality, adjustable hydrogen absorption and desorption, and extended lifespan continue to be strategic and advanced subjects.

The environmental and energy consequences of hydrogen fuel cell vehicles are greatly affected by the source of the hydrogen. Given the present industrial environment and technical breakthroughs, it is essential to highlight that improvements in hydrogen generation and fuel cell technology, as well as enhancements in manufacturing processes for fuel cell systems, are critical for the evolution of HFCVs. Future research on hydrogen energy storage should concentrate on electrolyzer technology, fuel cell technology, the study and development of hydrogen storage materials, and thorough performance evaluation. In the domain of hydrogen energy storage, all nations have established a comprehensive framework encompassing fundamental research, applied research, and demonstration projects.

References

[1] T.N. Veziroğlu, S. Şahi˙n, 21st century's energy: hydrogen energy system, Energy Conversion and Management 49 (2008) 1820.
[2] K. Sasaki, H-W. Li, A. Hayashi, J. Yamabe, T. Ogura, S.M. Lyth, Hydrogen Energy Engineering, Springer, Japan, 2016.
[3] I. Staffell, D. Scamman, A. Velazquez Abad, P. Balcombe, P.E. Dodds, P. Ekins, N. Shah, K.R. Ward, The role of hydrogen and fuel cells in the global energy system, Energy & Environmental Science 12 (2019) 463.
[4] O. Ehret, K. Bonhoff, Hydrogen as a fuel and energy storage: success factors for the German Energiewende, International Journal of Hydrogen Energy 40 (2015) 5526.
[5] F. Cuevas, J. Zhang, M. Latroche, The vision of France, Germany, and the European Union on future hydrogen energy research and innovation, Engineering 7 (2021) 715.
[6] F. Zhang, P. Zhao, M. Niu, J. Maddy, The survey of key technologies in hydrogen energy storage, International Journal of Hydrogen Energy 41 (2016) 14535.
[7] G.W. Crabtree, M.S. Dresselhaus, M.V. Buchanan, The hydrogen economy, Physics Today 57 (2004) 39.
[8] J. Zheng, X. Liu, P. Xu, P. Liu, Y. Zhao, J. Yang, Development of high pressure gaseous hydrogen storage technologies, International Journal of Hydrogen Energy 37 (2012) 1048.
[9] R. von Helmolt, U. Eberle, Fuel cell vehicles: status 2007, Journal of Power Sources 165 (2007) 833.
[10] X. Cheng, Z. Shi, N. Glass, L. Zhang, J. Zhang, D. Song, Z-S. Liu, H. Wang, J. Shen, A review of PEM hydrogen fuel cell contamination: impacts, mechanisms, and mitigation, Journal of Power Sources 165 (2007) 739.
[11] F. Ramadhani, M.A. Hussain, H. Mokhlis, S. Hajimolana, Optimization strategies for Solid Oxide Fuel Cell (SOFC) application: a literature survey, Renewable and Sustainable Energy Reviews 76 (2017) 460.
[12] M. Aasadnia, M. Mehrpooya, Large-scale liquid hydrogen production methods and approaches: a review, Applied Energy 212 (2018) 57.
[13] A. Lin, G. Bagnato, Revolutionising energy storage: the latest breakthrough in liquid organic hydrogen carriers, International Journal of Hydrogen Energy 63 (2024) 315.
[14] H. Yu, X. Li, J. Zheng, Beyond hydrogen storage: metal hydrides for catalysis, ACS Catalysis 14 (2024) 3139.
[15] P. Preuster, C. Papp, P. Wasserscheid, Liquid organic hydrogen carriers (LOHCs): toward a hydrogen-free hydrogen economy, Accounts of Chemical Research 50 (2017) 74.
[16] P. Jena, Materials for hydrogen storage: past, present, and future, Journal of Physical Chemistry Letters 2 (2011) 206.
[17] Z-X. Yang, X-G. Li, Q-L. Yao, Z-H. Lu, N. Zhang, J. Xia, K. Yang, Y-Q. Wang, K. Zhang, H-Z. Liu, L-T. Zhang, H-J. Lin, Q-J. Zhou, F. Wang, Z-M. Yu, J-M. Ma, 2022 roadmap on hydrogen energy from production to utilizations, Rare Metals 41 (2022) 3251.
[18] J.H.N. van Vucht, F.A. Kuijpers, H.C.A.M. Bruning, Reversible room-temperature absorption of large quantities of hydrogen by intermetallic compounds, Philips Research Reports (1970) 133.
[19] G. Sandrock, A panoramic overview of hydrogen storage alloys from a gas reaction point of view, Journal of Alloys and Compounds 293 (295) (1999) 877.
[20] M. Odysseos, P. De Rango, C.N. Christodoulou, E.K. Hlil, T. Steriotis, G. Karagiorgis, G. Charalambopoulou, T. Papapanagiotou, A. Ampoumogli, V. Psycharis, E. Koultoukis, D. Fruchart, A. Stubos, The effect of compositional changes on the structural and hydrogen storage properties of (La–Ce)Ni5 type intermetallics towards compounds suitable for metal hydride hydrogen compression, Journal of Alloys and Compounds 580 (2013) S268.
[21] M.H. Mendelsohn, D.M. Gruen, A.E. Dwight, $LaNi_{5-x}Al_x$ is a versatile alloy system for metal hydride applications, Nature 269 (1977) 45.
[22] J. Reilly, R. Wiswall Jr, Formation and properties of iron titanium hydride, Inorganic Chemistry 13 (1974) 218.

[23] K. Edalati, M. Matsuo, H. Emami, S. Itano, A. Alhamidi, A. Staykov, D.J. Smith, S.-i. Orimo, E. Akiba, Z. Horita, Impact of severe plastic deformation on microstructure and hydrogen storage of titanium-iron-manganese intermetallics, Scripta Materialia 124 (2016) 108.

[24] J.J. Reilly, R.H. Wiswall, Reaction of hydrogen with alloys of magnesium and nickel and the formation of Mg_2NiH_4, Inorganic Chemistry 7 (1968) 2254.

[25] K. Kadir, T. Sakai, I. Uehara, Synthesis and structure determination of a new series of hydrogen storage alloys; RMg2Ni9 (R = La, Ce, Pr, Nd, Sm and Gd) built from MgNi2 Laves-type layers alternating with AB5 layers, Journal of Alloys and Compounds 257 (1997) 115.

[26] W. Liu, C.J. Webb, E.M. Gray, Review of hydrogen storage in AB3 alloys targeting stationary fuel cell applications, International Journal of Hydrogen Energy 41 (2016) 3485.

[27] H. Yan, W. Xiong, L. Wang, B. Li, J. Li, X. Zhao, Investigations on AB3-, A2B7- and A5B19-type LaYNi system hydrogen storage alloys, International Journal of Hydrogen Energy 42 (2017) 2257.

[28] L. Ouyang, F. Liu, H. Wang, J. Liu, X-S. Yang, L. Sun, M. Zhu, Magnesium-based hydrogen storage compounds: a review, Journal of Alloys and Compounds 832 (2020) 154865.

[29] X. Zhang, Y. Liu, Z. Ren, X. Zhang, J. Hu, Z. Huang, Y. Lu, M. Gao, H. Pan, Realizing 6.7 wt% reversible storage of hydrogen at ambient temperature with non-confined ultrafine magnesium hydrides, Energy & Environmental Science 14 (2021) 2302.

[30] S. Orimo, H. Fujii, K. Ikeda, Notable hydriding properties of a nanostructured composite material of the Mg2Ni-H system synthesized by reactive mechanical grinding, Acta Materialia 45 (1997) 331.

[31] L. Zaluski, A. Zaluska, J.O. StromOlsen, Nanocrystalline metal hydrides, Journal of Alloys and Compounds 253 (1997) 70.

[32] H.J. Lin, L.Z. Ouyang, H. Wang, J.W. Liu, M. Zhu, Phase transition and hydrogen storage properties of melt-spun Mg3LaNi0.1 alloy, International Journal of Hydrogen Energy 37 (2012) 1145.

[33] H.J. Lin, L.Z. Ouyang, H. Wang, D.Q. Zhao, W.H. Wang, D.L. Sun, M. Zhu, Hydrogen storage properties of Mg–Ce–Ni nanocomposite induced from amorphous precursor with the highest Mg content, International Journal of Hydrogen Energy 37 (2012) 14329.

[34] H-J. Lin, C. Zhang, H. Wang, L. Ouyang, Y. Zhu, L. Li, W. Wang, M. Zhu, Controlling nanocrystallization and hydrogen storage property of Mg-based amorphous alloy via a gas-solid reaction, Journal of Alloys and Compounds 685 (2016) 272.

[35] J. Zhang, Y. Zhu, H. Lin, Y. Liu, Y. Zhang, S. Li, Z. Ma, L. Li Metal, Hydride nanoparticles with ultrahigh structural stability and hydrogen storage activity derived from microencapsulated nanoconfinement, Advanced Materials 29 (2017) 1700760.

[36] C.J. Chung, S-C. Lee, J.R. Groves, E.N. Brower, R. Sinclair, B.M. Clemens, Interfacial alloy hydride destabilization in Mg/Pd thin films, Physical Review Letters 108 (2012) 106102.

[37] A. Baldi, M. Gonzalez-Silveira, V. Palmisano, B. Dam, R. Griessen, Destabilization of the Mg-H system through elastic constraints, Physical Review Letters 102 (2009) 226102.

[38] L.Z. Ouyang, L. Yao, H.W. Dong, L.Q. Li, M. Zhu, Hydrogen storage properties of $LaMg_2Ni$ prepared by induction melting, Journal of Alloys and Compounds 485 (2009) 507.

[39] H.C. Zhong, H. Wang, L.Z. Ouyang, M. Zhu, Microstructure and hydrogen storage properties of Mg–Sn nanocomposite by mechanical milling, Journal of Alloys and Compounds 509 (2011) 4268.

[40] H.C. Zhong, H. Wang, J.W. Liu, D.L. Sun, M. Zhu, Altered desorption enthalpy of MgH2 by the reversible formation of Mg(In) solid solution, Scripta Materialia 65 (2011) 285.

[41] H. Gu, Y. Zhu, L. Li, Hydrogen storage properties of Mg–Ni–Cu prepared by hydriding combustion synthesis and mechanical milling (HCS + MM), International Journal of Hydrogen Energy 34 (2009) 2654.

[42] H. Shao, Y. Wang, H. Xu, X. Li, Preparation and hydrogen storage properties of nanostructured Mg2Cu alloy, Journal of Solid State Chemistry 178 (2005) 2211.

[43] H.Y. Shao, H.R. Xu, Y.T. Wang, X.G. Li, Preparation and hydrogen storage properties of Mg_2Ni intermetallic nanoparticles, Nanotechnology 15 (2004) 269.

[44] G. Barkhordarian, T. Klassen, R. Bormann, Fast hydrogen sorption kinetics of nanocrystalline Mg using Nb2O5 as catalyst, Scripta Materialia 49 (2003) 213.

[45] L.Z. Ouyang, X.S. Yang, M. Zhu, J.W. Liu, H.W. Dong, D.L. Sun, J. Zou, X.D. Yao, Enhanced hydrogen storage kinetics and stability by synergistic effects of in situ formed CeH2.73 and Ni in CeH2.73-MgH2-Ni nanocomposites, Journal of Physical Chemistry C 118 (2014) 7808.

[46] H-J. Lin, J-J. Tang, Q. Yu, H. Wang, L-Z. Ouyang, Y-J. Zhao, J-W. Liu, W-H. Wang, Zhu M. Symbiotic, $CeH_{2.73}/CeO_2$ catalyst: a novel hydrogen pump, Nano Energy 9 (2014) 80.

[47] J. Cui, J. Liu, H. Wang, L. Ouyang, D. Sun, M. Zhu, X. Yao, Mg-TM (TM: Ti, Nb, V, Co, Mo or Ni) core-shell like nanostructures: synthesis, hydrogen storage performance and catalytic mechanism, Journal of Materials Chemistry A 2 (2014) 9645.

[48] J.J. Vajo, S.L. Skeith, F. Mertens, Reversible storage of hydrogen in destabilized LiBH4, Journal of Physical Chemistry. B 109 (2005) 3719.

[49] M.A. Fetcenko, S.R. Ovshinsky, B. Reichman, K. Young, C. Fierro, J. Koch, A. Zallen, W. Mays, T. Ouchi, Recent advances in NiMH battery technology, Journal of Power Sources 165 (2007) 544.

[50] L. Ouyang, J. Huang, H. Wang, J. Liu, M. Zhu, Progress of hydrogen storage alloys for Ni-MH rechargeable power batteries in electric vehicles: a review, Materials Chemistry and Physics 200 (2017) 164.

[51] Y. Manoharan, S.E. Hosseini, B. Butler, H. Alzhahrani, B.T.F. Senior, T. Ashuri, J. Krohn, Hydrogen fuel cell vehicles; current status and future prospect, Applied Sciences 9 (2019) 2296.

[52] T. Yoshida, K. Kojima, Toyota MIRAI fuel cell vehicle and progress toward a future hydrogen society, The Electrochemical Society Interface 24 (2015) 45.

[53] F. Zhao, Z. Mu, H. Hao, Z. Liu, X. He, Victor Przesmitzki, S. Ahmad, A. Amer, Hydrogen fuel cell vehicle development in China: an industry chain perspective, Energy Technology 8 (2020) 2000179.

[54] Y. Matsuzaki, Residential applications: ENE-FARM, in: K. Sasaki, H-W. Li, A. Hayashi, J. Yamabe, T. Ogura, S.M. Lyth (Eds.), Hydrogen Energy Engineering: A Japanese Perspective, Springer Japan, Tokyo, 2016, p. 477.

[55] P. Colbertaldo, S.B. Agustin, S. Campanari, J. Brouwer, Impact of hydrogen energy storage on California electric power system: towards 100% renewable electricity, International Journal of Hydrogen Energy 44 (2019) 9558.

[56] E. López González, F. Isorna Llerena, M. Silva Pérez, F. Rosa Iglesias, J. Guerra Macho, Energy evaluation of a solar hydrogen storage facility: comparison with other electrical energy storage technologies, International Journal of Hydrogen Energy 40 (2015) 5518.

[57] J. Kurtz, G. Saur, S. Sprik, Hydrogen Fuel Cell Performance as Telecommunications Backup Power in the United States, National Renewable Energy Lab. (NREL), Golden, CO, United States, 2015.

[58] M.V. Lototskyy, V.A. Yartys, B.G. Pollet, R.C. Bowman Jr, Metal hydride hydrogen compressors: a review, International Journal of Hydrogen Energy 39 (2014) 5818.

[59] B.P. Tarasov, P.V. Fursikov, A.A. Volodin, M.S. Bocharnikov, Y.Y. Shimkus, A.M. Kashin, V.A. Yartys, S. Chidziva, S. Pasupathi, M.V. Lototskyy, Metal hydride hydrogen storage and compression systems for energy storage technologies, International Journal of Hydrogen Energy 46 (2021) 13647.

[60] K. Zhang, L. He, L. Jiang, S. Jiang, R. Yu, H.C. Lau, C. Xie, Z. Chen, The role of hydrogen in the energy transition of the oil and gas industry, Energy Reviews 3 (2024) 100090.

[61] H. Ishaq, C. Crawford, Review of ammonia production and utilization: enabling clean energy transition and net-zero climate targets, Energy Conversion and Management 300 (2024) 117869.

[62] Y.S.H. Najjar, Hydrogen safety: the road toward green technology, International Journal of Hydrogen Energy 38 (2013) 10716.

[63] L. Guo, J. Su, Z. Wang, J. Shi, X. Guan, W. Cao, Z. Ou, Hydrogen safety: an obstacle that must be overcome on the road towards future hydrogen economy, International Journal of Hydrogen Energy 51 (2024) 1055.

[64] S.K. Dwivedi, M. Vishwakarma, Hydrogen embrittlement in different materials: a review, International Journal of Hydrogen Energy 43 (2018) 21603.

[65] T. Hübert, L. Boon-Brett, G. Black, U. Banach, Hydrogen sensors – a review, Sensors and Actuators. B, Chemical 157 (2011) 329.

[66] S. Li, W. Li, L. Chen, H. Jiang, J. Zhang, D.W. Gao, Real-time robust state estimation for large-scale low-observability power-transportation system based on meta physics-informed graph timesnet, IEEE Transactions on Smart Grid 15 (6) (2024) 5500–5513.

[67] H. Ishaq, I. Dincer, C. Crawford, A review on hydrogen production and utilization: challenges and opportunities, International Journal of Hydrogen Energy 47 (2022) 26238.

[68] H. Xie, Z. Zhao, T. Liu, Y. Wu, C. Lan, W. Jiang, L. Zhu, Y. Wang, D. Yang, Z. Shao, A membrane-based seawater electrolyser for hydrogen generation, Nature 612 (2022) 673.

Chapter 6

Applications of energy storage systems in power systems and transportation systems

Shijie Li[a], Ziyun Liu[a], Haoqin Li[a], Huaiguang Jiang[a], and Xinhu Zheng[b,c,d]

[a]*School of Future Technology, South China University of Technology, Guangzhou, China,* [b]*The Hong Kong University of Science and Technology (Guangzhou), Guangzhou, China,* [c]*Thrust of Intelligent Transportation, Systems Hub, Guangzhou, China,* [d]*Thrust of Internet of Things, Information Hub, Guangzhou, China*

Contents

6.1 Applications of energy storage systems in power systems

In recent years, nearly half of the newly added global power generation capacity has come from renewable energy sources, which are essential in reducing carbon emissions from fossil fuels into the atmosphere [1]. As the world transitions towards a low-carbon power system, Energy Storage Systems (ESSs) are becoming increasingly essential [2]. On the generation side, Variable Energy Storage Systems (VESSs) can mitigate the intermittency caused by the large-scale integration of Variable Renewable Energy (VRE), thereby reducing the uncertainty of VRE and enhancing the reliability and stability of the power system [3]. VESSs also provide frequency and voltage support [4]. On the grid side, VESSs can defer investments in transmission systems, reduce the need for costly spinning reserves traditionally provided by

Modern Energy Storage. https://doi.org/10.1016/B978-0-32-395419-8.00012-X

fossil-fuel power generation, and enhance power quality [5]. On the consumer side, VESSs can smooth power fluctuations, enable economic dispatch of the system, and provide both emergency backup power and protection for sensitive equipment [6]. Coordinated regulation at different levels can reduce the peak-to-valley gap between generation and load, thereby significantly improving the overall stability of the power system [7], as illustrated in Fig. 6.1.

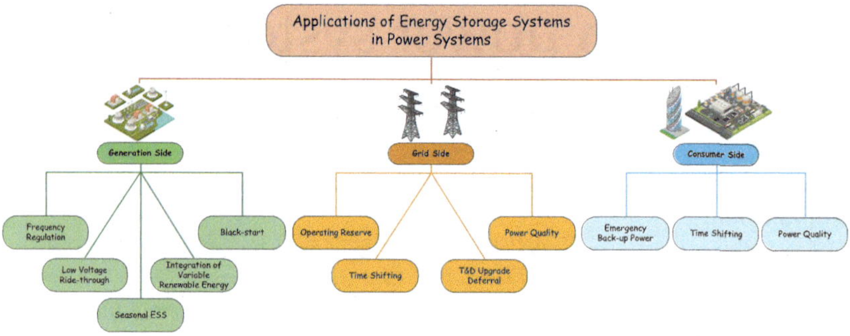

FIGURE 6.1 Overall framework for the application of ESS in power systems.

6.1.1 Applications of energy storage systems on the generation side

a) Frequency Regulation

Frequency regulation is a critical task in power system operation, aimed at maintaining the system frequency within a specified range. When there is an imbalance between the generation output and the load, this imbalance can cause the grid frequency to deviate from its standard value, thereby affecting the stability and security of the power system. To address this, ESSs help maintain grid frequency stability by rapidly responding to changes in supply and demand within the power system [8].

- ESSs possess excellent fast response capabilities. When a frequency deviation exceeds a predefined threshold, ESSs can be utilized for primary frequency control to restore the frequency to its normal range within a very short time. Furthermore, ESS can also participate in secondary frequency control over longer time scales, adjusting the output levels of various resources, including the ESS itself, to achieve overall balance and address sudden power fluctuations.
- ESSs can perform bidirectional frequency regulation. Specifically, whether the system requires an increase or decrease in power supply to maintain balance, ESSs can flexibly adjust the direction of their output power, enabling both positive and negative regulation.
- ESSs can enhance system efficiency. In comparison with traditional generating units, using advanced energy storage solutions such as batteries for frequency regulation is more efficient and energy-saving, reducing unnecessary fuel consumption and emissions.

- Some countries and regions already allow qualified ESS projects to provide frequency regulation services, thereby generating economic benefits. Consequently, these services contribute to the stability and reliability of the power system while also offering financial incentives for ESS operators.

b) Low-Voltage Ride Through

Low-Voltage Ride Through (LVRT) refers to the capability of generation equipment to continue operating and maintain connection to the power system during voltage sags caused by faults, until the voltage returns to normal levels. This is particularly important in scenarios where a large amount of VRE is integrated into the power system, as energy storage can significantly enhance system stability and resilience to sudden disturbances.

- When a voltage sag occurs, energy storage can rapidly release stored energy to provide the necessary active and reactive power support to loads or generating equipment, helping them to maintain operation. This rapid response capability helps protect the generating equipment from disconnecting, thereby reducing the risk of large-scale power outages.
- ESS can smooth voltage fluctuations through charging and discharging operations, thereby mitigating the impact of voltage sags on equipment. By absorbing excess reactive power or injecting deficient reactive power, energy storage can effectively stabilize voltage levels, ensuring continuous system operation during faults.
- By reducing frequent start-stop operations, ESSs can alleviate mechanical stress on generating equipment, thereby extending their operational lifespan.

c) Generation Side Black Start

Black start refers to the process of restoring a power system from a complete shutdown to normal operation after a widespread power outage. ESS can markedly enhance the recovery speed and reliability of the power system, thereby effectively improving the efficiency of black-start operations.

- In the event of a complete grid shutdown, traditional generating units require an external power source to start. ESS, such as battery storage, can provide the initial energy needed to start critical generating units without the need for an external power source. This capability can significantly reduce the recovery time, especially in remote areas or regions with limited backup power resources.
- During the initial stages of a black start, ESS can provide temporary power to critical loads, such as communication equipment and emergency lighting. This ensures that essential infrastructure remains operational until the main power supply is restored.
- As the generation side gradually returns to operation, ESS can smooth out power fluctuations during the transition. For example, when gas turbines are starting up, ESS can supplement their output power, ensuring a stable power supply.

- Traditional black start methods often rely on external diesel generators or other backup power sources, which are not only costly but also subject to fuel supply and maintenance issues. ESS can reduce dependence on these external power sources by quickly charging and discharging, and adjusting output power according to demand. This enhances the flexibility, autonomy, and reliability of the power system.

d) Integration of Variable Renewable Energy

As the penetration of VRE increases, the large-scale integration of VRE into the power system presents significant challenges. ESSs support the large-scale integration of VRE by smoothing output, enhancing grid stability, optimizing energy dispatch, and improving economic efficiency [9].

- VRE, such as wind and solar energy, are characterized by high intermittency and variability. ESS can absorb these fluctuations and provide a stable power output. Specifically, short-term smoothing by ESS, achieved through rapid charging and discharging, can mitigate instantaneous fluctuations in wind or photovoltaic generation within seconds to minutes. Long-term smoothing by ESS involves storing excess energy and releasing it during peak demand periods, thereby alleviating daily or seasonal variations and providing a more consistent power supply.
- Since not all the energy generated by VRE is utilized, ESS can mitigate the curtailment of VRE, reducing energy waste caused by grid capacity limitations.
- By rapidly responding to changes in grid demand, ESS enables more flexible and efficient energy distribution strategies, facilitates the creation of virtual power generation, alleviates grid congestion, and enhances the overall flexibility and reliability of the power system.

e) Seasonal Energy Storage Systems

Seasonal ESSs store energy during certain periods of the year and release it during other periods to address seasonal energy supply and demand imbalances, thereby balancing seasonal variations and simultaneously reducing the curtailment of VRE [10]. Each of these storage methods has its unique characteristics and applicable scenarios:

- Pumped hydro storage utilizes excess electrical energy to pump water from a lower reservoir to a higher one. When electricity is needed, the stored water is released to generate power.
- Compressed air energy storage compresses air using excess electrical energy during low-demand periods and stores it in underground caverns. When electricity is required, the compressed air is released to drive a gas turbine for power generation.
- Liquid air energy storage involves cooling air to a liquid state and storing it. When electricity is needed, the liquefied air is heated to expand and drive a turbine for power generation.

- Chemical energy storage (such as hydrogen storage) involves producing hydrogen through electrolysis of water and storing it. When electricity is needed, the stored hydrogen can be used in fuel cells or combusted to generate power.
- Thermal energy storage converts excess electrical energy into thermal energy, which is stored in high-temperature materials or phase-change materials. When electricity is needed, the stored thermal energy is used in a thermodynamic cycle to generate power.

However, these seasonal ESS methods often face challenges in commercialization due to technological immaturity and high construction difficulties.

6.1.2 Applications of energy storage systems on the grid side

a) Grid Side Operating Reserve

Operating reserve refers to the additional generating capacity reserved in a power system to address sudden contingencies, such as generator failures, ensuring stable operation during short-term fluctuations.

- ESSs can serve as part of the spinning reserve due to their extremely fast response times, ready to provide power on demand. Similarly, ESS can function as a nonspinning reserve, remaining in a standby state without generating power and capable of quickly starting up and providing power when needed.
- ESS can also perform load following, where they adjust their output power rapidly to match changes in load, thereby balancing supply and demand.

Several countries and regions have already incorporated ESS into their operational reserves, or have integrated distributed ESS to form a virtual power generation.

b) Grid Side Power Quality

Power quality refers to the degree to which the electricity supplied by the power system meets the required standards, including aspects such as voltage stability, frequency stability, and harmonic content. These characteristics significantly impact the overall stability and reliability of the power system [11]. In addition to their roles in frequency control, voltage stabilization, and large-scale integration of VRE, ESS play a crucial role in improving power quality on the grid side through the following ways:

- Harmonic Suppression: New types of ESS, such as supercapacitors, have excellent high-frequency response characteristics. They can be used to filter out high-order harmonic components in the grid, reducing harmonic pollution caused by nonlinear electrical equipment. This helps prevent harmonic oscillations and other faults in the power system.
- Enhancing System Reliability: When a fault in the main grid causes a power outage, ESS installed at critical nodes can immediately switch to an independent power source mode, continuing to supply power to essential

loads until the main grid is restored. This prevents widespread oscillations or even grid islanding, thereby enhancing the overall reliability and resilience of the power system.

c) Grid Side Time Shifting

ESSs store energy during periods of low demand and release it during peak demand, thereby smoothing the power supply, reducing the gap between peak and off-peak loads, and minimizing curtailment of VRE, which enhances the economic efficiency of the grid.

- By storing energy during periods of low electricity prices (e.g., at night) and using the stored energy during periods of high prices (e.g., during the day), ESS can reduce electricity costs and improve the operational efficiency of the grid.
- During peak demand periods, ESS can discharge stored energy to meet additional demand, reducing reliance on traditional power plants and alleviating pressure on the grid.

Many countries and regions have already widely adopted ESS for time shifting, employing batteries and pumped hydro storage to enhance the utilization efficiency of VRE, reduce peak load on the grid, and contribute to the development of more efficient, reliable, and sustainable power systems.

d) T&D upgrade Deferral

Deferring transmission and distribution upgrades refers to the deployment of ESS to alleviate bottlenecks in the grid, thereby postponing or avoiding costly upgrades to transmission lines, substations, or other grid infrastructure. ESS can provide additional power during peak demand periods, reducing the strain on existing infrastructure [12].

- During peak demand periods, certain transmission lines may reach their maximum capacity, leading to grid congestion. By deploying ESS at these critical nodes, stored energy can be released during peak demand, thereby reducing the load on transmission lines.
- In some regions, growing electricity demand may exceed the capacity of existing substations. By deploying ESS, additional power can be provided during peak demand periods, alleviating grid congestion and thereby deferring the expansion or construction of new substations.
- In regions with weak or fault-prone grids, ESS can serve as a backup power source, providing emergency power during grid failures and enhancing the reliability of the power supply.

Some countries have deployed battery ESS in key regions to alleviate grid congestion. These ESSs provide additional power during peak demand or emergency failures, thereby deferring the need for upgrades to transmission lines and substations and reducing the occurrence of power outages.

6.1.3 Applications of energy storage systems on the consumer side

Certain industrial users, such as chemical plants and data centers, have high requirements for power quality. ESS can help these users maintain the operation of critical equipment during voltage sags, thus avoiding production interruptions and economic losses [13].

a) Consumer Side Power Quality

Power quality directly affects the performance and lifespan of electrical equipment, as well as the overall stability and reliability of the power system. For critical equipment on the consumer side, ESS can filter out noise and disturbances from the grid, providing clean power and protecting sensitive equipment from damage [5].

- Issues such as voltage sags can cause damage to critical equipment on the consumer side, ranging from minor disturbances to severe outages and equipment failure. ESS can provide rapid, seamless power switching, offering necessary inertia support and a stable power supply. This is particularly effective in handling transient faults, ensuring the continuous operation of critical equipment.
- Nonlinear loads generate harmonic currents, which can reduce the efficiency of the power system and potentially damage sensitive equipment. By combining appropriately configured filters with ESS, the impact of these disturbances can be mitigated on the consumer side, thereby improving the overall power quality.

b) Emergency Backup Power

Emergency backup power refers to a power source that can rapidly provide electricity to maintain the operation of critical equipment and systems in the event of a failure or interruption of the main power supply. Uninterruptible Power Supply (UPS) is a key power protection device used to provide continuous, uninterrupted power to critical loads when the main power supply fails. The applications of emergency backup power, based on BESS, include but are not limited to the following scenarios:

- Data centers require a highly reliable power supply to ensure the continuity of data processing and storage. BESS can immediately provide power in the event of a main power failure, ensuring seamless power transfer and preventing data loss and service interruptions.
- Hospitals and medical facilities, especially operating rooms, intensive care units, and emergency equipment, have a high dependency on power. BESS can ensure that these critical devices continue to operate with stable voltage during power outages, safeguarding patient safety.
- Telecommunication base stations require continuous power to maintain the stability of communication networks. BESS can provide backup power during grid failures, ensuring that communication services remain uninterrupted.

- Certain industrial facilities, such as chemical plants, refineries, and manufacturing plants, need a stable power supply to prevent production interruptions and safety incidents. BESS can provide emergency backup power, ensuring the normal operation of critical processes and equipment.
- Data processing centers for banks and financial institutions require a highly reliable power supply to ensure the continuity of transactions and data processing.
- In residential and commercial buildings, BESS can serve as an emergency backup power, ensuring the continuity of basic living and business activities during power outages.

Several countries and regions have already adopted BESS as the emergency backup power and have achieved significant benefits from their applications.

c) Consumer Side Time Shifting

ESSs also play a crucial role in time shifting on the consumer side [1]. Specifically, ESSs can be utilized to:

- Reduce Peak Demand and Lower Electricity Costs: ESS can charge during off-peak hours when electricity prices are lower and discharge during peak hours to supply power to the load. This reduces the electricity consumption of users during high-price or high-demand periods, thereby alleviating grid pressure and lowering the electricity bills of users.
- Manage Electric Vehicle (EV) Charging: With the increasing adoption of EVs, the concentrated and highly variable demand from fast-charging stations can cause significant stress on local grids. Distributed ESS can be used to regulate charging power, preventing grid overloads caused by a large number of EVs charging simultaneously. This not only meets the needs of users for fast charging but also ensures the safe and stable operation of the grid.

Through these applications, time shifting with ESS helps to reduce the electricity costs of users while maintaining the balance between supply and demand in the grid. This, in turn, contributes to the overall safety and reliability of the power system, as illustrated in Fig. 6.2.

6.2 Applications of energy storage systems in transportation systems

The application of ESS in transportation systems is extensive and diverse. As illustrated in Fig. 6.3, it covers multiple fields such as EVs, rail transit, Unmanned Aerial Vehicles (UAVs), and ships. In EVs, innovations in lithium-ion batteries, charging technology, and battery-swapping technology have significantly enhanced vehicle range and charging efficiency, while Vehicle-to-Grid (V2G) technology enables bidirectional interaction between vehicles and the power grid. In rail transit, ESS such as batteries, flywheels, and supercapacitors effectively recover regenerative braking energy, balance power supply and demand, and support catenary-free operation, improving energy

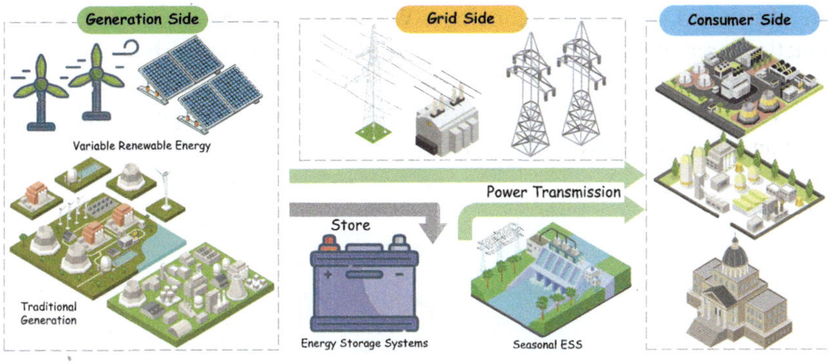

FIGURE 6.2 Role of ESSs in power systems.

utilization efficiency and system stability. In the fields of UAVs and ships, the application of ESS and hybrid power systems has significantly boosted energy efficiency, extended operational duration, and substantially reduced carbon emissions, providing crucial momentum for the sustainable development of green transportation. The continuous advancement of ESS offers vital support for the intelligent, efficient, and sustainable development of transportation systems.

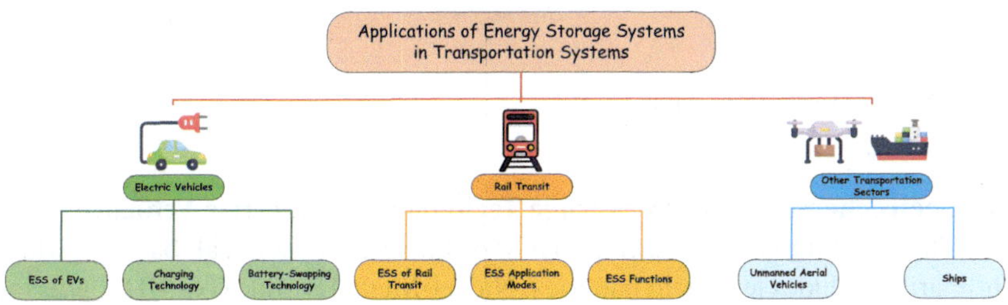

FIGURE 6.3 Overall framework for the application of ESSs in transportation systems.

6.2.1 Applications of energy storage systems in electric vehicles

EVs can be categorized into several categories based on their powertrain systems and intended applications, including Pure Electric Vehicles (PEVs), Hybrid Electric Vehicles (HEVs), Plug-in Hybrid Electric Vehicles (PHEVs), Extended-Range Electric Vehicles (EREVs), and Fuel Cell Electric Vehicles (FCEVs) [14]. PEVs, also known as Battery Electric Vehicles (BEVs), are fully powered by batteries, while FCEVs use fuel cells or a combination of fuel cells and batteries for power. Although HEVs and PHEVs combine internal combustion engines with electric propulsion systems, which distinguishes them from BEVs in terms of power sources and application scenarios, their core ESS are also based on battery technology. With the variation in vehicle types and functional requirements, the battery capacity, energy density, driving range, and safety design also differ. The following sections analyze the key technologies of EVs, including ESS, charging technology, and battery-swapping technology, with an overall framework illustrated in Fig. 6.4.

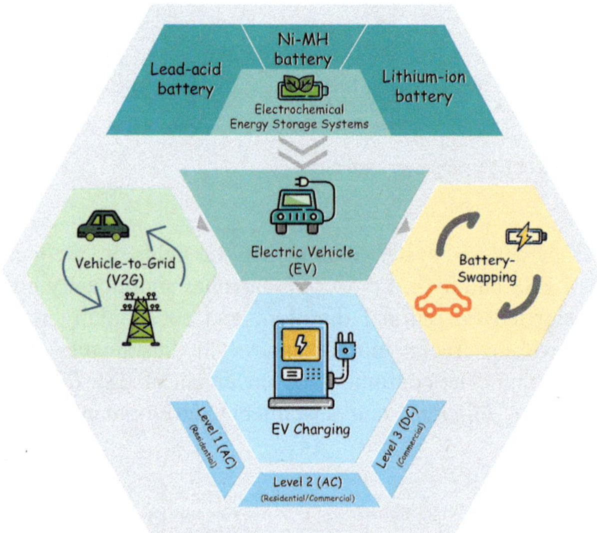

FIGURE 6.4 Applications of ESS in EVs.

a) Electrochemical Energy Storage Technology for Electric Vehicles

The most important ESS in EVs is the electrochemical battery, which primarily includes lead-acid batteries, Nickel-Metal Hydride (Ni-MH) batteries, and Lithium-ion (Li-ion) batteries.

- Lead-acid batteries, due to their maturity, low cost, and high reliability, were widely used in EVs in the past. They are particularly common in low-speed EVs such as forklifts and golf carts, as well as in some HEVs serving as auxiliary power sources. However, their relatively low specific energy and energy density have constrained their application in high-performance EVs [14].
- Compared to lead-acid batteries, Ni-MH batteries exhibit higher specific energy and energy density, along with superior cycle life, low-temperature performance, and safety, thereby being widely employed in HEVs. However, with the growing demand for higher performance and longer driving range, the use of Ni-MH batteries has gradually decreased, with Li-ion batteries becoming the preferred choice thanks to their superior performance attributes. [15].

- Li-ion batteries, recognized for their superior energy density, extended cycle life, and exceptional power output, have emerged as the dominant battery technology in the field of EVs. Depending on the cathode material, Li-ion batteries can be classified into various types, with the most commonly used in EVs being Lithium Iron Phosphate (LFP), Nickel Manganese Cobalt (NMC), and Nickel Cobalt and Aluminum (NCA). LFP is renowned for its exceptional safety, long cycle life, and cost-effectiveness, although it has a relatively low energy density (120–160 Wh/kg) and poor low-temperature performance. It is typically used in vehicles that prioritize higher safety, extended service life, and strict cost control. LFP is particularly well-suited for heavy-duty applications, such as buses and trucks [16], especially on short-distance routes that require frequent recharging, as it meets the demands for range and rapid charging. NMC offers a good balance between energy density (about 200–300 Wh/kg) and power density, with flexible formulation adjustments to optimize specific performance metrics. The relatively low self-heating rate ensures its suitability for automotive applications that require extended periods of stable performance [16]. NCA provides exceptionally high energy density and a long lifecycle, making it ideal for applications demanding long driving ranges and high performance. Despite its higher cost, the performance advantages NCA offers render it the preferred choice for high-end EVs.

Although significant progress has been made in the application of Li-ion batteries in EVs, challenges and limitations remain, including limited energy density improvements, long charging times, high costs, and safety risks. Solid-state batteries, as an emerging technology, offer higher energy densities, enhanced safety, and longer lifespans, but currently face challenges such as high production costs, poor low-temperature performance, and interface instability [14]. As technology advances, ESSs for future EVs are expected to diversify, leading to improved battery performance, cost, and safety, thus promoting the broader adoption and development of EVs.

b) Electric Vehicle Charging Technology

EV charging technology is a critical component for ensuring the proper operation of EVs. Currently, EV charging technologies primarily include slow charging, fast charging, and V2G.

- EV charging can be categorized into different modes based on the charging power source and charging efficiency. According to the SAE J1772 standards, slow charging mainly includes AC Level 1 and AC Level 2 charging [17]. AC Level 1 charging is typically suitable for household AC power and utilizes a standard household outlet (120 V), providing lower power (up to 1.92 kW). It is ideal for overnight or long-term parking situations. AC Level 2 charging usually operates at 240 V and requires a dedicated charging station. Compared to AC Level 1, it offers faster charging speeds with a maximum power output of up to 19.2 kW. It is widely used in office buildings, shopping malls, public charging stations, and other environments where higher charging efficiency is required [18].

- DC fast charging, also known as Level 3 charging, uses high-power DC sources, with power ratings ranging from 50 kW to 350 kW. High-power charging stations directly supply DC power to the EV battery, enabling rapid charging in a short period. It is suitable for highway service areas and locations with high-speed charging demands [18]. Supercharging stations now provide DC fast charging capabilities, supporting high-power charging with a maximum charging power of up to 250 kW [19]. At peak power, they can replenish up to approximately 250 kilometers of range in as little as 15 minutes.
- V2G technology enables bi-directional interaction between EVs and the grid, enabling EVs to both draw energy from the grid and supply stored energy back to it. V2G technology optimizes the charging and discharging times of EVs through smart charging and discharging strategies, alleviating peak load pressures and boosting the stability and reliability of the grid [22]. For EV owners, V2G helps optimize charging costs while providing valuable energy storage support to the grid. Through intelligent scheduling, V2G can distribute the charging activity more evenly, preventing vehicle congestion at specific charging stations or areas, and reducing traffic during peak charging times.

c) Electric Vehicle Battery-Swapping Technology

The EV battery-swapping technology achieves rapid charging by replacing drained batteries with fully charged ones at Battery-Swapping Stations (BSSs) for vehicles whose State of Charge (SoC) falls below a predetermined threshold.

BSS operates under two main models: centralized and distributed [20][21].

- Centralized BSS employs a method of off-site charging and swapping, where battery charging and replacement services are conducted at separate facilities. In this model, the EV battery is first swapped at the battery swapping station for a fully charged battery, and the depleted battery is then transferred to a centralized battery charging station for charging.
- Distributed BSS combines both charging and battery swapping at the same location, with the charging system and battery swapping facilities working in coordination [20]. This model reduces the complexity of battery logistics but faces challenges such as increased local grid load and high costs for charging modules.

Battery-swapping technology decouples the charging process from vehicle usage, significantly reducing charging wait times and alleviating range anxiety for users. Additionally, BSS enables centralized management and maintenance of batteries, allowing better control over battery charging states, temperature, and other parameters. This contributes to lowering battery failure rates, prolonging battery lifespan, and improving the overall reliability of the system. From the grid's perspective, BSS offers significant flexibility in demand management. BSS can adjust charging schedules based on grid load conditions, avoiding peak-time grid stress. Furthermore, they can return surplus energy stored in batteries to the grid, supporting load balancing and demand management.

6.2.2 Applications of energy storage systems in rail transit

a) Common Energy Storage Technologies in Rail Transit

Common energy storage technologies used in rail transit systems include batteries, flywheels, and supercapacitors. Specifically:

- Batteries, such as Li-ion batteries and Ni-MH batteries, offer high energy density and are commercially mature, making them suitable for long-duration power supply needs. They are also capable of responding quickly to changes in power demand. However, they come with challenges such as high cost, thermal management issues, and limited cycle life.
- Flywheel Energy Storage Systems (FESS) are electromechanical ESS that store kinetic energy through a rapidly rotating rotor. Compared to chemical energy storage, FESS offers superior cycling stability and durability, as well as high energy efficiency, high power density, high energy density, ease of maintenance, and fast response times. These characteristics make FESS particularly well-suited for handling the instantaneous high-power demands during train start-up and braking. They are ideal for metro or light rail systems with frequent stops and starts, helping stabilize the power supply network and reduce power fluctuations. Additionally, FESS has minimal environmental impact [24].
- Supercapacitors, such as Electric Double-Layer Capacitors (EDLCs), provide high power density, fast charging and discharging, long lifespan, low heat generation, and high safety. These features make them particularly suitable for addressing the high instantaneous power demands during train start-up and braking. In rail transit, EDLCs can be used to recover kinetic energy during train braking and rapidly release the stored energy during acceleration, significantly improving the efficiency of regenerative braking and greatly enhancing the overall energy utilization of the train [23].

b) Application modes of Energy Storage systems in Rail Transit

ESS in railway systems can be implemented in two distinct modes: Stationary Energy Storage Systems (SESSs) and Onboard Energy Storage Systems (OESSs) [23]. The two types of ESS each have their own emphasis, collaborating to improve the overall performance of the railway system.

- SESSs are typically installed in fixed railway infrastructure, such as stations and power substations. They are primarily used to regulate the power grid load, store excess electrical energy, and release it during peak demand, thus decreasing dependence on the external grid. In addition, SESS can recover regenerative energy generated during train braking, providing backup power for the system and supporting the integration of green energy, thus improving the utilization of renewable energy. To address various power fluctuations and unforeseen events during rail transit operations, SESS must have high power output and large energy storage capacity. Additionally, SESS should feature long charge-discharge lifecycles to ensure efficient performance during frequent charging and discharging cycles, reducing operational costs and maintenance frequency, thus enhancing the system's economic viability and long-term stability.

- OESSs are installed onboard trains and are specifically designed to capture regenerative energy generated during braking and provide power to the train when necessary, optimizing the train's energy efficiency. OESS can support catenary-free operation, making it particularly suitable for short-distance sections, offering an additional power source to handle emergencies or enhance operational flexibility. Trains can rely solely on their onboard energy storage and regenerated energy to store, release, and utilize the energy available in their OESS, reducing dependence on the external grid. Since OESSs operate directly with the train, their size and weight are subject to more stringent limitations [24]. Compared to SESS, OESSs have lower power and energy capacity as they are designed only to store the recovered energy of a single train, focusing primarily on short-term energy storage and release, which is ideal for balancing the electrical demand during train operations.

c) Functions of Energy Storage Systems in Rail Transit

The application of ESS in rail transit can be classified into three main functions: regenerative braking energy recovery, power supply-demand balancing, and operation without overhead contact systems.

- Regenerative Braking Energy Recovery is an important application of ESS. Both SESS and OESS can effectively recover Regenerative Braking Energy (RBE) generated during train braking. When the train brakes, the traction motor converts kinetic energy into electrical energy, and the RBE is transmitted and stored via the power system [24]. In charging mode, the energy is absorbed by the ESS, and charging stops once the voltage or SoC reaches the preset value. The stored energy is then used by the traction system in discharge mode until the voltage or SoC drops to the minimum value. Through RBE recovery, ESS can convert the kinetic energy generated during deceleration into electrical energy and store it for future use, significantly boosting energy utilization efficiency.
- ESSs help balance power supply and demand, reducing the effects of transient power fluctuations on the grid [25]. For example, seasonal ESSs in the London Underground, Japan, and Western Europe effectively reduce voltage fluctuations and peak currents by utilizing flywheel and EDLC technologies, improving the stability of the power system [26]. ESSs also reduce power losses and mitigate the risk of faults by compensating for excess power and absorbing power fluctuations.

- In the case of operation without overhead contact systems, ESSs allow some sections of the track to be operated without the need for traditional overhead catenary systems. These ESSs charge during train operation and provide power in sections without overhead contact, and are widely used in light rail and tram systems [24]. These systems typically combine EDLC and Li-ion batteries to support fast charging and efficient operation. Not only do these systems reduce construction costs, but they also enhance the flexibility and scalability of the rail transit network.

6.2.3 Applications of energy storage systems in other transportation sectors

a) Applications of Energy Storage in Unmanned Aerial Vehicles

The flight time and endurance of UAVs are often limited by the battery capacity and energy density. Therefore efficient ESSs are crucial for enhancing UAV performance.

- Lithium batteries, especially Lithium Polymer (LiPo) and Li-ion batteries, are widely used in consumer-grade and small commercial UAVs owing to their excellent energy density, lightweight characteristics, and relatively low cost. They support a variety of tasks such as agricultural monitoring, building inspections, and logistics transportation. Lithium batteries, through the use of a battery management system BMS, ensure stable voltage output and prolonged lifespan, ensuring reliable UAV operation.
- Fuel cells, due to their long endurance and environmentally friendly characteristics, are gradually becoming the core ESS for industrial UAVs. Hydrogen fuel cells, with their high energy density and rapid refueling capabilities, can provide longer flight times, making them especially suitable for tasks requiring prolonged cruising, surveillance, and emergency response.
- Solar power technology is also being applied to long-endurance UAVs, especially in fields such as environmental monitoring, meteorological research, and remote sensing, where it offers unique advantages. For instance, solar-powered UAVs have already demonstrated the ability to achieve uninterrupted flight for several weeks, offering new possibilities for continuous operations of UAVs in extreme conditions [27].

b) Applications of Energy Storage in Ships

Currently, ESSs applied in ships mainly include batteries, supercapacitors, and flywheels. Battery-powered electric ships, especially for short-distance voyages and port operations, have become an ideal choice for zero-emission and low-noise operations. Li-ion batteries, with their high energy density and long lifecycle, are commonly employed in short-distance ferries and port service vessels, effectively enhancing range and economic efficiency. In scenarios requiring rapid charging and discharging, supercapacitors and flywheels, due to their high power density characteristics, serve as ideal ESS. These systems are particularly suitable for applications requiring an uninterrupted power supply, such as UPS and black start, ensuring that ships can quickly start and restore power during critical moments.

Hybrid propulsion technology in vessels typically combines batteries, supercapacitors, and flywheels with internal combustion engines or fuel cells as auxiliary power systems [28]. By intelligently integrating these energy storage devices, ships can flexibly switch between different operating conditions, ensuring that the internal combustion engine always operates within its optimal efficiency range, thus minimizing fuel consumption and harmful emissions, and improving overall energy utilization efficiency. Hybrid vessels fully leverage the advantages of traditional internal combustion engines and electric propulsion systems, ensuring sufficient range while significantly reducing pollutant emissions, aligning with the future trend of green shipping development.

6.3 Conclusion and future prospective

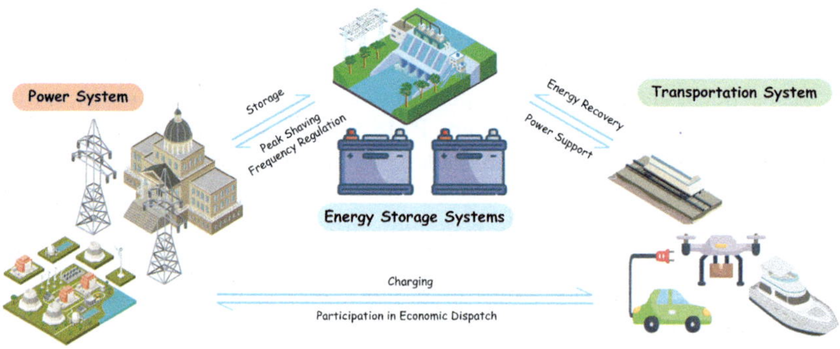

FIGURE 6.5 ESS, power system, and transportation system interact with and support each other.

Applications of ESS in power systems and transportation systems not only play a crucial role in improving energy utilization efficiency and promoting environmental sustainability but also provide critical support for the advancement of intelligent green development. The mutual interactions and support among ESS, power systems, and transportation systems are illustrated in Fig. 6.5.

In power systems, ESSs significantly bolster grid stability and reliability. In the short term, they address the intermittency and volatility of renewable energy, smoothing the output of wind and solar power, which enhances grid stability and overall energy efficiency. In the long term, ESS help balance seasonal energy supply and demand, lower costs related to prolonged energy storage, and enhance the effective use of renewable energy resources.

In transportation systems, the evolution of ESS has driven the green and intelligent transformation of EVs, rail transit, drones, ships, and other transportation sectors, significantly enhancing energy utilization efficiency and system performance. By integrating ESS technologies, such as batteries, supercapacitors, and flywheels, transportation systems can achieve more efficient energy management, enhance the stability and responsiveness of power systems, diminish reliance on conventional fossil fuels, and advance the development of low-carbon transportation networks.

Despite numerous advantages, ESSs face challenges like high costs and limited lifespans. However, technological advancements and economies of scale are expected to address these issues. Establishing robust policies and market mechanisms is crucial to encourage ESS participation in power and transportation systems, supported by incentives, technical standards, and regulations. Additionally, the progression of intelligent algorithms will enable ESS to refine operational strategies under increasingly complex conditions, thereby enhancing the flexibility and reliability of smart grids and transportation systems. In the future, ESS will play an increasingly vital role in developing green smart grids and intelligent transportation systems.

References

[1] A.G. Olabi, M.A. Abdelkareem, T. Wilberforce, E.T. Sayed, Application of graphene in energy storage device–a review, Renewable and Sustainable Energy Reviews 135 (2021) 110026.

[2] A. Baroutaji, T. Wilberforce, M. Ramadan, A.G. Olabi, Comprehensive investigation on hydrogen and fuel cell technology in the aviation and aerospace sectors, Renewable & Sustainable Energy Reviews 106 (2019) 31–40.

[3] K. Xie, H. Hui, Y. Ding, Review of modeling and control strategy of thermostatically controlled loads for virtual energy storage system, Protection and Control of Modern Power Systems 4 (4) (2019) 1–13.

[4] M. Cheng, S.S. Sami, J. Wu, Benefits of using virtual energy storage system for power system frequency response, Applied Energy 194 (2017) 376–385.

[5] A.G. Olabi, C. Onumaegbu, T. Wilberforce, M. Ramadan, M.A. Abdelkareem, A.H. Al-Alami, Critical review of energy storage systems, Energy 214 (2021) 118987.

[6] Z. Zhang, T. Ding, Q. Zhou, Y. Sun, M. Qu, Z. Zeng, Y. Ju, L. Li, K. Wang, F. Chi, A review of technologies and applications on versatile energy storage systems, Renewable and Sustainable Energy Reviews 148 (2021) 111263.

[7] K. Meng, Z.Y. Dong, Z. Xu, Y. Zheng, D.J. Hill, Coordinated dispatch of virtual energy storage systems in smart distribution networks for loading management, IEEE Transactions on Systems, Man and Cybernetics: Systems 49 (4) (2019) 776–786.

[8] M. Jafari, A. Botterud, A. Sakti, Decarbonizing power systems: a critical review of the role of energy storage, Renewable and Sustainable Energy Reviews 158 (2022) 112077.

[9] A.Z.A. Shaqsi, K. Sopian, A. Al-Hinai, Review of energy storage services, applications, limitations, and benefits, Energy Reports 6 (2020) 288–306.

[10] N.A. Sepulveda, J.D. Jenkins, A. Edington, D.S. Mallapragada, R.K. Lester, The design space for long-duration energy storage in decarbonized power systems, Nature Energy 6 (5) (2021) 506–516.

[11] S. Li, W. Li, L. Chen, H. Jiang, J. Zhang, D.W. Gao, Real-time robust state estimation for large-scale low-observability power-transportation system based on meta physics-informed graph timesnet, IEEE Transactions on Smart Grid 15 (6) (2024) 5500–5513.

[12] S. Li, W. Li, L. Chen, H. Li, Y. Zhang, H. Jiang, A short-term load forecasting method for large-scale power distribution systems based on a novel spatio-temporal neural network, in: 2023 China Automation Congress (CAC), 2023, pp. 9235–9241.

[13] Q. Zhong, Q. He, G. Wang, L. Wang, Q. Li, Optimal sizing and placement method for dynamic voltage restorers with mitigation expectation index, IEEE Transactions on Power Delivery 36 (6) (2021) 3561–3569.

[14] M. Amer, J. Masri, U. Sajjad, K. Hamid, et al., Electric vehicles: battery technologies, charging standards, ai communications, challenges, and future directions, Energy Conversion and Management: X (2024) 100751.

[15] Z.P. Cano, D. Banham, S. Ye, A. Hintennach, J. Lu, M. Fowler, Z. Chen, Batteries and fuel cells for emerging electric vehicle markets, Nature Energy 3 (4) (2018) 279–289.

[16] Y. Ding, Z.P. Cano, A. Yu, J. Lu, Z. Chen, Automotive li-ion batteries: current status and future perspectives, Electrochemical Energy Reviews 2 (2019) 1–28.

[17] G. Rajendran, C.A. Vaithilingam, N. Misron, K. Naidu, M.R. Ahmed, A comprehensive review on system architecture and international standards for electric vehicle charging stations, Journal of Energy Storage 42 (2021) 103099.

[18] I. Rahman, P.M. Vasant, B.S.M. Singh, M. Abdullah-Al-Wadud, N. Adnan, Review of recent trends in optimization techniques for plug-in hybrid, and electric vehicle charging infrastructures, Renewable and Sustainable Energy Reviews 58 (2016) 1039–1047.

[19] S. Hemavathi, A. Shinisha, A study on trends and developments in electric vehicle charging technologies, Journal of Energy Storage 52 (2022) 105013.

[20] X. Chen, K. Xing, F. Ni, Y. Wu, Y. Xia, An electric vehicle battery-swapping system: concept, architectures, and implementations, IEEE Intelligent Transportation Systems Magazine 14 (5) (2021) 175–194.

[21] D. Cui, Z. Wang, P. Liu, S. Wang, D.G. Dorrell, X. Li, W. Zhan, Operation optimization approaches of electric vehicle battery swapping and charging station: a literature review, Energy 263 (2023) 126095.

[22] B.K. Sovacool, L. Noel, J. Axsen, W. Kempton, The neglected social dimensions to a vehicle-to-grid (v2g) transition: a critical and systematic review, Environmental Research Letters 13 (1) (2018) 013001.

[23] N. Ghaviha, J. Campillo, M. Bohlin, E. Dahlquist, Review of application of energy storage devices in railway transportation, Energy Procedia 105 (2017) 4561–4568.

[24] M. Domínguez, A. Fernández-Cardador, A. Fernández-Rodríguez, A. Cucala, R. Pecharromán, P.U. Sánchez, I.V. Cortázar, Review on the use of energy storage systems in railway applications, Renewable and Sustainable Energy Reviews 207 (2025) 114904.

[25] S. Vazquez, S.M. Lukic, E. Galvan, L.G. Franquelo, J.M. Carrasco, Energy storage systems for transport and grid applications, IEEE Transactions on Industrial Electronics 57 (12) (2010) 3881–3895.

[26] P. Radcliffe, J.S. Wallace, L.H. Shu, Stationary applications of energy storage technologies for transit systems, in: 2010 IEEE Electrical Power & Energy Conference, IEEE, 2010, pp. 1–7.

[27] M.N. Boukoberine, Z. Zhou, M. Benbouzid, A critical review on unmanned aerial vehicles power supply and energy management: solutions, strategies, and prospects, Applied Energy 255 (2019) 113823.

[28] A. Haxhiu, A. Abdelhakim, S. Kanerva, J. Bogen, Electric power integration schemes of the hybrid fuel cells and batteries-fed marine vessels—an overview, IEEE Transactions on Transportation Electrification 8 (2) (2021) 1885–1905.

Chapter 7

Aqueous ion batteries: fundamentals, innovations, and applications

Jiayan Zhu[a], **Haobo Dong**[b], **and Huaiguang Jiang**[b]
[a]*College of Physics, Jilin University, Changchun, China,* [b]*School of Future Technology, South China University of Technology, Guangzhou, China*

Contents

Modern Energy Storage. https://doi.org/10.1016/B978-0-32-395419-8.00013-1

7.1 Introduction to aqueous ion batteries

7.1.1 Overview of energy storage technologies and the role of AIBs

Energy storage technologies are critical to the global energy transition, with the primary goal of enabling efficient energy storage and flexible dispatch to accommodate renewable energy fluctuations, balance power demand, and facilitate electrification in sectors like transportation, grid systems, and portable electronics. Currently, mainstream technologies include pumped hydro storage, electrochemical storage, thermal storage, and mechanical storage. Among these, electrochemical storage has attracted considerable attention due to its high energy density, rapid response, and versatility in deployment. Within this category, lithium-ion batteries dominate, yet their drawbacks, such as flammability, high cost, resource reliance, and environmental concerns, limit their use in large-scale storage and high-safety applications.

Aqueous Ion Batteries (AIBs) have gained attention as a promising energy storage solution because of their enhanced safety, affordability, and environmental sustainability. Unlike conventional lithium-ion batteries, AIBs utilize water-based electrolytes, which provide advantages such as higher ionic conductivity, simpler manufacturing processes, and a reduced risk of fire. These features make AIBs particularly suitable for large-scale energy storage, grid frequency regulation, and portable electronic devices.

7.1.2 Historical development and the basic principles of AIBs

In 1986 Japanese scientist Yamamoto first developed the $Zn/ZnSO_4/MnO_2$ battery system, which is an aqueous zinc-manganese battery (also known as Aqueous Zinc Ion Battery, abbreviated as AZIBs) using a neutral or weakly acidic aqueous solution as the electrolyte [1]. This system offers benefits such as high energy density, a broad electrochemical window, and high ionic conductivity, making it a promising alternative to alkaline zinc-manganese batteries. In 1994 Prof. J. R. Dahn's research group at Dalhousie University in Canada first reported an aqueous lithium-ion battery using an aqueous solution as the electrolyte in *Science* [2]. The cathode used $LiMnO_2$, the anode used

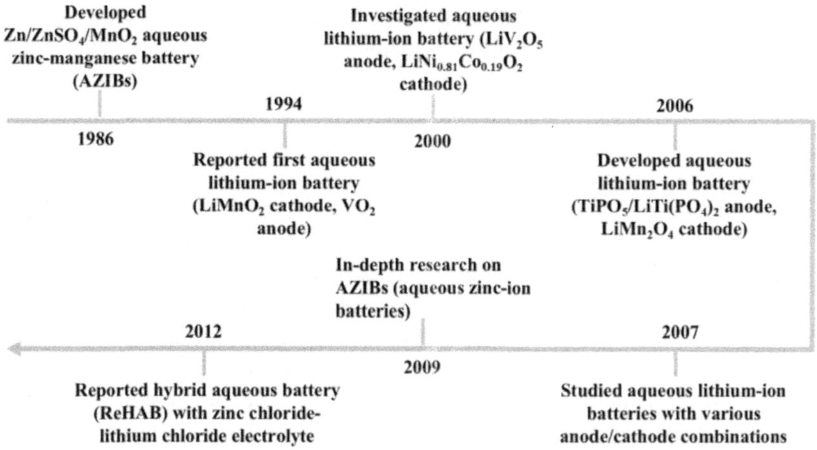

FIGURE 7.1 Historical development of AIBs.

VO_2, and the electrolyte solution was a neutral Li_2SO_4 solution. Operating at an average voltage of 1.5 V, the battery delivered an energy density of 75 Wh/kg, with a practical energy density around 40 Wh/kg, surpassing lead-acid batteries (30 Wh/kg) and being on par with Ni-Cd batteries. In 2000 the Toki group in Japan investigated an aqueous lithium-ion battery with LiV_2O_5 as the anode, $LiNi_{0.81}Co_{0.19}O_2$ as the cathode, and Li_2SO_4 as the electrolyte [3]. In 2006 Prof. Liquan Chen's research group at the Institute of Physics, Chinese Academy of Sciences, studied novel anode materials and reported an aqueous lithium-ion battery based on $TiPO_5$ or $LiTi(PO_4)_2/LiNO_3$ and $LiMn_2O_4$ as the cathode [4]. In 2007 Professor Zhiping Wu's group at Fudan University and Professor Yongyao Xia's group separately studied aqueous lithium-ion battery systems based on $LiV_2O_5/LiNO_3/LiCoO_2$ and $LiTi(PO_4)_2@C/Li_2SO_4/LiMn_2O_4$ [5]. In 2009 Prof. Feiyu Kang's research team at Tsinghua University again conducted in-depth studies on AZIBs [6]. In 2012 Prof. Pu Chen's group at the University of Waterloo in Canada first reported a Hybrid Aqueous Battery (ReHAB) with zinc chloride and lithium chloride as the mixed electrolyte, $LiMn_2O_4$ as the cathode, and zinc as the anode [7]. Since then, various aqueous battery systems based on different electrolytes and electrode materials have been reported and extensively researched, see Fig. 7.1.

Based on the electrochemical reaction mechanisms of the positive and negative electrodes, four types of aqueous batteries under neutral or weak acidic conditions are most commonly studied. The first category includes aqueous lithium-ion batteries, such as $LiMn_2O_4/LiTi(PO_4)_2$ and $LiNi_{0.81}Co_{0.19}O_2/LiVO_3$, where lithium ions reversibly intercalate and deintercalate between the positive and negative electrodes. This process gives rise to the characteristic rocking-chair lithium-ion battery mechanism in an aqueous system during charging and discharging. The second type is aqueous sodium-ion batteries, such as $NaMnO_2/NaTi(PO_4)_2$ and $Na_3V_2(PO_4)_3/NaTi(PO_4)_2$, where sodium ions undergo reversible intercalation and deintercalation, following the "aqueous rocking-chair sodium-ion battery" mechanism. The third type is hybrid aqueous batteries, such as $LiMn_2O_4/ZnNa_4MnO_2/Zn$ and $LiMn_2O_4/Na_{0.22}MnO_4$, which use a mixture of salts as the electrolyte. In these batteries, metal ions intercalate and de-intercalate between the electrodes, or the negative electrode undergoes deposition and dissolution reactions. The fourth type is aqueous zinc-ion batteries, such as MnO_2/Zn and $ZnMn_2O_4/Zn$, where the negative electrode is composed of zinc metal or electrodes that mainly consist of zinc-intercalated compounds. The energy storage and release in these batteries are achieved through the intercalation and deintercalation of zinc ions.

Aqueous rechargeable batteries, while offering enhanced safety and environmental advantages, still encounter numerous challenges. In aqueous electrolyte systems, the ion intercalation process is more intricate compared to organic electrolytes, which may result in several unwanted side reactions. These can include the interaction of electrode materials with water or oxygen, cointercalation of protons and metal ions, hydrogen/oxygen evolution, and the dissolution of electrode materials into the electrolyte. Such reactions notably impede the progress and practical use of aqueous batteries.

7.2 Fundamentals of aqueous ion batteries

7.2.1 Electrochemistry of AIBs

The working principle and reaction mechanism of aqueous batteries are grounded in electrochemical processes, which include ion intercalation and deintercalation, electron migration, and electrolyte ionic conductivity. In these systems, electrochemical reactions occur as ions transfer between the positive and negative electrodes via the electrolyte, allowing energy to be stored and released. Each battery setup consists of a negative electrode, a positive electrode, and an electrolyte. During both charging and discharging, ions continually intercalate and deintercalate between the electrodes, while electrons flow through the external circuit, facilitating energy conversion. The reaction mechanisms of aqueous batteries can vary depending on the specific type (e.g., lithium-ion, sodium-ion, or zinc-ion batteries), as discussed in Section 7.2.3. Furthermore, the performance of aqueous batteries is not only shaped by electrochemical reaction mechanisms but also by factors like the choice of electrolyte, electrode material properties, ionic conductivity, and the mitigation of side reactions. Therefore optimizing these parameters is essential to enhancing battery performance and prolonging its lifespan [8,9].

7.2.2 Components of AIBs: electrodes, electrolytes, separators, and other components

AIBs consist of several key components, each contributing to the performance and efficiency of the battery. These components include electrodes (anode and

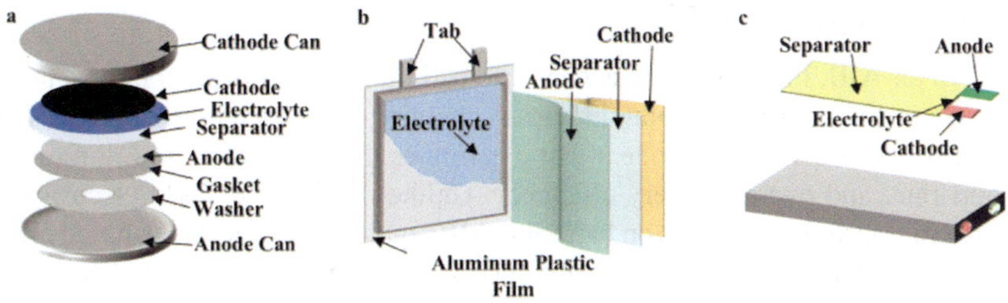

(a) button cell, (b) pouch cell, (c) prismatic cell

FIGURE 7.2 Cross-sectional view of aqueous batteries.

cathode), electrolytes, separators, and other auxiliary components. Based on different application requirements, aqueous ion batteries come in various shapes and can be categorized into three types: button cells, pouch cells, and prismatic cells, as shown in Fig. 7.2. Button cells use stainless-steel casings, making them suitable for small or miniature energy storage devices such as calculators and digital watches. Pouch cells are packaged with aluminum-plastic film, while prismatic cells typically use ABS or aluminum casings. Both types can serve as consumer and power batteries for applications such as mobile phones, digital cameras, and electric vehicles.

- Electrodes: Anode and Cathode

The electrodes are fundamental to the electrochemical performance of AIBs. These components are where the charge and discharge processes take place, involving ion intercalation and deintercalation. The selection of materials and their electrochemical properties has a significant impact on the battery's efficiency, capacity, cycle stability, and energy density.

The anode functions as the site for ion storage during the discharge cycle. Common anode materials for AIBs include zinc, lithium, sodium, and various transition metals. Zinc-based anodes are especially favored in aqueous Zinc-Ion Batteries (ZIBs) due to their high energy density, cost-effectiveness, and nontoxicity. However, challenges such as dendrite formation and corrosion can compromise their long-term cycling stability. To overcome these challenges, modifications like coating with polymers (e.g., PAA, alginic acid) or the use of nanomaterials are being explored [10].

The cathode is responsible for ion release during the charging cycle and the opposite process during discharging. For AIBs, cathode materials often include manganese dioxide, vanadium oxide, and Prussian blue analogs [11]. The cathode's performance is dictated by factors such as its capacity, cycling stability, and compatibility with the electrolyte. Both electrodes must exhibit high electrical conductivity, stability under electrochemical cycling, and minimal degradation over time.

- Electrolytes

In an AIB, the electrolyte acts as the medium that facilitates ion movement between the anode and cathode during charging and discharging cycles. The electrolyte is a vital component that determines the voltage window, ion transport rate, and overall battery efficiency. Unlike the organic-based electrolytes used in traditional lithium-ion batteries, AIBs typically utilize water-based electrolytes. These electrolytes are safer due to their nonflammability and are environmentally friendly. Common aqueous electrolytes include Zinc Sulfate (ZS), sodium sulfate, lithium sulfate, and mixtures with organic or inorganic additives to improve performance. The pH, ionic conductivity, and stability of the electrolyte are essential factors in designing efficient AIBs. Modifying the electrolyte can enhance the battery's overall performance. Additives like alginic acid, phosphoric acid, and TriFluoroMethaneSulfonic acid (TFMS) are being explored to improve ionic conductivity, suppress dendrite formation, and enhance cycle stability. These modifications can also help adjust the electrochemical stability window, which is particularly important for higher-voltage systems. In aqueous systems, ion solvation plays a significant role in determining the movement of ions within the electrolyte [12–14]. The structure and solvation of metal ions such as Zn^{2+}, Na^{+}, or Li^{+} are key to understanding the battery performance, as they affect ion diffusion rates and interfacial reactions at the electrode-electrolyte interface.

- Separators

The separator is a critical component that physically isolates the anode and cathode to prevent direct contact, while still allowing ions to pass through. The separator ensures the smooth operation of the electrochemical reaction by preventing internal short circuits and maintaining ion flow between electrodes. Separators in AIBs are typically made of porous materials such as polyolefin (PE, PP), PolyVinyliDene Fluoride (PVDF), or cellulose-based materials. The material must have high ionic conductivity, chemical stability, and mechanical strength. It should also be resistant to swelling or decomposition in aqueous environments [15]. Besides preventing short circuits, the separator also plays a role in controlling the ion diffusion rate, which can influence the battery's efficiency and cycling life. A high-quality separator should be able to endure the chemical environment of the electrolyte without significant degradation over time.

- Other Components

Current Collectors: These are materials that conduct electrical current from the electrodes to the external circuit. The choice of current collector material impacts the battery's electrical conductivity and stability. For positive electrodes, materials like aluminum foil are often used, while copper foil or corrosion-resistant metals are employed for negative electrodes.

Battery Casings: The casing provides mechanical protection and ensures the battery's safety by preventing leaks and external damage. Materials like

Acrylonitrile Butadiene Styrene (ABS) plastic, aluminum alloy, or stainless steel are used depending on the application and requirements for strength and durability. The casing also serves to maintain battery integrity by preventing moisture or air from compromising the electrolyte.

Protective Circuit Boards (PCBs): These are crucial for monitoring and ensuring the safe operation of the battery. The PCB prevents overcharging, overdischarging, and short circuits, and manages current flow to avoid damage. For AIBs, the PCB helps optimize the charge/discharge process, protecting the cells from excessive stress and ensuring that the battery operates within safe parameters [16].

Battery Management Systems (BMS): The BMS is a sophisticated electronic system responsible for overseeing the performance of the battery pack. It monitors the health of individual cells, adjusts the charge/discharge cycles, balances the cells, and predicts the overall lifetime of the battery. A properly designed BMS enhances the safety, efficiency, and lifespan of AIBs, especially when dealing with the complex electrochemical systems found in aqueous batteries.

7.2.3 Types of ions used in AIBs (e.g., Li^+, Na^+, K^+, Zn^{2+}, Mg^{2+}) and their transport mechanisms

In AIBs ion type is one of the key factors determining battery performance. Common ions in these systems include sodium (Na^+), lithium (Li^+), zinc (Zn^{2+}), potassium (K^+), and magnesium (Mg^{2+}) ions. Each ion exhibits distinct behaviors and transport mechanisms in aqueous media, and understanding these mechanisms is crucial for optimizing battery performance.

- Aqueous Lithium-Ion Batteries

Aqueous Lithium-Ion Batteries (ALIBs) utilize water-based electrolytes, providing benefits such as enhanced safety, environmental sustainability, and cost-effectiveness. The high ionic conductivity of these batteries allows for fast charge and discharge rates. However, their relatively narrow electrochemical stability window ($\sim$1.23 V) restricts energy density, and side reactions due to water decomposition can reduce the battery's cycle life [17].

In aqueous media, ion transport is primarily governed by diffusion and migration, with their behaviors influenced by solvation effects, salt concentration, and solution structure. Lithium ions (Li^+) typically exist in the form of hydrated ions (e.g., $[Li(H_2O)_4]^+$), where solvation increases their effective radius and reduces mobility. Increasing the salt concentration in the electrolyte, such as adopting "water-in-salt" electrolytes, reduces free water decomposition, optimizes ion transport properties, and broadens the electrochemical stability window. Additionally, the hydrogen-bonded network formed by water molecules is disrupted in highly concentrated solutions, enhancing solution stability and interfacial performance [18].

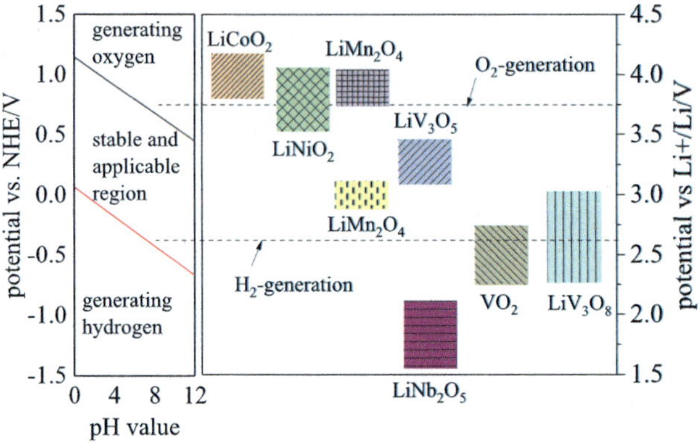

FIGURE 7.3 H_2O stability region for the entire pH range (left) and the potentials for Li^+ deintercalation for selected oxide compounds relative to Li metal (right).

At the electrode/electrolyte interface, lithium ions undergo desolvation before entering the solid-phase electrode. This process directly affects ion transport efficiency at the interface and overall battery stability. High-concentration electrolytes can form a stable Solid-Electrolyte Interphase (SEI) at the interface, suppressing side reactions and improving cycle life. Optimizing electrode interfaces and electrolyte composition is essential for enhancing the performance of aqueous lithium-ion batteries.

To address the limitations of ALIBs, researchers have proposed various optimization strategies. Using high-concentration lithium salt electrolytes, such as LiTFSI, significantly reduces water decomposition and expands the electrochemical window. As shown in Fig. 7.3, the electrochemical stability region of H_2O and the Li^+ deintercalation potentials of representative oxide compounds jointly determine the feasible operating potential range of aqueous lithium-ion batteries. Introducing additives can regulate the hydrogen-bond network in water improving transport properties. Coupling these advancements with the development of electrode materials that exhibit low hydrogen evolution and high oxygen evolution potentials, such as $LiMn_2O_4$ and $Li_4Ti_5O_{12}$ [19], can further enhance battery performance. These strategies collectively provide directions for the development of ALIBs and lay a foundation for their applications in portable energy storage and large-scale storage systems [20].

- Aqueous Sodium-Ion Batteries

In Aqueous Sodium-Ion Batteries (ASIBs), ion transport is governed by diffusion and migration, influenced by solvation structures, electrolyte concentration, and the behavior of water molecules. Sodium ions typically exist as hydrated ions with a large effective radius, reducing their mobility. Strategies such as high-concentration electrolytes (e.g., "water-in-salt") mitigate these issues by minimizing free water activity, expanding the electrochemical stability window, and enhancing ion transport properties. These approaches also disrupt water's hydrogen-bond network, improving solution stability and interfacial performance, which are critical for optimizing the battery's overall efficiency and durability [21].

- Aqueous Zinc-Ion Batteries

Aqueous zinc-ion batteries offer high energy density due to the high theoretical capacity and divalent nature of zinc, as well as enhanced safety and environmental sustainability. However, challenges like dendrite growth and side reactions limit their cycle life. Zinc ions exist as strongly solvated species (e.g., $[Zn(H_2O)_6]^{2+}$), and high-concentration electrolytes reduce free water content and stabilize the interface, suppressing hydrogen evolution and improving ion transport efficiency. The transport of zinc ions is influenced by desolvation behavior and electrode surface structures, and optimizing electrolyte composition and interface design significantly enhances performance [22].

- Aqueous Potassium-Ion Batteries

Aqueous Potassium-Ion Batteries (AKIBs) hold great potential for energy storage due to the low cost, abundant availability of potassium, and the high mobility of potassium ions. Potassium ions have weaker solvation effects and a larger radius, resulting in higher mobility compared to lithium and sodium ions, but they suffer from lower stability. High-concentration electrolytes can reduce free water decomposition, optimize ionic interactions, and expand the electrochemical window, thereby improving battery performance. However, the transport of potassium ions is still limited by the stability of electrode materials and interfacial reactions, necessitating further improvements in electrode design and interface regulation [23].

- Aqueous Magnesium-Ion Batteries

Aqueous Magnesium-Ion Batteries (AMIBs) leverage the high capacity and abundant resources of magnesium, but their slow reaction kinetics limit practical applications. Magnesium ions have strong solvation abilities, with highly stable hydrated ions (e.g., $[Mg(H_2O)_6]^{2+}$), which result in low mobility. High-concentration electrolytes and functional additives can alleviate desolvation energy barriers and enhance magnesium ion transport efficiency. Additionally, interface modification and electrode material optimization are essential to improving cycling performance, advancing the development of AMIBs for aqueous energy storage applications [24].

In aqueous batteries, the ion transport mechanism is a core factor determining battery performance. Due to differences in solvation structures, mobility, and interfacial behavior, various ions contribute differently to battery performance. By employing high-concentration electrolytes, additive engineering, and interface optimization, transport performance can be significantly enhanced, offering vast potential for the application of aqueous batteries in the energy storage field.

- Hybrid Aqueous Batteries

In current research, neutral or mildly acidic aqueous electrolyte-based hybrid battery systems can be categorized into two types. The first type

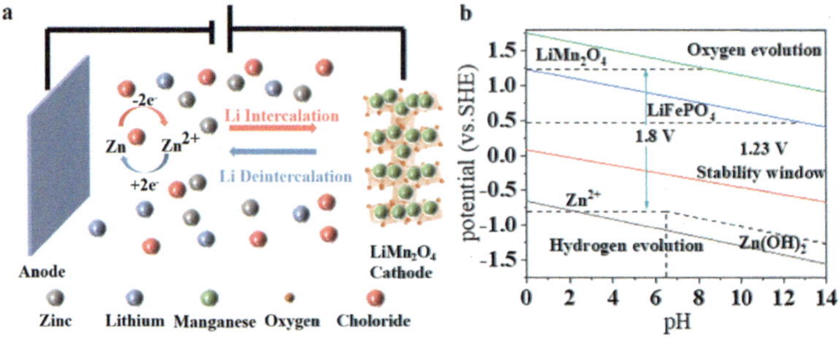

FIGURE 7.4 (a) The schematic diagram of the working principle of hybrid aqueous batteries, and (b) the direct relationship between the H_2/O_2 evolution potential and the solution pH value.

uses metallic zinc as the anode, a mixed salt electrolyte containing zinc and lithium, sodium, or potassium ions, and cathode materials containing lithium, sodium, or potassium compounds. This configuration constructs systems such as $Zn/LiMn_2O_4$, $Zn/Li_3V_2(PO_4)_3$, $Zn/Na_3V_2(PO_4)_3$, $Zn/Na_{0.44}MnO_2$, and $Zn/K_{0.19}MnO_2$, corresponding to the Li^+/Zn^{2+}, Na^+/Zn^{2+}, and K^+/Zn^{2+} hybrid ion systems. The second type includes battery systems such as $NaTi_2(PO_4)_3/LiMn_2O_4$.

Taking Hybrid Aqueous Batteries (ReHABs) as an example, these batteries use lithium intercalation compound $LiMn_2O_4$ as the cathode and metallic zinc as the anode. The electrolyte design deviates from conventional neutral or alkaline systems, adopting an aqueous solution with a pH of 4.0 containing 3M LiCl and 4M $ZnCl_2$. This configuration achieves a high energy density (50–80 $Wh \cdot kg^{-1}$) and excellent cycling performance, with a capacity retention of 90% after 1000 cycles [25]. During charge-discharge processes, the anode undergoes reversible zinc ion deposition and dissolution, while the cathode experiences reversible lithium-ion intercalation and deintercalation, as illustrated in Fig. 7.4a.

Since the concept of hybrid aqueous batteries was introduced, research has focused on cathode materials. Researchers have explored various cathode materials to pair with metallic zinc, with $LiMn_2O_4$ and $LiFePO_4$ gaining attention due to their low cost and abundance [26]. However, these materials exhibit poor conductivity, which has been improved by compositing them with carbon materials. Additionally, in 2016, Prof. Xianwen Wu from Jishou University synthesized $Na_{0.44}MnO_2$ using a sol-gel method and applied it to hybrid aqueous battery systems. The theoretical basis for cathode material selection is shown in Fig. 7.4b [27].

7.3 Materials for aqueous ion batteries

7.3.1 Advances in electrode materials for high performance and safety

In aqueous sodium-ion batteries, $Na_3V_2(PO_4)_3$ as a cathode material demonstrates excellent rate performance and cycling stability. Research shows that

$Na_3V_2(PO_4)_3$, synthesized via a hydrothermal-assisted sol-gel method, achieves a competitive capacity at a current density of 10 $mA \cdot g^{-1}$. By applying carbon coating to the electrode material's surface, side reactions between the active materials and water can be prevented, thereby enhancing the cycling stability of the electrode. For instance, carbon-coated $LiTi_2(PO_4)_3$ exhibits outstanding cycling performance in aqueous lithium-ion batteries, with a capacity retention of over 90% after 1000 charge-discharge cycles [28].

By engineering directional ion transport channels within the electrode, the rate performance can be greatly enhanced. Studies have shown that VO_2 (B) nanobelts, with their oriented tunnel structures, exhibit exceptional rate performance and cycling stability in aqueous zinc-ion batteries [29].

In aqueous lithium-ion batteries, using high-molecular-weight polymers capable of doping and dedoping as the anode material effectively improves the battery's cycling life and safety. For example, a novel high-performance aqueous rechargeable lithium battery employs materials that allow lithium-ion intercalation and deintercalation as the cathode, and high-molecular-weight polymers as the anode, exhibiting high energy and power densities.

In recent years, with the continuous development of battery design and architecture, traditional electrode preparation methods have revealed some drawbacks. The conventional grinding and coating method typically involves mixing the materials, binders, and conductive agents in certain proportions, grinding them together, and then coating the slurry onto the current collector to form the electrode. However, this method often leads to cracking and detachment of the active materials during bending and cycling, resulting in insufficient electrode flexibility. As a result, many researchers are turning to the design of self-supported electrodes to overcome these issues.

Self-supported electrodes maintain the morphology of the material after preparation, preventing negative impacts such as material aggregation and structural damage. They offer better structural stability and higher flexibility. Unlike traditional electrodes, self-supported electrodes are created by in-situ growth or deposition methods, directly growing binder-free composite electrodes on a three-dimensional substrate. Commonly used three-dimensional substrates include Carbon Fiber Paper (CFP), Carbon Fiber Cloth (CFC), Carbon Fiber Felt (CFF), and Stainless Steel Welded Mesh (SSWM) [30]. These substrates not only improve the electrode's conductivity but also provide support for the active material, forming a stable porous structure that further enhances electrode performance. Some three-dimensional substrates can even be used in the design of flexible batteries. Furthermore, the use of binder-free electrodes significantly reduces the material's impedance, thereby improving battery performance.

For example, Tang et al. synthesized a nanoflower-shaped SSWM@Mn_3O_4 composite material by in-situ growth of Mn_3O_4 on stainless steel mesh using a hydrothermal method (as shown in Fig. 7.5a) [31]. This material, composed of ultra-thin nanosheets forming a multilayered nanoflower structure,

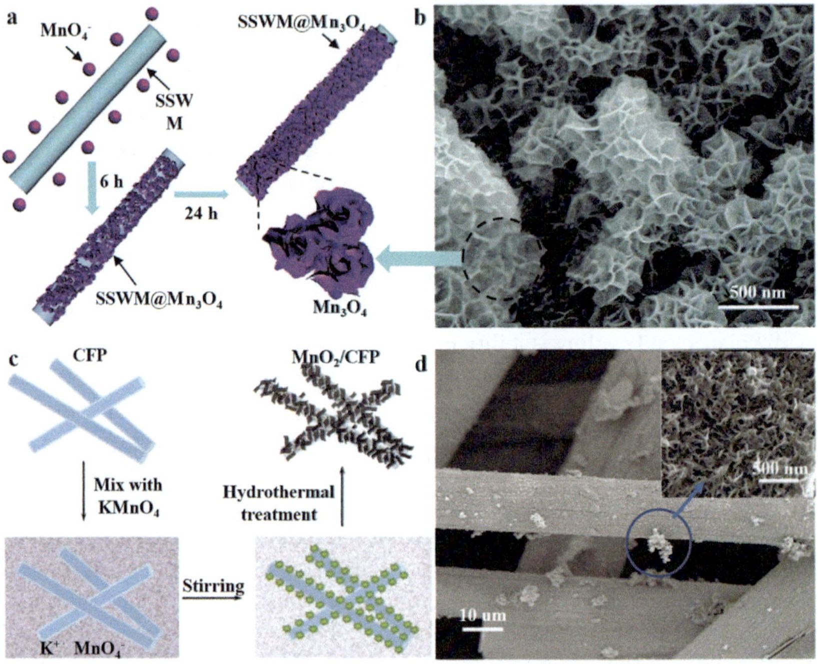

FIGURE 7.5 $SSWM@Mn_3O_4$ (a) Synthesis scheme, (b) SEM image of $MnO_2@CFP$ (c) Schematic of the electrodeposition process of Nano-MnO_2 on CFP, and (d) SEM images of $MnO_2@CFP$ after electrodeposition.

is beneficial for the migration and diffusion of Zn^{2+}. Wang et al. employed in-situ electrodeposition to fabricate $MnO_2@CFP$ cathodes on carbon fiber paper (Fig. 7.5c). In the electrodeposition process, Mn^{2+} from $MnSO_4$ solution was oxidized, resulting in the uniform deposition of MnO_2 onto the surface of CFP [32]. The SEM images (Fig. 7.5d) show that the MnO_2 consists of interconnected nanosheets, forming a highly porous structure that increases the electrode/electrolyte contact area, reduces the ion diffusion path, and enables faster electrochemical kinetics.

7.3.2 Development of electrolytes for wider voltage windows and improved ion conductivity

By increasing the concentration of salt in the electrolyte, the electrochemical window of aqueous batteries can be effectively expanded. High-concentration "water-in-salt" electrolytes can suppress water decomposition, enabling higher operating voltages and energy densities. In aqueous zinc-ion batteries, adding organic molecules such as serine as electrolyte additives can regulate the deposition behavior of zinc ions, inhibit dendrite formation, and improve the cycling stability of the battery. Studies have shown that the introduction of serine can induce preferential growth of zinc on the (100) crystal plane, reducing the occurrence of side reactions [33].

7.3.3 Innovations in separators and current collectors for enhanced battery efficiency

Using 3D printing to fabricate current collectors with specific structures can enhance the conductivity and mechanical strength of the electrodes, thereby improving the overall performance of the battery. For example, 3D-printed current collectors exhibit excellent performance in rechargeable batteries, including higher energy density and longer cycle life [34].

In aqueous lithium-ion batteries, employing lithium-ion-containing polymers as protective layers on the separator can prevent direct contact between the electrode active materials and the aqueous electrolyte, reduce side reactions, and enhance the battery's safety and cycling stability.

In conclusion, the ongoing advancements in electrode materials, electrolytes, and separators for aqueous ion batteries are driving significant improvements in their performance, safety, and efficiency. The integration of novel materials, such as self-supported electrodes and high-concentration electrolytes, alongside innovations in 3D-printed current collectors and protective separators, promises to overcome the traditional limitations of aqueous battery systems. These developments not only expand the electrochemical windows and enhance ion conductivity but also improve the overall stability and longevity of the batteries, positioning them as viable alternatives for sustainable and high-performance energy storage solutions.

7.4 Challenges in aqueous ion battery development

7.4.1 Technical and material challenges

AIBs have made significant progress in battery structure design and architecture optimization. Traditional AIBs typically rely on simple electrolytes and electrode materials, but in recent years, as demand for battery performance has increased, structural design and architecture optimization have become key areas of research.

To enhance battery performance, researchers have been actively investigating new electrode materials. Currently, innovative materials such as carbon-based compounds, transition metal oxides, sulfides, and phosphides are being utilized to improve the energy density and cycling stability of AIBs. In particular, modifying the microstructure of these materials, such as increasing the specific surface area and controlling the pore structure, has proven to be an effective way to boost the electrochemical performance of the batteries. Research has also focused on improving the conductivity and stability of aqueous battery electrolytes. For example, fluorine-containing electrolytes with high conductivity and stability have been developed, and the compatibility of various aqueous

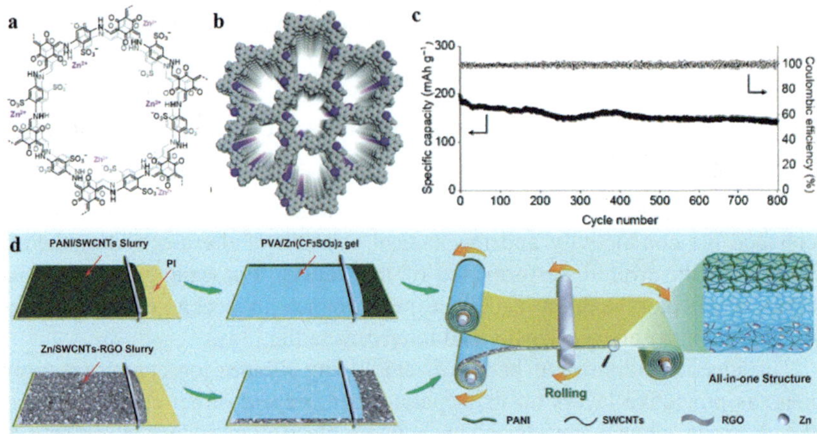

FIGURE 7.6 (a) Chemical structure; (b) Structural model (Zn^{2+}: purple, TpPa-SO^{3-}: gray); (c) Design of ultrathin all-in-one ZIBs: schematic process.

salt solutions has been studied [35,36]. To optimize interface reactions, modified electrolytes play an important role in suppressing side reactions and improving battery efficiency. Ion-conducting Polymer Electrolytes (PEs) are also considered effective at suppressing the growth of zinc dendrites and, due to their limited water content, can also inhibit the dissolution of active materials. For instance, Park et al. proposed a novel single-ion conducting hydrogel electrolyte based on sulfonated zinc covalent organic frameworks (TpPa-$SO_3Zn_{0.5}$), as shown in Figs. 7.6(a) and (b) [37]. Driven by the structure and physicochemical properties, the TpPa-$SO_3Zn_{0.5}$ membrane enhances the redox reliability of the Zn metal anode, stabilizes the cathode material structure, and enables fast ion transport, as depicted in Fig. 7.6(c). As a result, the modified Zn-MnO_2 full battery retains a capacity of 144 mAh·g^{-1} with a capacity retention rate of 73% after 800 cycles at a current density of 0.6 A·g^{-1}. Yao et al. designed a scalable method for assembling flexible battery components, as shown in Fig. 7.6(d) [38]. After preparing the positive and negative electrodes using a one-piece coating method, the electrodes were closely bonded together using a roller-pressing technique, applying mechanical pressure to both electrode materials, thus creating a thin, integrated flexible battery.

The battery interface, particularly the interface between the electrode and the electrolyte, directly impacts the battery's performance. As a result, in recent years, the stability and reaction mechanisms of the electrode-electrolyte interface have become a hot research topic. The application of nanostructures has led to breakthroughs in interface engineering, effectively reducing interface instability and extending the battery's lifespan.

7.4.2 Safety, scalability, and environmental impact

Traditional aqueous ion batteries generally have lower energy density, limiting their application in high-energy-demand scenarios. However, by incorporating high-performance electrode materials, such as transition metal oxides with high specific capacity and modified carbon materials, alongside optimizing electrolytes, researchers have achieved notable advancements in energy density [39–41]. For example, with the use of composite electrode materials and electrolyte systems, the energy density has reached levels that are comparable to or exceed those of some organic batteries. The charging speed of AIBs has long been limited by ion migration rates and electrode material reaction kinetics. By enhancing electrode conductivity with nanostructured materials, improving ion conductivity, and employing advanced electrolyte formulations to reduce internal resistance, the charging speed has been significantly improved. In recent years, studies have shown that some optimized AIBs can achieve rapid charging within a few minutes, greatly reducing charging time. Improving cycle life is a key direction in the development of AIB technology. By employing interface engineering, optimizing the stability of electrode materials, and developing new electrolytes, the cycle life of AIBs has been significantly enhanced. Particularly in terms of suppressing electrode corrosion, reducing solubility, and improving surface stability, techniques such as coating and electrode surface modification have significantly reduced battery degradation during repeated charge and discharge cycles [42,43].

7.4.3 Economic considerations and market competitiveness

Water electrolysis in traditional AIBs often leads to a decline in battery performance. To overcome this issue, researchers have proposed several innovative solutions, such as developing high-voltage-tolerant electrolyte systems, using ion-selective membranes to prevent water electrolysis side reactions, and designing highly stable electrode materials. For example, certain nanostructured electrode materials demonstrate excellent resistance to electrolysis and can operate at higher voltages without causing electrolyte hydrolysis [44].

The stability of electrodes directly affects the overall performance of the battery, especially during cycling. To address this, surface modification, coating technologies, and alloying methods have been employed to effectively enhance the stability of electrode materials [43]. These approaches improve the electrodes' corrosion resistance during repeated charge and discharge cycles, reduce electrode material dissolution or expansion, and ensure long-term stable operation.

In solving water electrolysis and electrode stability issues, modified electrolytes play a crucial role. For example, the use of fluorine compounds or other highly efficient and stable electrolytes can improve electrolyte stability and reduce side reactions. Additives in electrolytes, such as metal ions and polymer molecules, are also used to regulate the conductivity and stability of the electrolyte, effectively suppressing water electrolysis and electrode degradation.

Through these technological advancements, AIBs have made significant breakthroughs in various performance metrics, offering greater potential for applications in energy storage. With continuous technological development, aqueous ion batteries are expected to become a competitive green energy storage solution in the future.

Through these technological advancements, aqueous ion batteries have made significant breakthroughs in various performance metrics, offering greater potential for applications in the energy storage field. With continuous technological development, aqueous ion batteries are expected to become a competitive green energy storage solution in the future.

7.5 Applications of aqueous ion batteries

7.5.1 Integration with renewable energy sources

- Stability and Performance of Electrode Materials

In aqueous batteries, transition metal oxides used as cathode materials often face challenges of dissolution, leading to capacity decay and reduced cycle life. For instance, in aqueous lithium-ion batteries, the solubility of these oxides negatively impacts stability. Researchers are developing more stable electrode materials using techniques such as doping and surface coating to enhance their performance in aqueous environments.

Metal sulfides, despite their high capacity, suffer from poor cycling stability. For example, aqueous metal sulfide batteries encounter significant issues with cathode stability. Utilizing materials like silver sulfide (Ag_2S), which exhibit excellent conductivity and low solubility [45], has significantly improved cycling stability.

- Stability of Electrolytes and Electrochemical Window

At high voltages, aqueous batteries may experience water electrolysis, leading to performance degradation. For example, aqueous zinc-ion batteries face water electrolysis under high-voltage conditions. Optimizing electrolyte formulations and developing high-voltage-tolerant electrolytes can expand the electrochemical stability window, mitigating this issue.

Traditional aqueous electrolytes show performance limitations at low temperatures due to solvent freezing or reduced ionic mobility. For instance, aqueous zinc-ion batteries face challenges in low-temperature environments [46]. The development of composite aqueous electrolytes without organic solvents can address desolvation issues and enhance performance in low-temperature conditions.

- Interface Stability and Dendrite Growth

During charge and discharge cycles, metal anodes may develop dendrites, leading to short circuits and safety risks. For example, zinc dendrite growth is a significant issue in aqueous zinc-ion batteries [47]. The application of integrated Janus hydrogel membranes effectively regulates water molecule activity at electrode interfaces, suppressing dendrite formation and improving both safety and cycle life.

- Operating Voltage and Energy Density

The working voltage of aqueous batteries is constrained by the electrochemical stability window of water, limiting their energy density. For example, the working voltage of aqueous zinc batteries presents a major challenge for practical applications. However, by using bipolar membranes to separate electrolytes with different pH levels, the high oxygen evolution overpotential of acidic electrolytes and the high hydrogen evolution overpotential of alkaline electrolytes can be leveraged to increase the operating voltage.

7.5.2 Use in portable electronics, electric vehicles, and grid storage

- Safety Concerns

Aqueous batteries may generate hydrogen and oxygen gases under high-voltage conditions, posing explosion risks. Researchers are addressing this by optimizing electrolyte formulations to develop high-voltage-tolerant electrolytes, expanding the electrochemical stability window, and reducing gas evolution.

Dendrite growth in metal anodes can cause short circuits and compromise safety. The adoption of integrated Janus hydrogel membranes helps regulate water molecule activity at interfaces, mitigating corrosion and dendrite formation while improving safety and longevity.

- Challenges in Scaled Production

The production of aqueous batteries requires precise control over electrolyte composition and assembly processes, which increases complexity and cost. Researchers are working to simplify production through automation and streamlined processes to reduce costs and improve efficiency.

The consistent supply of raw materials like high-purity water and inorganic salts is essential for large-scale production. Strengthening partnerships with raw material suppliers ensures stability in the supply chain, supporting continuous production.

- Environmental Impact

The production and recycling of aqueous batteries can consume significant water resources, raising environmental concerns. Developing water recycling technologies reduces consumption and minimizes the ecological footprint.

Battery production and usage generate waste, such as wastewater and off-gases, which must be effectively managed. Establishing comprehensive waste treatment and recycling systems ensures proper handling, reducing environmental pollution. Future potential applications and the impact on energy systems.

Through continued research and innovation, addressing safety, scalability, and environmental challenges will enable aqueous batteries to become a competitive and sustainable energy storage solution in the future.

7.5.3 Economic considerations and market competitiveness

- Economic Analysis of Aqueous Ion Batteries

Aqueous ion batteries often rely on readily available and cost-effective materials, such as water-based electrolytes and metal salts [48]. Compared to organic solvent-based systems, aqueous batteries benefit from lower raw material costs. For instance, zinc, a commonly used anode material, is abundant and inexpensive, making it economically viable. However, some cathode materials, such as transition metal oxides and sulfides, remain costly due to complex synthesis processes and limited availability. Efforts to utilize more affordable alternatives, such as manganese-based or iron-based compounds, are underway to further reduce costs.

The production of aqueous ion batteries typically requires simpler assembly processes compared to lithium-ion batteries due to their nonflammable and less toxic components. This reduces the need for stringent safety measures during manufacturing. However, the challenge lies in ensuring consistency in electrolyte composition and electrode material quality, which can increase costs in high-precision production. Automation and advanced manufacturing techniques are being developed to scale production while minimizing costs.

Aqueous batteries generally have lower maintenance costs due to their inherent safety and stability. For example, the reduced risk of thermal runaway eliminates the need for complex monitoring systems [49]. However, issues like capacity fading and electrode corrosion over time can lead to higher replacement and maintenance expenses. Research into durable materials and efficient recycling processes aims to mitigate these costs, ensuring a longer lifecycle and lower total cost of ownership.

- Market Competitiveness of Aqueous Ion Batteries

Aqueous ion batteries are inherently safer than lithium-ion batteries because water-based electrolytes are nonflammable. This makes them an ideal choice for applications that demand high safety standards, such as grid storage and consumer electronics.

The use of nontoxic materials and water as the solvent reduces the environmental impact, giving aqueous batteries a competitive edge in markets where sustainability is a priority. For instance, government incentives for green energy solutions could further boost adoption. Lower raw material and manufacturing costs make aqueous batteries an attractive alternative for large-scale applications, such as renewable energy storage and backup power systems.

Although there has been notable advancement, the energy density of aqueous ion batteries remains inferior to that of lithium-ion batteries, restricting their use in applications that require high energy output, such as electric vehicles. Continued research is essential to improve energy density through innovations in electrode and electrolyte materials.

The dominance of lithium-ion batteries in the market presents a significant barrier to entry for aqueous batteries. Their established supply chains and economies of scale make it challenging for aqueous systems to compete, particularly in cost-sensitive markets.

Aqueous ion battery technologies are diverse, and the lack of standardization can hinder mass adoption. Developing industry standards and certifications will be critical to gaining market acceptance.

- Future Prospects and Strategies

Aqueous ion batteries are particularly well-suited for niche markets, such as stationary energy storage, portable electronics, and low-speed electric vehicles, where safety and cost take precedence over energy density. Collaborations with renewable energy companies and government agencies can help integrate aqueous batteries into green energy initiatives, increasing their visibility and market share. Establishing efficient recycling systems for aqueous batteries can further reduce costs and environmental impact, enhancing their appeal to environmentally conscious consumers and businesses. Continuous investment in research and development to improve energy density, cycle life, and manufacturing efficiency will be crucial in making aqueous ion batteries more competitive in the broader energy storage market.

By addressing economic and competitive challenges through innovation, standardization, and strategic partnerships, aqueous ion batteries have the potential to carve out a significant niche in the energy storage market, contributing to a greener and more sustainable future.

7.6 Future directions and research opportunities

7.6.1 Emerging trends in material science and electrochemistry for AIBs

As the world transitions to green energy, the proportion of renewable energy sources, such as solar and wind power, continues to grow in the global energy mix. However, due to the intermittent and unstable nature of these energy sources, there is an increasing demand for efficient, cost-effective, and environmentally friendly energy storage technologies. Aqueous ion batteries, with their high safety, low cost, and green characteristics, have emerged as an ideal choice for integration with renewable energy systems.

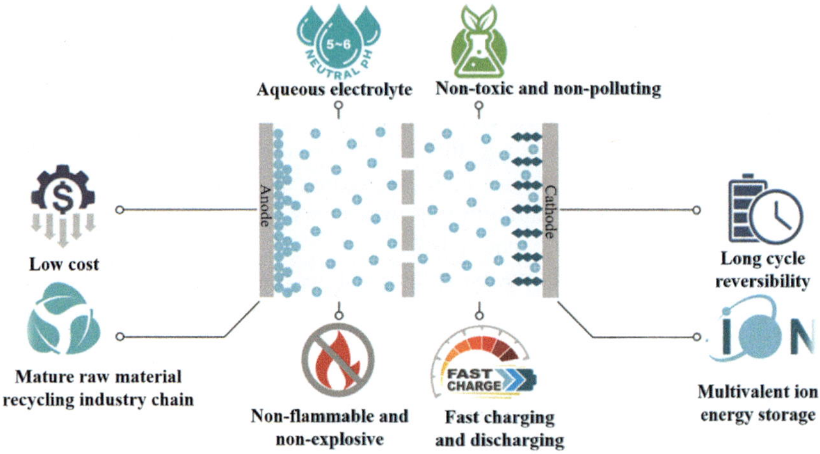

FIGURE 7.7 Advantages of aqueous batteries.

The advantages of aqueous ion batteries in renewable energy systems are evident, as shown in Fig. 7.7. First, their use of water-based electrolytes makes them nonflammable and nonexplosive, making them particularly suitable for high-safety applications, such as wind farms or residential solar systems. This reduces the risk of fires or explosions in energy storage facilities. Second, aqueous batteries feature low internal resistance and high ion mobility, enabling them to respond rapidly to grid demands. They can quickly release energy during short interruptions in photovoltaic generation or fluctuations in wind speed, ensuring system stability. Additionally, compared to traditional energy storage technologies, aqueous batteries have a smaller environmental footprint during manufacturing, usage, and recycling, aligning with the low-carbon development goals promoted by renewable energy. Potential for breakthroughs in energy storage capacity and efficiency.

7.6.2 Use in portable electronics, electric vehicles, and grid storage

- Portable Electronics

The safety and eco-friendly nature of aqueous ion batteries make them a promising alternative energy storage technology, especially in scenarios demanding high safety standards. Examples include:

Traditional lithium-ion batteries pose risks of overheating and combustion, leading some manufacturers to explore aqueous zinc-ion batteries for children's toys and smartwatches. These batteries are nonflammable, nonexplosive, and low-cost, making them suitable for large-scale production. Portable medical devices like ECG monitors and infusion pumps require high stability and safety.

Research institutions have tested aqueous zinc-ion batteries to provide a durable and secure power source for such applications.

- Electric Vehicles

While lithium-ion batteries dominate the Electric Vehicle (EV) market, AIBs are showing potential in specific areas: In a pilot project in a Chinese city, electric buses were equipped with aqueous zinc-iron batteries. These batteries charge quickly, meeting the high-frequency operational needs of public transportation. Their low cost and eco-friendly properties make them an ideal choice for public transit. For community service vehicles and park shuttles, which require less energy density but prioritize safety and cost, aqueous zinc-ion batteries have been adopted by some manufacturers due to their high safety, long cycle life, and affordability.

- Grid Storage

Aqueous ion batteries have demonstrated the most significant impact in grid storage, particularly in solar and wind energy systems. In California, United States, a solar power plant uses a storage system based on aqueous vanadium batteries. These batteries offer over 10,000 cycles, enabling the long-term storage of solar energy and nighttime energy release, thereby enhancing solar energy utilization efficiency. In a Danish wind farm, an aqueous zinc-air battery system was deployed to manage wind power fluctuations. This system provides fast energy release and storage, ensuring grid stability. In a remote Indian village, a collaborative project between the government and research institutions installed an aqueous battery storage system integrated with solar panels to provide 24/7 reliable electricity. This solution significantly reduced grid construction costs and minimized diesel generator pollution.

As technology advances, the application scenarios for aqueous ion batteries are expected to diversify further. For instance, large data centers are exploring aqueous batteries as Uninterrupted Power Supply (UPS) solutions to ensure continuous operations. Additionally, high-energy industrial sectors, such as steel and chemical industries, are experimenting with aqueous battery storage systems to reduce carbon emissions and improve energy efficiency. Collaborations, funding, and policy implications for accelerating AIB development. These examples demonstrate that aqueous ion batteries have already showcased unique advantages in portable devices, electric vehicles, and grid storage. Their future market potential is highly promising.

7.6.3 Future potential applications and the impact on energy systems

As the demand for sustainable and efficient energy storage grows, aqueous batteries have begun to demonstrate their versatility in various emerging application scenarios. Their unique combination of safety, environmental friendliness, and cost-effectiveness makes them increasingly relevant in areas that require robust and scalable energy solutions.

- Emerging Application Scenarios

With the proliferation of rooftop solar panels and small wind turbines, the need for safe and reliable energy storage for residential and commercial users has surged. Aqueous batteries, with their nonflammable water-based electrolytes, provide a safer alternative to lithium-ion batteries. They can store excess energy generated during the day and release it during peak hours, optimizing energy use and reducing dependence on the grid.

Remote and off-grid locations often rely on diesel generators, which are expensive and environmentally damaging. Aqueous batteries integrated with renewable energy systems, such as small-scale solar or wind installations, offer a sustainable and cost-effective solution for powering remote villages, disaster relief centers, or military outposts. Their high safety and low maintenance requirements make them ideal for these settings.

In maritime industries and underwater operations, safety and resistance to harsh environments are critical. Aqueous batteries, with their stability and environmental compatibility, can be used in electric ships, underwater drones, and autonomous submersibles. Their nontoxic components also minimize environmental risks in case of leakage.

Aqueous batteries are well-suited for powering the energy demands of smart cities, where dynamic energy storage and distribution systems are needed. They can be used to balance supply and demand in urban grids, store energy from distributed renewable sources, and even provide backup power for critical infrastructure such as hospitals, transportation networks, and communication systems.

- Potential Impact on Future Energy Systems

The integration of aqueous batteries with renewable energy sources can address the intermittency issues of solar and wind power, making these energy sources more reliable. By enabling the storage and smooth dispatch of renewable energy, aqueous batteries could significantly reduce dependence on fossil fuels, supporting the global transition to a low-carbon energy system.

The rapid response capability of aqueous batteries makes them valuable in grid stabilization. They can handle frequency regulation, voltage control, and peak shaving, ensuring a more resilient energy infrastructure. In regions prone to extreme weather events, aqueous batteries can provide backup power and enhance grid reliability.

With their scalability and compatibility with renewable energy, aqueous batteries are ideal for decentralized energy systems. This approach reduces transmission losses, empowers local communities with energy independence, and fosters a more equitable energy distribution system.

The development of aqueous batteries could inspire further innovations in energy storage technologies, particularly in the search for nontoxic, abundant, and affordable materials. This progress could have a cascading effect on the broader energy storage industry, driving down costs and increasing adoption rates.

In conclusion, aqueous batteries hold immense promise for transforming emerging energy applications and reshaping the future energy landscape. Their unique characteristics align closely with the goals of sustainable development, making them a cornerstone technology in building a greener and more resilient energy system.

7.7 Future directions and research opportunities

7.7.1 Emerging trends in material science and electrochemistry for AIBs

Water-based ion batteries are in a phase of rapid development. At the forefront of materials science and electrochemistry research, scientists are exploring novel materials and innovative designs to further enhance the performance, stability, and application potential of these batteries. Researchers are actively developing new cathode and anode materials to improve energy density and cycle stability. For instance, two-dimensional materials like MXene and amorphous structure materials have demonstrated superior ion storage capabilities and structural stability [50,51]. Additionally, the development of low-cost biomass-based carbon materials offers promising potential for large-scale production.

Recent studies focus on unconventional aqueous electrolytes, such as water-ionic liquid mixtures and superconcentrated saline electrolytes. These electrolytes exhibit a broader electrochemical window and lower volatility. Furthermore, dynamic ionic liquid electrolytes with selective conductivity are being designed to achieve higher stability and adaptability in batteries. In terms of interface stability, innovative solutions such as dynamic adaptive interfacial layers and in-situ generated protective films have been proposed in recent years. These approaches not only enhance interface stability but also repair defects, offering new strategies for designing long-life batteries. Moreover, nanoscale patterned interface technologies are being utilized to further reduce the risk of dendrite growth.

Intelligent aqueous battery systems have become a research focus. By integrating AI algorithms with sensing technologies, precise monitoring and real-time adjustment of battery operating conditions are achieved. Concurrently, research on adaptive battery materials is advancing rapidly. Shape-memory materials and dynamically reconfigurable gel electrolytes, in particular, provide new solutions for high-safety batteries. Progress in green manufacturing processes has centered on reducing the use of toxic chemicals and developing biodegradable battery components. For example, the application of pollution-free hydrothermal synthesis to produce nanomaterials and the acceleration of research into materials derived from natural resources, such as chitosan and alginates, have significantly reduced the carbon footprint of battery production.

Advancements in materials science and electrochemistry have driven the rapid evolution of aqueous batteries toward high performance, enhanced safety, and environmental friendliness. Looking ahead, with the integration of more advanced technologies, aqueous batteries are expected to play a pivotal role in the energy storage sector, offering competitive solutions to support the global transition to a green energy system.

7.7.2 Potential for breakthroughs in energy storage capacity and efficiency

Future research will focus on developing new electrode materials with higher specific capacities, such as nanostructured oxides, highly conductive polymers, and bifunctional composite materials (combining conductivity and catalytic activity). These materials can not only enhance energy storage capacity but also significantly reduce the internal resistance of the battery, thus improving energy conversion efficiency. Researchers are working to expand the electrochemical window by introducing high-concentration electrolytes, ionic liquids, or other mixed solution systems to support higher operating voltages. This approach effectively addresses the limitations of water splitting, enhancing the total energy density of aqueous batteries and providing a solid foundation for high-performance energy storage systems.

In terms of interface optimization, the use of novel surface modification techniques (such as in-situ film formation and electrode surface coatings) can reduce side reactions and minimize energy loss. Additionally, dynamic interface optimization strategies, such as adaptive interface films, can continuously repair interface damage during cycling, further improving energy efficiency and cycle life. Multifunctional separators that effectively suppress dendrite growth and control ion flow and distribution, have been shown to enhance charging and discharging efficiency. For example, nanofiber-enhanced separator materials and polymer separators with ion-selective functions have demonstrated significant improvements in battery efficiency, especially under high-rate charging and discharging conditions.

By combining advanced thermal management systems and intelligent control systems, researchers are optimizing the overall efficiency of aqueous batteries. For instance, by monitoring the battery state (SOC) in real-time and adjusting operating parameters, energy loss can be minimized. Meanwhile, modular battery system designs further enhance overall energy storage efficiency and simplify the deployment process for large-scale storage.

Enhancing energy storage capacity and efficiency is a key focus for the future advancement of aqueous batteries. Through a collaborative effort involving material innovation, electrolyte optimization, interface engineering, and system integration design, breakthroughs in energy storage capacity and efficiency will provide more sustainable and efficient solutions for energy storage technologies.

7.7.3 Collaborations, funding, and policy implications for accelerating AIB development

AIBs, emerging as a novel energy storage technology, have garnered considerable attention for their high safety, cost-effectiveness, and environmental sustainability. To accelerate their development, collaboration from various sectors, financial support, and policy guidance are essential. Industry-academia-research cooperation is a crucial pathway to advancing the technology of aqueous ion batteries. Universities, research institutions, and enterprises should collaborate to drive technological development and commercialization. For example, the Energy Innovation Center at the University of California, San Diego, has received funding from the U.S. Department of Energy and is dedicated to developing next-generation energy storage technologies. Additionally, forming industry alliances for aqueous batteries can integrate resources from upstream and downstream companies, creating synergies and enhancing market competitiveness. On the international front, collaborating with advanced research institutions and enterprises globally will help introduce cutting-edge technologies and experiences, accelerating technological breakthroughs and market applications for aqueous batteries.

Governments should establish special funds to support fundamental research, technological development, and demonstration applications for aqueous ion batteries. For instance, Guangdong Province has released an implementation plan to promote the high-quality development of new energy storage industries, supporting the demonstration of sodium-ion batteries in storage and other sectors. Financial institutions should increase their financing support for aqueous battery companies, offering loans, equity investments, and other diversified financial services. Additionally, through policy guidance, governments can attract social capital to invest in the aqueous battery industry, creating a diversified investment and financing system.

Governments should also formulate development plans for the aqueous ion battery industry, clarifying development goals, key tasks, and support measures to guide the orderly development of the industry. Efforts should be made to accelerate the establishment of technical standards and testing and certification systems for aqueous batteries, standardizing market order and improving product quality. Tax reductions should be offered to companies engaged in the research and production of aqueous batteries, lowering operating costs and incentivizing technological innovation. A reasonable electricity pricing mechanism and energy storage compensation system should be established to allow energy storage entities to benefit commercially from electricity price fluctuations, promoting the large-scale application of new energy storage technologies.

Through multiparty collaboration, increased financial investment, and enhanced policy support, aqueous ion batteries are expected to develop rapidly and become an essential force in driving energy transition and achieving carbon neutrality goals.

7.8 Productization and commercialization

7.8.1 Potential product

With continuous technological advancements, the productization and commercialization of aqueous batteries are developing in a diversified direction, presenting various potential product forms and business models.

- Potential Product Forms

The inherent safety of aqueous batteries makes them a suitable option for portable electronics like smartphones and tablets. Their affordability and eco-friendly nature also cater to the growing consumer preference for green products. While their energy density currently falls short of that of lithium-ion batteries, their safety and cost-effectiveness make them a viable candidate for the EV market [52]. As technology advances, aqueous batteries are expected to emerge as a dominant power source for electric vehicles. Furthermore, their low cost and environmentally friendly properties give them an edge in large-scale energy storage applications. For example, Guangdong Province has introduced a plan to advance the high-quality development of new energy storage industries, including supporting the use of sodium-ion batteries in energy storage projects.

- Business Models

Companies directly produce and sell aqueous battery products to consumers and enterprise customers. For instance, Guangdong Kaijin New Energy Technology Co., Ltd. generates profits primarily through product sales in its business model. For large-scale energy storage systems, companies can offer battery leasing and maintenance services, charging rent and service fees. This model can reduce the initial investment cost for customers and promote the widespread application of aqueous batteries. Companies collaborate with universities and research institutions to jointly develop aqueous battery technologies and license the technological results to other companies, charging licensing fees. This model can accelerate the promotion and application of the technology. Companies integrate upstream and downstream resources to form a complete industry chain, from raw material supply to product sales, enhancing market competitiveness. For example, Guangdong Kaijin New Energy Technology Co., Ltd. has integrated graphite processing and other links into a complete industry chain in its business cooperation model.

7.8.2 Potential companies and application scenarios

Leading companies have emerged both domestically and internationally, achieving significant progress in multiple practical application scenarios. In China, in 2020, Zheneng Zhongke Energy Storage announced an investment of 60 million yuan to establish a 1 MWh pilot production line for aqueous zinc energy storage batteries. After more than three years of research and development, the project passed its completion acceptance in July 2023. In 2022, Hanwei Technology established China's first automated aqueous zinc-ion battery R&D pilot production line in Zhejiang. Currently, their aqueous zinc-ion batteries have entered the commercialization phase, with order values nearing 100 million yuan. In July 2023, Jiaotou Zhuoyue New Energy launched a high-tech zone production base project for aqueous zinc-ion batteries in Yangjiang City, with plans to build a 1 GWh production and packaging factory for zinc-ion power and energy storage cells, with an expected annual output value of approximately 1.2 billion yuan once fully operational.

Internationally, Aquion Energy in the United States focuses on the development of AIBs, with products primarily used in residential and commercial energy storage systems. The company uses eco-friendly materials and is committed to providing safe and sustainable energy solutions. Japan's Sumitomo Chemical has conducted extensive research in aqueous batteries and developed various aqueous battery products widely used in power tools, home appliances, and other sectors. These products are known for their high safety and long lifespan.

Aqueous batteries, with their high safety and affordability, are ideal for home energy storage systems. For instance, Zheneng Zhongke Energy Storage has deployed aqueous zinc-ion batteries in residential energy storage, addressing consumer needs for safety and environmental sustainability. Hanwei Technology's aqueous zinc-ion batteries have entered the commercialization phase, with order values approaching 100 million yuan, mainly applied in commercial energy storage systems, helping enterprises reduce energy costs and improve energy efficiency. The high safety of aqueous batteries also presents potential application value in the electric vehicle sector. As technology advances, aqueous batteries are expected to become the mainstream power battery for electric vehicles in the future.

In summary, leading companies in the domestic and international aqueous battery industry have made significant progress in various practical application scenarios. As technology continues to mature, aqueous batteries are expected to play a larger role in energy storage and the EV market.

■ Chapter Abstracts

AIBs have gained significant attention as a safer and more environmentally sustainable alternative to traditional energy storage systems, especially in light of the growing demand for renewable energy solutions. Characterized by their use of water-based electrolytes, AIBs offer distinct advantages over traditional batteries, including nonflammability, low toxicity, and high ionic conductivity, which contribute to enhanced safety and potentially lower manufacturing costs. This review delves into the fundamental aspects of AIBs, covering the variety of ions utilized (e.g., lithium, sodium, potassium, magnesium, and zinc ions) and their unique transport mechanisms within the aqueous medium. It also explores the diverse materials and innovative designs employed in electrodes and electrolytes to overcome challenges associated with narrow electrochemical windows and the stability of aqueous systems. Significant advancements in electrode material science and electrolyte formulation have led to breakthroughs in the energy density, charging rates, and cycle life of AIBs, pushing the boundaries of their performance closer to, and in some cases surpassing, that of their nonaqueous counterparts. The review highlights the wide range of applications for AIBs, from grid storage and portable electronics to electric vehicles, underscoring their role in the transition towards sustainable energy systems. Despite their promising features, AIBs face challenges related to ion mobility, electrode degradation, and energy efficiency that require ongoing research. The review concludes with a perspective on future directions in AIB research, emphasizing the importance of interdisciplinary approaches in material science, electrochemistry, and engineering to realize the full potential of aqueous-ion batteries in modern energy storage. A ranking index, combining both electrochemical and mechanical properties, is introduced as a quantitative factor to review the feasibility of the AIBs to be commercialized.

References

[1] T. Shoji, M. Hishinuma, T. Yamamoto, Zinc-manganese dioxide galvanic cell using zinc sulphate as electrolyte. Rechargeability of the cell, Journal of Applied Electrochemistry 18 (1988) 521–526.

[2] W. Li, J.R. Dahn, D.S. Wainwright, Rechargeable lithium batteries with aqueous electrolytes, Science 264 (1994) 1115–1118.

[3] J. Köhler, et al., LiV3O8: characterization as anode material for an aqueous rechargeable Li-ion battery system, Electrochimica Acta 46 (2000) 59–65.

[4] H. Wang, et al., Electrochemical properties of TiP2O7 and LiTi2(PO4)3 as anode material for lithium ion battery with aqueous solution electrolyte, Electrochimica Acta 52 (2007) 3280–3285.

[5] G. Wang, et al., An aqueous rechargeable lithium battery with good cycling performance, Angewandte Chemie 46 (2007) 295–297.

[6] C. Xu, et al., Reversible insertion properties of zinc ion into manganese dioxide and its application for energy storage, Electrochemical and Solid-State Letters 12 (2009) A61.

[7] P. Chen, B.A. Powell, M. Mortimer, P.C. Ke, Adaptive interactions between zinc oxide nanoparticles and Chlorella sp, Environmental Science & Technology 46 (2012) 12178–12185.

[8] G. Fang, J. Zhou, A. Pan, S. Liang, Recent advances in aqueous zinc-ion batteries, ACS Energy Letters 3 (2018) 2480–2501.

[9] G. Li, et al., Developing cathode materials for aqueous zinc ion batteries: challenges and practical prospects, Advanced Functional Materials 34 (2024) 2301291.

[10] H. Dong, et al., Bio-inspired polyanionic electrolytes for highly stable zinc-ion batteries, Angewandte Chemie 135 (2023) e202311268.
[11] Y. Xu, et al., Recent advances on challenges and strategies of manganese dioxide cathodes for aqueous zinc-ion batteries, Energy & Environmental Materials 6 (2023) e12575.
[12] X. Wu, et al., Reverse dual-ion battery via a ZnCl2 water-in-salt electrolyte, Journal of the American Chemical Society 141 (2019) 6338–6344.
[13] L. Suo, et al., Advanced high-voltage aqueous lithium-ion battery enabled by "water-in-bisalt" electrolyte, Angewandte Chemie 128 (2016) 7252–7257.
[14] L. Jiang, et al., High-voltage aqueous Na-ion battery enabled by inert-cation-assisted water-in-salt electrolyte, Advanced Materials 32 (2020) 1904427.
[15] P. Arora, Z. Zhang, Battery separators, Chemical Reviews 104 (2004) 4419–4462.
[16] K. Huang, J. Guo, Z. Xu, Recycling of waste printed circuit boards: a review of current technologies and treatment status in China, Journal of Hazardous Materials 164 (2009) 399–408.
[17] L. Suo, et al., "Water-in-salt" electrolyte enables high-voltage aqueous lithium-ion chemistries, Science 350 (2015) 938–943.
[18] J. Luo, W. Cui, P. He, Y. Xia, Raising the cycling stability of aqueous lithium-ion batteries by eliminating oxygen in the electrolyte, Nature Chemistry 2 (2010) 760–765.
[19] S. Liu, et al., Rechargeable aqueous lithium-ion battery of TiO2/LiMn2O4 with a high voltage, Journal of the Electrochemical Society 158 (2011) A1490.
[20] J.Y. Luo, Y.Y. Xia, Aqueous lithium-ion battery LiTi2 (PO4) 3/LiMn2O4 with high power and energy densities as well as superior cycling stability, Advanced Functional Materials 17 (2007) 3877–3884.
[21] L. Suo, et al., "Water-in-salt" electrolyte makes aqueous sodium-ion battery safe, green, and long-lasting, Advanced Energy Materials 7 (2017) 1701189.
[22] H. Yu, D. Chen, L. Zhang, S. Huang, L. Zhou, G. Kuang, W. Wei, L. Chen, Y. Chen, Electrolyte engineering for optimizing anode/electrolyte interface towards superior aqueous zinc-ion batteries: a review, Transactions of Nonferrous Metals Society of China 34 (2024) 3118–3150.
[23] X. Wu, D.P. Leonard, X. Ji, Emerging non-aqueous potassium-ion batteries: challenges and opportunities, Chemistry of Materials 29 (2017) 5031–5042.
[24] F. Wang, et al., High-voltage aqueous magnesium ion batteries, ACS Central Science 3 (2017) 1121–1128.
[25] J. Yan, et al., Rechargeable hybrid aqueous batteries, Journal of Power Sources 216 (2012) 222–226.
[26] B. Sadeghi, M.R. Sarraf, H.R. Shahverdi, The effect of $LiFePO_4$ coating on electrochemical performance of $LiMn_2O_4$ cathode material, Iranian Journal of Chemistry & Chemical Engineering 31 (4) (2012) 29–33.
[27] X. Wu, et al., The electrochemical performance of aqueous rechargeable battery of Zn/Na0.44MnO2 based on hybrid electrolyte, Journal of Power Sources 336 (2016) 35–39.
[28] Y. Wang, et al., Carbon-coated $LiTi_2$ $(PO_4)_3$ composites synthesized through tannic acid with high rate performance for aqueous lithium-ion batteries, Journal of Alloys and Compounds 939 (2023) 168704.
[29] L. Kou, et al., One-dimensional tunnel VO_2 (B) cathode materials for high-performance aqueous zinc ion batteries: a mini review of recent advances and future perspectives, Green Chemistry (2024).
[30] J. Wang, et al., Integrated three-dimensional carbon paper/carbon tubes/cobalt-sulfide sheets as an efficient electrode for overall water splitting, ACS Nano 10 (2016) 2342–2348.
[31] C. Zhu, et al., Binder-free stainless Steel@Mn_3O_4 nanoflower composite: a high-activity aqueous zinc-ion battery cathode with high-capacity and long-cycle-life, Journal of Materials Chemistry A 6 (20) (2018) 9677–9683, https://doi.org/10.1039/C8TA01198B.
[32] W. Yang, et al., Photo-assisted charging of carbon fiber paper-supported CeO2/MnO2hetero-junction and its long-lasting capacitance enhancement in dark, Journal of Advanced Ceramics 11 (2022) 1735–1750.

[33] M. Zhou, et al., Surface-preferred crystal plane for a stable and reversible zinc anode, Advanced Materials 33 (2021) 2100187.
[34] F. Zhang, et al., 3D printing technologies for electrochemical energy storage, Nano Energy 40 (2017) 418–431.
[35] N. Von Aspern, G.V. Röschenthaler, M. Winter, I. Cekic Laskovic, Fluorine and lithium: ideal partners for high-performance rechargeable battery electrolytes, Angewandte Chemie 58 (2019) 15978–16000.
[36] X. Zhao, Z. Zhao Karger, M. Fichtner, X. Shen, Halide-based materials and chemistry for rechargeable batteries, Angewandte Chemie 59 (2020) 5902–5949.
[37] S. Park, et al., A single-ion conducting covalent organic framework for aqueous rechargeable Zn-ion batteries, Chemical Science 11 (2020).
[38] M. Yao, et al., Scalable assembly of flexible ultrathin all-in-one zinc-ion batteries with highly stretchable, editable, and customizable functions, Advanced Materials 33 (2021) 2008140.
[39] J. Hassoun, K. Lee, Y. Sun, B. Scrosati, An advanced lithium ion battery based on high performance electrode materials, Journal of the American Chemical Society 133 (2011) 3139–3143.
[40] K.K. Patel, et al., Evolution and recent developments of high performance electrode material for supercapacitors: a review, Journal of Energy Storage 44 (2021) 103366.
[41] F. Liu, S. Song, D. Xue, H. Zhang, Folded structured graphene paper for high performance electrode materials, Advanced Materials (Deerfield Beach, Fla.) 24 (2012) 1089–1094.
[42] A. Merz, A.J. Bard, A stable surface modified platinum electrode prepared by coating with electroactive polymer, Journal of the American Chemical Society 100 (1978) 3222–3223.
[43] L.J. Fu, et al., Surface modifications of electrode materials for lithium ion batteries, Solid State Sciences 8 (2006) 113–128.
[44] A. Manthiram, A.V. Murugan, A. Sarkar, T. Muraliganth, Nanostructured electrode materials for electrochemical energy storage and conversion, Energy & Environmental Science 1 (2008) 621–638.
[45] M.H. Hebb, Electrical conductivity of silver sulfide, Journal of Chemical Physics 20 (1952) 185–190.
[46] M. Chen, et al., Aqueous zinc-ion batteries at extreme temperature: mechanisms, challenges, and strategies, Energy Storage Materials 51 (2022) 683–718.
[47] Q. Zhang, et al., Interfacial design of dendrite-free zinc anodes for aqueous zinc-ion batteries, Angewandte Chemie 59 (2020) 13180–13191.
[48] H. Zhang, et al., Challenges and strategies for high-energy aqueous electrolyte rechargeable batteries, Angewandte Chemie 60 (2021) 598–616.
[49] Y. Wang, et al., A review on research status and key technologies of battery thermal management and its enhanced safety, International Journal of Energy Research 42 (2018) 4008–4033.
[50] Y. Zhang, et al., Adjustable MXene-based materials in metal-ion batteries: progress, prospects, and challenges, Small Structures 5 (2024) 2300255.
[51] W. Fu, M. Aizudin, P.S. Lee, E.H. Ang, Recent progress in the applications of MXene-based materials in multivalent ion batteries, Small (2024) 2404093.
[52] S. Li, et al., Real-time robust state estimation for large-scale low-observability power-transportation system based on meta physics-informed graph timesnet, IEEE Transactions on Smart Grid 15 (2024) 5500–5513.

Chapter 8

The role of energy storage in achieving carbon peak and carbon neutrality goals

Ce Liang[a], Lin Wu[a], Haoqin Li[a], Shijie Li[a], Huaiguang Jiang[a], Lin Shu[a], and Yi Guo[b]

[a]*School of Future Technology, South China University of Technology, Guangzhou, China,* [b]*School of Automation, Beijing Institute of Technology, Beijing, China*

Contents

Modern Energy Storage. https://doi.org/10.1016/B978-0-32-395419-8.00014-3

8.1 The role of energy storage in achieving carbon peak and carbon neutrality goals

8.1.1 Carbon peaking and carbon neutrality goals and challenges

8.1.1.1 Current status of global carbon emissions

The upward trend in global carbon emissions persists, reflecting the growing demand for energy driven by industrialization and population growth.

According to the Global Carbon Budget 2024, global carbon emissions reached 11.1±0.9 Gt C/yr in 2023, with fossil fuel and cement emissions increasing by 1.3% to 10.1±0.5 Gt C/yr, while land-use change emissions declined slightly to 1.0±0.7 Gt C/yr [1]. As shown in Fig. 8.1, among the various sources of emissions, those resulting from fossil fuel combustion and cement production (EFOS) dominate, contributing approximately 10.1±0.5 Gt C/yr, which accounts for 91% of total global emissions. Breaking these emissions into their components, the contributions related to energy highlight a stark reliance on traditional fossil fuels. Coal, the major contributor, accounts for 41% of emissions, followed by oil (32%), natural gas (21%), and cement production (4%). Gas flaring and other minor sources together contribute less than 1%. Additionally, emissions linked to land use changes (ELUC) and deforestation stand at 1.0±0.7 Gt C/yr, representing a smaller share but one with significant ecological consequences, particularly in terms of diminishing the natural capacity of ecosystems to absorb carbon. The sectoral distribution of carbon emissions further underscores the influence of energy consumption. Globally, energy-related emissions account for 73.2% of all carbon emissions [2], making it the primary contributor to the current climate challenge. Within this category:

- **Industrial energy use** is responsible for 24.2%, with significant contributions from heavy industries of high energy such as steel production (7.2%) and chemical and petrochemical manufacturing (3.6%). Other sectors, including paper production, nonferrous metals, and machinery manufacturing, collectively add 10.6%.
- **Transportation** accounts for 16.2%, with road transport leading at 11.9%, while aviation (1.9%), maritime shipping (1.7%), and rail transport (0.4%) represent smaller but still notable shares.
- **Building energy use** contributes 17.5%, with 10.9% stemming from residential buildings and 6.6% from commercial buildings. Key emission drivers in this sector include heating, lighting, and the use of household appliances, see Figs. 8.1 and 8.2.

As illustrated in Fig. 8.2, the sub-sectoral breakdown reveals that industrial processes and energy conversion dominate within the industrial category, while transportation and building energy use contribute significantly to the overall footprint. Understanding these sub-sectoral distributions is essential for designing targeted mitigation strategies.

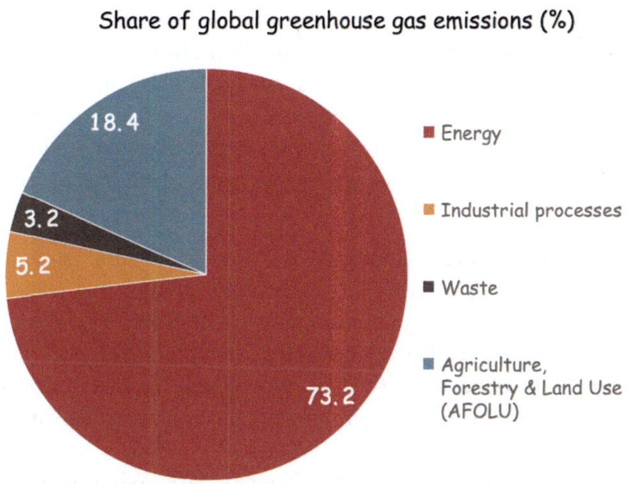

FIGURE 8.1 Global carbon dioxide emissions by sector.

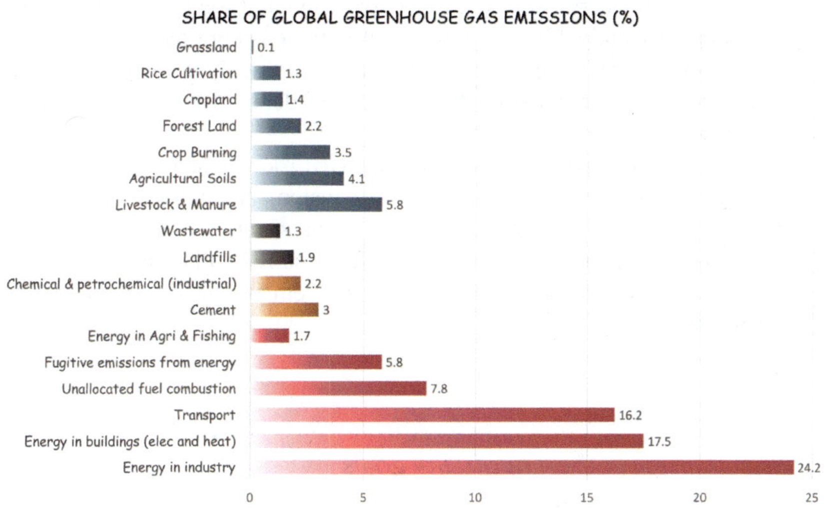

FIGURE 8.2 Global carbon dioxide emissions by subsector.

Furthermore, emissions from industrial processes, distinct from energy consumption, contribute 5.2%, with cement production alone accounting for 3%, while chemical and petrochemical processes contribute 2.2%. Other emission sources include waste management, such as landfill emissions and wastewater treatment (3.2%), as well as Agriculture, Forestry, and Other Land Use (AFOLU) activities, which together account for a significant 18.4% of total emissions.

These sectoral contributions highlight the diverse and interlinked sources of emissions that make global carbon reduction efforts increasingly complex. Addressing these challenges requires a comprehensive understanding of the driving forces behind persistent growth in emissions, which are rooted in both structural economic factors and systemic dependence on traditional energy sources. Three primary factors underlie the continued rise in global carbon emissions:

1. **Rising Global Energy Demand**. As industrialization and urbanization expand globally, especially in developing regions, the energy demand continues to grow at an unprecedented pace. Rapid economic growth in South Asia and Africa has significantly driven this increase. These regions remain heavily dependent on coal and oil due to their affordability and accessibility, further bolstering the dominance of fossil fuels in their energy systems.
2. **Dominance of Fossil Fuels in the Energy Mix**. While renewable energy technologies such as solar and wind have witnessed rapid advancements in recent years, their overall contribution remains insufficient to meet the surging global energy demand. Currently, wind and solar energy contribute only approximately 12% of global electricity consumption, which is much less than the scale required to offset fossil fuel use. This imbalance underscores the urgent need for accelerated deployment of clean energy solutions.
3. **Land Use Changes and Ecosystem Disruption**. Changes in land use, including deforestation, agricultural expansion, and urban development, have significantly disrupted natural carbon sinks, such as forests and wetlands. These disruptions not only release stored carbon but also reduce the ecosystem's capacity to absorb future emissions, compounding the problem.

8.1.1.2 Global carbon neutrality goals

- **The Paris Agreement**

  Adopted in 2015, serves as the cornerstone of global efforts to combat climate change and achieve carbon neutrality [3]. It sets the overarching goal of limiting the increase in global average temperature to below 2 °C above preindustrial levels, with a more ambitious aim of limiting it to 1.5 °C. To achieve this, developed countries are required to take the lead in reducing emissions while providing financial and technological support to developing nations. This collaborative framework has spurred widespread commitments, with over 140 countries pledging to achieve carbon neutrality. Collectively, these nations account for more than 90% of global GDP and approximately 80% of global emissions.

Each country has tailored its approach to carbon neutrality based on its unique circumstances, including economic structure, resource endowments, and technological capabilities, see Table 8.1.

TABLE 8.1 Summary of global carbon neutrality goals and measures.

Country	Target Year	Key Goals	Main Measures
China	2030 2060	Peak emissions by 2030; Carbon neutrality by 2060	– 25% nonfossil energy by 2030, 80%+ by 2060 – 1.2 billion kW wind/solar capacity by 2030 – 13.5% energy intensity cut by 2025
United States	2050	Achieve net-zero emissions	– Invest in renewable energy (solar, wind) – Phase out internal combustion vehicles by 2035 – Promote carbon capture (CCS) technologies
European Union	2050	Carbon neutrality	– 75%+ renewable energy share – Implement Carbon Border Adjustment Mechanism (CBAM) – Advance circular economy
Japan	2050	Carbon neutrality	– Build hydrogen production, utilization network – Increase reliance on nuclear energy – Promote energy-efficient buildings
India	2070	Carbon neutrality	– 50% nonfossil fuel capacity by 2030 – Expand solar investments

- **China's "Dual Carbon" Goals**
  China has set ambitious targets to peak carbon emissions by 2030 and achieve carbon neutrality by 2060 [4]. Key measures include:
  Increasing the share of nonfossil fuels in primary energy consumption to 25% by 2030 and over 80% by 2060. Expanding wind and solar installed capacity to 1.2 billion kilowatts by 2030. Reducing energy intensity by 13.5% by 2025, thereby improving energy efficiency.
- **United States' Net-Zero Emissions by 2050**
  The United States aims to achieve net-zero emissions by 2050 through a combination of clean energy deployment and technological innovation [5]:
  Large-scale investment in renewable energy, particularly solar and wind. Accelerated development of electric vehicles, to phase out internal combustion engines by 2035. Deployment of Carbon Capture and Storage (CCS) technologies to mitigate emissions from industrial and agricultural sectors.
- **European Union's European Green Deal**
  The EU has committed to carbon neutrality by 2050 [6], emphasizing:

Increase the share of renewable energy in total consumption by over 75%.
Introducing the Carbon Border Adjustment Mechanism (CBAM) to prevent carbon leakage from high-emission imports.
Advancing the circular economy to reduce resource consumption and industrial waste.

- **Japan's 2050 Carbon Neutrality Plan**
  Japan's strategy revolves around hydrogen and nuclear energy [7]:
  Establishing a nationwide hydrogen production and utilization network. Increasing reliance on nuclear energy to ensure stable, clean energy supplies. Promoting energy-efficient building designs.
- **India's 2070 Carbon Neutrality Goal**
  India, one of the largest developing economies [8], aims to balance rapid growth with environmental sustainability by:
  Raising the share of nonfossil fuel capacity to 50% by 2030. Expanding solar energy investments to meet rising electricity demand.

8.1.1.3 Challenges in carbon peaking and carbon neutrality

Achieving carbon peaking and carbon neutrality is a complex endeavor fraught with challenges at the technological, economic, social, and environmental levels. These intertwined issues pose significant hurdles for governments in formulating and implementing effective policies. Below, the challenges are analyzed in detail, focusing on technological, economic, and environmental aspects, with real-world examples to illustrate the difficulties.

- **Technological Challenges**
  Technological development is one of the primary bottlenecks in achieving carbon neutrality, encompassing areas such as the standardization of low-carbon technologies, the advancement of energy storage systems, and the establishment of supporting infrastructure [9]. These limitations not only hinder the pace of renewable energy adoption but also restrict the systemic transition to low-carbon economies.
  Many low-carbon technologies remain in the research or early application stages, and significant discrepancies exist in technical standards across countries and regions. This lack of standardization creates compatibility issues, hampering the large-scale deployment of these technologies. For example, CCS, a critical technology to reduce industrial carbon emissions, has progressed slowly due to the absence of unified technical standards and regulatory frameworks worldwide. Even in countries with pilot projects, economic and safety concerns persist. The Boundary Dam project in Canada, one of the world's first large-scale CCS projects, faced significant challenges such as high costs and technical complexity, limiting its scalability and further implementation.
  The intermittency and variability of renewable energy sources like wind and solar necessitate efficient and reliable energy storage systems. However, the

high costs, limited efficiency, and insufficient lifespan of current storage technologies hinder their widespread adoption. Lithium-ion batteries, for example, are a dominant storage technology, but their production remains expensive, and the global supply chain for lithium is unstable. According to the International Energy Agency (IEA) [10], global lithium demand is expected to increase 40-fold by 2030, further straining supply and raising economic barriers. Additionally, the recycling and reuse technologies for large-scale storage batteries remain immature, potentially causing new environmental pollution when batteries fail. For example, in China, many photovoltaic power plants struggle to supply electricity at night due to inadequate storage capacity, undermining the reliability of renewable energy [11].

The transition to a low-carbon economy requires substantial investments in new infrastructure, including power grids, transport systems, and industrial facilities. For instance, integrating decentralized renewable energy sources into existing power grids often faces technical and infrastructure-related challenges. Without the necessary upgrades, the capacity to handle surges or shortages in renewable energy remains limited, slowing the overall transition process.

– **Environmental Pressure on Ecosystems and Resources**

The environmental implications of achieving carbon neutrality are significant, particularly in terms of land use, critical resource extraction, and waste management [15]. These factors place substantial stress on ecosystems and natural resources, potentially offsetting the benefits of decarbonization.

Land Use Conflicts and Ecosystem Disruption. The construction of renewable energy infrastructure, such as photovoltaic power stations and wind farms, requires vast amounts of land, leading to increasing pressure on land resources. In western China, large-scale photovoltaic installations have encroached upon grasslands and desert regions. While these areas may appear suitable for renewable energy development, long-term land occupation could lead to soil degradation and a decline in biodiversity. Similarly, wind farms in several EU member states have faced opposition from environmental organizations due to their impact on bird migration routes and ecosystem balance. These conflicts highlight the need for more precise land-use planning and ecological protection measures to mitigate the adverse effects of renewable energy expansion.

OverExploitation of Critical Mineral Resources. Clean energy technologies rely heavily on critical minerals such as lithium, cobalt, nickel, and rare earth metals. The surge in global demand for clean energy has driven rapid expansion in the extraction of these resources, with severe environmental consequences. For example, in South America's "lithium triangle" (Bolivia, Chile, and Argentina), lithium mining has led to the depletion of groundwater resources, threatening local agriculture and ecosystems. Similarly, in the Democratic Republic of Congo, widespread cobalt mining has not only destroyed forest ecosystems but also polluted rivers and soils due to poor regulatory oversight. The environmental damage from such resource extraction could rival that caused by fossil fuels if not managed sustainably.

Waste Management From Renewable Energy Technologies. As clean energy technologies proliferate, the disposal of retired equipment and waste has become a growing environmental challenge [12]. For example, end-of-life photovoltaic panels are generating increasing amounts of waste, but recycling technologies and policies for these materials remain underdeveloped. Many panels are simply landfilled or incinerated, potentially causing soil and air pollution. Similarly, the low recycling rates of lithium-ion batteries from electric vehicles pose significant risks. Improper disposal of these batteries can release toxic substances, threatening both the environment and public health. Moreover, nuclear energy, a key component of many carbon neutrality strategies, faces persistent challenges in safely storing radioactive waste over the long term.

Conflict Between Renewable Energy Expansion and Carbon Sinks. The expansion of renewable energy infrastructure, while reducing carbon emissions from fossil fuels, can inadvertently compromise natural carbon sinks through land-use changes. For example, in India, wind farms have encroached on agricultural land, creating livelihood challenges for local farmers. In other cases, forest clearing for renewable energy projects has reduced natural carbon sequestration capacity, contradicting the broader goals of carbon neutrality. Addressing these issues requires integrating ecological preservation into renewable energy development planning.

- **Economic Challenges: Funding and Market Mechanisms**

  Achieving carbon neutrality demands massive financial resources, encompassing the development of clean energy technologies, infrastructure construction, and industrial restructuring [16]. However, funding gaps, market inefficiencies, and fossil fuel subsidies present significant economic obstacles.

  Funding Gaps and Financing Challenges. The sheer scale of investment required for carbon neutrality is staggering. According to the IEA [13], achieving global carbon neutrality by 2050 will require over $1 trillion in annual investments in the energy sector alone. This poses a significant burden, particularly for developing countries. For example, despite India's commitment to achieving carbon neutrality by 2070, its renewable energy projects face a funding gap of hundreds of billions of dollars, slowing the construction of solar and wind energy facilities. Additionally, many developing countries struggle to attract external capital for low-carbon projects due to low credit ratings, forcing them to continue relying on high-carbon fossil fuels.

  Market Mechanism Inefficiencies. Current market mechanisms often fail to provide sufficient incentives for low-carbon technologies to achieve commercial viability. For instance, wind, solar, and green hydrogen technologies require substantial upfront investment and have long payback periods. At the same time, carbon pricing and trading systems remain underdeveloped in many countries. As of 2022, World Bank data indicates that carbon markets cover only 23% of global emissions [14], with carbon prices in most countries far below the levels needed to drive meaningful technological transitions. For example, while the European Union's carbon price of €90 per ton has begun to pressure high-emission industries, carbon prices in many countries remain below $10 per ton, providing insufficient economic signals for change.

Economic Disruption from Industrial Transition. The transition to a low-carbon economy poses significant challenges for high-emission industries such as coal, oil, and steel. For regions heavily reliant on traditional energy industries, this shift threatens jobs and economic stability. Balancing short-term economic costs with long-term benefits requires careful policy design and targeted support for affected communities.

These multifaceted challenges—spanning technological limitations, economic constraints, and environmental trade-offs—underscore the urgent need for innovative solutions. Energy storage systems emerge as a critical enabler in addressing these issues by enhancing grid flexibility, enabling higher renewable energy penetration, and reducing reliance on fossil fuel backup. By bridging the gap between intermittent renewable generation and consistent energy demand, storage technologies can accelerate the transition toward carbon neutrality while maintaining economic stability and energy security.

8.1.2 Current status and trends of energy storage development

8.1.2.1 Global energy storage market status and trends

Different economies have varying approaches to energy storage, but common themes include the integration of renewable energy, enhancing grid reliability, and supporting energy policy objectives [17]. The fundamental trends in the development of Energy Storage Technologies can be observed from Fig. 8.3. The global energy storage capacity has shown continuous growth, led by pumped hydro and electrochemical technologies, with the latter emerging as the fastest-growing segment. China, the United States, and the EU account for more than 80% of the global increase [18]. By 2021, the global cumulative energy storage capacity reached 209.4 GW, up 9.6% year-on-year, with 18.3 GW of new installations marking a 181.5% increase.

From the perspective of energy storage types, pumped hydro remains dominant with 180.5 GW (86.2% of total capacity) by 2021. Electrochemical storage grew significantly, from 3.0 GW in 2017 to 24.5 GW in 2021 (11.7% share), with a CAGR of 69.3%. By 2025, the global electrochemical storage capacity is expected to reach 148 GW, comprising 40% of total energy storage, with a Compound Annual Growth Rate (CAGR) of 52.5% from 2021 to 2025 [19].

The U.S. energy storage market is currently experiencing rapid growth, driven by multiple factors such as policy support, technological innovation, and market demand [21]. From a policy perspective, both federal and state

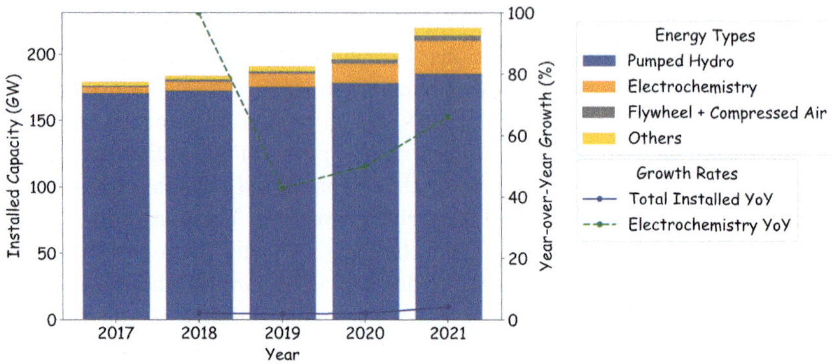

FIGURE 8.3 Energy storage technologies from 2017 to 2021.

regulations, along with economic incentives, have provided strong support for investment in storage systems. The Inflation Reduction Act (IRA) includes subsidies for energy storage, enabling it to be an independent beneficiary of tax incentives, with a 10-year investment tax credit (ITC) [22]. This has significantly reduced the initial investment costs of storage projects, fueling the market's rapid expansion. Fig. 8.4 illustrates the regional impact of the IRA on energy storage deployment across the United States. As shown, states with favorable renewable energy policies and higher solar and wind penetration—such as California, Texas, and Florida—have experienced the most significant capacity growth. The IRA's 10-year investment tax credit has particularly accelerated deployments in the Southwest and Northeast, reducing project payback periods by approximately 30% and attracting substantial private investment. This regional disparity highlights how federal policy can catalyze localized energy storage markets, creating a model for other nations seeking to scale their storage infrastructure.

As shown in, the energy storage capacity across various regions of the United States has significantly increased.

On the technology front, the application of new electrolyte technology has significantly improved the energy density and stability of batteries. Meanwhile, advancements in production processes and the scaling of manufacturing have effectively lowered unit storage costs. Additionally, the development of battery and new material technologies has accelerated the iteration of storage products.

From a market demand perspective, the trend in the United States is shifting toward large-scale storage systems, reflecting the growing need for grid stability. As more large storage projects come online, storage technology is helping to balance supply and demand fluctuations, enhancing the flexibility and reliability of the grid. According to forecasts, by 2025, the cumulative installed capacity of energy storage in the United States will exceed 75 GW, with an annual growth rate of 88.4%. Meanwhile, the North American battery storage system market is expected to grow from $3.91 billion in 2024 to $15.28 billion by 2029, with an annual growth rate of 31.28%[20].

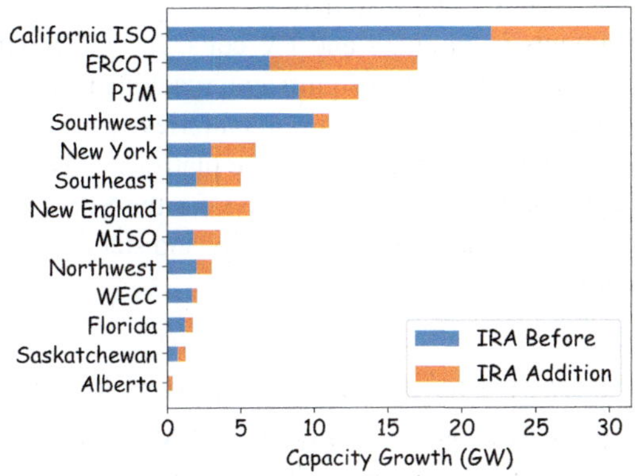

FIGURE 8.4 IRA ACT droves U.S. capacity growth.

The European Union considers energy storage critical for renewable targets and energy security, with policies like the "Clean Energy for All Europeans" package driving progress. Germany's energy storage market is expected to grow at a CAGR of over 10% from 2024 to 2029. By September 2024, Germany's grid storage capacity increased by nearly a third to 1.8 GWh [23][24].

To study the future development of the Swiss electricity system and its flexibility needs and supply, the Energy Science Center at ETH Zurich developed the Nexus-e integrated energy system modeling platform [25][26]. Through extensive simulations, the platform demonstrates that Switzerland can maintain system flexibility and security while meeting its nuclear phase-out and renewable energy targets. According to the simulation results, new photovoltaic (PV) installations largely replaced nuclear capacity, with some investments in biomass and PV cells. In high-flexibility scenarios, PV cell investment was particularly significant, reaching nearly 10 GW by 2050. However, there was little investment in grid batteries or wind energy.

Nexus-e provides a comprehensive representation of Switzerland's energy-economic system, aiming to identify cost-optimal investments and operations for both centralized and distributed energy resources. The platform combines both bottom-up and top-down energy modeling approaches, covering a broad range of the energy-economy system, from regional to national levels, with high temporal resolution. Its modular structure and well-defined interfaces allow the integration of different domains (such as electrical engineering, power markets, and economics), network levels (centralized and distributed), and timescales (from subhourly to annual), creating a unified platform. This enables the platform to capture the interdependencies across various components, sectors, and levels of the Swiss electricity system.

Through this comprehensive representation, the platform helps demonstrate that Switzerland can maintain system flexibility and security while meeting its nuclear phase-out and renewable energy targets. According to the simulation results, new PhotoVoltaic (PV) installations largely replaced nuclear capacity, with some investments in biomass and PV cells. In high-flexibility scenarios, PV cell investment was particularly significant, reaching nearly 10 GW by 2050. However, there was little investment in grid batteries or wind energy.

By running these simulations, the Nexus-e platform aids decision-makers in identifying the optimal development pathways for the Swiss electricity system, ensuring that the system can maintain flexibility and stability amid rapid growth in renewable energy.

8.1.2.2 The international energy storage development trend

1. **Declining Battery Storage Costs and Rapid Market Expansion.** Technological advancements and economies of scale have significantly reduced the cost of battery storage. According to the IEA, costs are expected to drop by another 40% by 2030, which will drive broader adoption of energy storage solutions [27].

2. **International Trade Boosting Energy Storage Growth.** As clean technologies gradually replace traditional fossil fuel-based trade, batteries and related storage products are becoming a key part of the clean technology trade. With rising demand for battery and energy storage products in regions like the European Union, the global energy storage market is expanding at a rapid pace [28].

3. **Lithium-Ion Batteries** dominate the market, while emerging alternative technologies are rapidly developing. Lithium-ion batteries continue to dominate the global energy storage market due to their high energy density and falling costs. Notably, Lithium Iron (Fe) Phosphate (LFP) batteries became the dominant type in new storage installations in 2022, offering a cost-effective solution for large-scale energy storage. Sodium-ion batteries, with advantages in cost and material availability, are expected to capture an increasing share of the energy storage market by 2030 [19].

4. **China's Leadership in Energy Storage Manufacturing.** Companies like CATL have secured a significant share of the market, positioning China as a global leader in energy storage manufacturing. China is also leading the transformation of global renewable energy storage. It is expanding into emerging markets in Southeast Asia and Latin America, where it serves as a low-cost production hub for energy storage devices, especially solar photovoltaic components and batteries. These regions will enhance the flexibility of global energy storage supply chains and contribute to the global energy transition [29].

5. **Policy Measures Driving Market Growth.** Governments worldwide are introducing policies to accelerate energy storage deployment. For example, the G7 group plans to increase global electricity storage capacity by six times by

2030, emphasizing the strategic role of storage in achieving energy transition goals [29].

6. **Integration of Energy Storage with Renewable Energy.** Energy storage is crucial for integrating renewable energy into the grid, enabling the storage of excess energy during peak production and releasing it when demand spikes. This will enhance grid stability and support the transition to cleaner energy sources [30].

8.1.2.3 Status and outlook of China's energy storage market

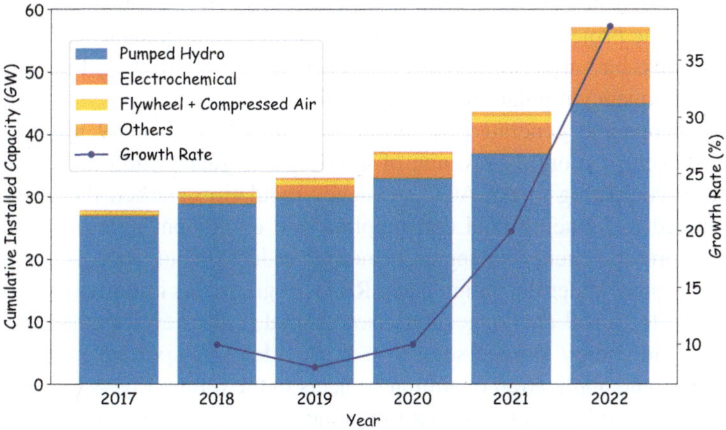

FIGURE 8.5 China's emergy storage scale and growth rate.

China's energy storage market is growing rapidly, with electrochemical energy storage expected to reach 70 GW by 2025, growing at a CAGR of 88.9%. Electrochemical storage now accounts for nearly 90% of total energy storage capacity. Policies promoting energy storage integration with renewable energy are driving this growth, particularly in the 14th Five-Year Plan period, which aims to add over 50 GW of new capacity [31]. While global renewable energy expansion creates similar challenges worldwide, China's rapid deployment makes grid stabilization particularly critical within its domestic market. Fig. 8.5 shows China's Energy Storage Scale and Growth Rate.

In Guangdong Province, the new energy storage industry generated 150 billion yuan in 2022, with a total installed capacity of 710,000 kW. By the end of 2023, installed capacity surpassed 1.6 million kW, a 125% increase [32].

In November 2023, a project [30] was launched in Zhuhai, Guangdong Province: a 16 MW lithium iron phosphate ($LiFePO_4$) battery and 4 MW supercapacitor hybrid energy storage system for AGC (Automatic Generation Control) frequency regulation at Zhuhai Jinwan Power Plant. The project is designed to enhance the frequency regulation efficiency of the power system while demonstrating the potential of new energy storage technologies in energy dispatch, particularly through the combined use of supercapacitors and $LiFePO_4$ batteries in a hybrid storage solution. The system utilizes a hybrid energy storage technology that integrates $LiFePO_4$ batteries with supercapacitors, primarily for AGC frequency regulation in the power grid. Compared to traditional $LiFePO_4$ battery-based frequency regulation systems, the supercapacitors offer significant advantages, including high power density, fast charge and discharge rates, long lifespan, resistance to extreme environments, and safety and environmental benefits.

During operation, the system achieved an impressive composite K-value of 1.3 to 1.6, notably higher than traditional $LiFePO_4$ battery-based projects. Monthly frequency regulation revenues exceeded 5 million yuan. The performance of the project has been outstanding, with the supercapacitors achieving 7200 cycles, far surpassing the 1200 cycles typical of $LiFePO_4$ batteries. The average frequency regulation K-value reached 1.381, with a daily peak of 1.92. The total revenue generated by the project has reached 30.69 million yuan, with an average monthly revenue of 4.38 million yuan. Additionally, by efficiently participating in the AGC frequency regulation market, the project has contributed to reducing coal consumption and equipment aging of coal-fired units, improving energy-saving and emission reduction outcomes. As a demonstration project under the national key R&D program, this initiative has not only addressed technical challenges but also resulted in the development of proprietary hybrid energy storage solutions. It provides an important model for the promotion and application of similar projects in the future. Fig. 8.6 presents the forecast of cumulative installed capacity and the share of new energy storage in China.

8.1.2.4 Outlook for energy storage development

1. **Chinese Market Potential.** The energy storage industry has vast growth prospects. In China, the sector's output value surpassed 300 billion RMB, and it

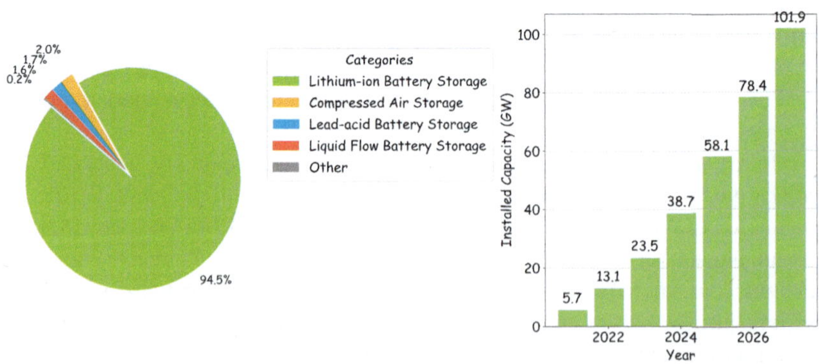

FIGURE 8.6 (left) Share of energy storage technologies (2021–2027) and (right) cumulative installed capacity (GW) in China [30].

is projected that by 2030, the cumulative installed capacity will reach 220 GW, with the market value exceeding 3 trillion RMB.

2. **Technological Diversification.** The trend toward technological diversification in energy storage will continue. Innovations such as gravity-based storage, liquid air storage, and carbon dioxide storage are set to expand. In the future, the industry will need to balance safety, cost-efficiency, and performance. Breakthroughs in intelligent systems and AI-assisted grid management will further accelerate technological progress and adoption.

3. **Investment and Financing.** The energy storage industry will continue to attract substantial investments, particularly in new battery technologies and hydrogen storage. Financing models will evolve with mechanisms like inertial compensation and long-term power purchase agreements, providing stable revenue streams. Green finance, including venture capital and green loans, will offer sustainable funding to support industry growth.

4. **Application Expansion.** The range of applications for energy storage technologies will keep expanding. New models are emerging in renewable energy grid integration, shared storage systems, and electricity spot markets. Significant potential exists across various operational layers and tasks within power systems, encompassing power generation, grid management, and end-user applications.

5. **Regional Coordination and Resource Sharing.** Regional differences in energy storage efficiency and system maturity remain significant. In some areas, grid management and security need improvement. Solutions such as shared storage and leasing policies will help address regional imbalances and improve the utilization of storage technologies.

6. **International Collaboration.** International cooperation will play a key role in advancing energy storage technology and optimizing supply chains. Global collaboration and knowledge-sharing will accelerate technological progress and cost reduction, while integrating new information technologies and the renewable energy sector will help upgrade the industry toward greater intelligence, sustainability, and innovation [33].

8.1.3 Strategic position and role of energy storage

8.1.3.1 The role of energy storage in renewable energy integration

With the rapid growth of renewable energy worldwide, energy storage technologies have become pivotal in achieving the energy transition and meeting carbon neutrality targets. The intermittent and variable nature of renewable energy sources makes it challenging to maintain a stable energy supply on their own [34]. By bridging the temporal gap between energy production and consumption, energy storage technologies enhance the flexibility and stability of power systems, becoming a cornerstone of future electricity systems. Energy storage systems can balance load fluctuations on the generation side and optimize grid operations on the transmission side. Furthermore, the introduction of hybrid energy storage systems expands the application scenarios for storage technologies, offering comprehensive solutions. The characteristics and application scenarios of common energy storage devices are summarized in Table 8.2.

The following sections elaborate on the critical roles of energy storage in renewable energy integration from three perspectives: the generation side, the transmission side, and hybrid energy storage systems.

1. **Balancing Load Fluctuations in Energy Generation**
   On the generation side, the intermittency and variability of renewable energy sources are primary challenges that energy storage technologies are designed to address. For instance, solar and wind power generation heavily depend on natural conditions. Daytime solar irradiance, the absence of sunlight at night, and fluctuations in wind speeds can result in significant variations in power generation capacity, which often do not align with electricity demand. Energy storage technologies mitigate this issue by storing excess electricity during periods of high generation and releasing it during periods of low generation, thus balancing supply and demand at the generation level. For example:
   Pumped Hydro Storage (PHS). This mature energy storage technology, primarily used for daily energy arbitrage and peak shaving (typically 4–12 hours), has been widely implemented globally [35]. Its working principle involves pumping water to a higher elevation reservoir during off-peak electricity demand and releasing it to generate electricity during peak demand. In China's Yunnan power grid, the use of pumped hydro storage has increased the combined utilization efficiency of wind and hydro power by 30%, significantly reducing wasted wind and hydro energy.
   Electrochemical Storage (Batteries). Lithium-ion and flow batteries have played critical roles. in addressing the variability of renewable energy generation [36]. In Japan, for example, lithium-ion batteries are used in solar power stations to store excess electricity generated during the day, which is then released at night to meet electricity demands. This approach not only improves the efficiency of power generation but also reduces the regulatory burden on traditional power grids.
2. **Enhancing Grid Flexibility and Stability**
   On the transmission system side, energy storage technologies effectively alleviate the operational pressures posed by the integration of renewable energy into the grid. Traditional power grids, designed around fossil fuel-based generation, offer stable outputs and are relatively easy to regulate. However, the variability of wind and solar energy introduces frequency and voltage instabilities into the system. Energy storage systems address these challenges by rapidly responding to grid demands, thereby enhancing grid flexibility and stability.

TABLE 8.2 Characteristics and application scenarios of common energy storage devices.

Storage Device	Energy Density	Power Density	Efficiency	Charge/Discharge Time	Cost	Main Applications	Advantages	Disadvantages	References
Li-ion Battery	High	Moderate	High (~90–95%)	Hours	Moderate	Portable electronics, EVs, grid storage	High energy density, well-established	Degrades with use, temperature sensitive	[47]
Lead-acid Battery	Low	Low	Moderate (~70–80%)	Hours	Low	UPS, grid storage, automotive	Low cost, reliable, mature	Low energy density, heavy, environmental issues	[48]
Supercapacitor	Low	Very High	Very High (>95%)	Seconds–min	High	Regenerative braking, power quality	Very fast charge/discharge, very long life	Low energy density, high cost per unit energy	[49]
Flywheel	Moderate	High	High (~85–95%)	Seconds–min	High	Frequency regulation, UPS	No chemical degradation, fast response	High mechanical maintenance, energy loss over time	[50,51]
Flow Battery	Moderate	Low–Med	Moderate (~70–85%)	Min–hours	Moderate	Long-term grid storage	Scalable energy capacity, long discharge	Complex plumbing and maintenance	[52]
Compressed Air (CAES)	Low	Low	Moderate (~40–75%)	Hours–days	Moderate	Grid storage, load balancing	Large-scale storage, long life	Requires specific geological formations	[53,54]
Thermal Energy Storage (TES)	Variable	Variable	Variable	Hours–days	Variable	Building heating/cooling, industrial processes	Flexible application range	Efficiency varies by installation	[55,56]
Hydrogen Storage	Low	Low	Low (~35–55%)	Hours	High	Vehicle fuel, grid storage	Large capacity, suitable for massive storage	High cost, requires advanced infrastructure	[57–59]
SMES	Low	Very High	Very High (>95%)	Instant	Very High	Power quality, grid stability	Extremely fast response, very high efficiency	High cost, complex low-temperature requirements	[60]
Pumped Hydro Storage	Low	Low	High (~70–85%)	Hours–days	Low	Grid storage, load balancing	Long life, large-scale storage	Geographic and environmental limitations	[61,62]

Maintaining grid frequency stability requires real-time balancing of generation and load. However, the rapid fluctuations in wind and solar power outputs can cause significant frequency deviations. Superconducting Magnetic Energy Storage (SMES) systems, with their millisecond-level response times, are particularly effective in frequency regulation [37]. For example, in California, SMES systems have been successfully deployed to stabilize frequency fluctuations caused by wind power [38] variability, preventing grid instability.

Energy storage technologies also provide voltage support during transmission. Flywheel Energy Storage (FES) systems, which store energy in the form of mechanical motion, can quickly release energy to stabilize voltage drops in transmission lines [38]. In remote regions of Australia, flywheel systems have significantly reduced transmission losses [38] and minimized the risk of power outages.

Energy storage systems can serve as backup power sources to ensure the reliability of the electricity supply during grid failures or extreme weather events. In typhoon-prone Japan, battery storage systems are widely integrated into transmission networks as emergency backup power sources [36], mitigating the risk of widespread outages caused by extreme weather.

3. **Hybrid Energy Storage Systems**

Single energy storage technologies often face limitations in terms of power output and storage duration, making it difficult to simultaneously meet diverse grid demands. Hybrid Energy Storage Systems (HESSs), which combine the strengths of multiple storage technologies, offer solutions for both short-term high-power output and long-duration storage needs [39]. HESS is a critical development direction for modern grid storage technologies.

HESSs refer to coordinated systems comprising two or more energy storage technologies that complement each other's strengths to meet a variety of storage demands. As shown in Fig. 8.7, a typical HESS integrates multiple technologies: lithium-ion batteries provide short-term high-power output for instantaneous grid demand adjustments; flywheels ensure frequency regulation and voltage stability for high-power, short-duration needs; compressed air energy storage (CAES) offers long-duration bulk storage; and thermal energy storage converts surplus electricity into heat for industrial applications. This multi-layered architecture enables HESS to address both transient grid disturbances and long-term energy imbalances. The schematic below illustrates a typical HESS structure (see Fig. 8.7):

Key Components of HESS: Battery Storage: Provides short-term high-power output for instantaneous grid demand adjustments, such as using lithium-ion batteries to address peak loads. Flywheel Storage: Stores energy as mechanical motion, ensuring frequency regulation and voltage

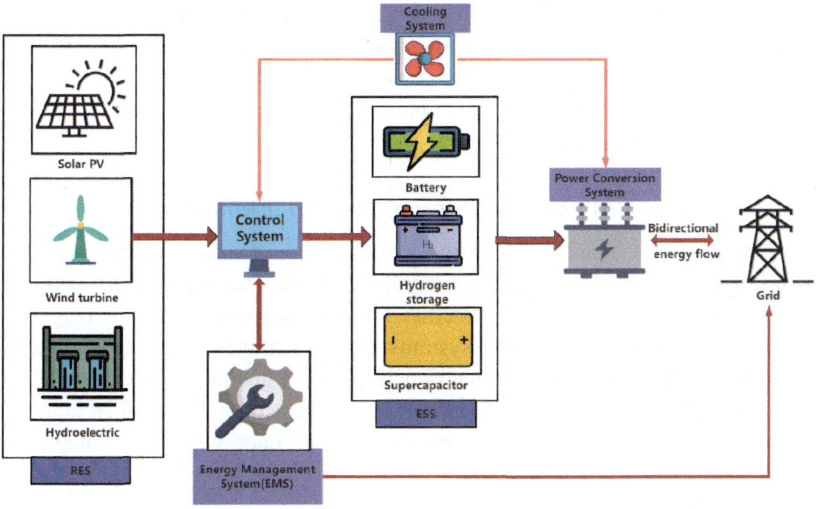

FIGURE 8.7 Structure of the hybrid energy storage system.

stability, particularly for high-power, short-duration needs. Compressed Air Energy Storage (CAES): Offers long-duration, large-scale energy storage by compressing and storing air during off-peak periods and releasing it during peak demand. Thermal Energy Storage (TES): Converts surplus electricity into heat for industrial applications or grid load balancing. Advantages and Applications of HESS The multifunctional nature of HESS enables it to address diverse challenges in renewable energy integration. For instance:

In Germany, wind farms utilize a combination of lithium-ion batteries for short-term frequency regulation and compressed air storage systems for long-term energy balancing. This setup ensures a stable 24-hour power supply and reduces reliance on traditional fossil fuel power plants.

Industrial Applications of HESS: Beyond grid-scale deployments, HESS offers significant value in industrial settings. In energy-intensive industries, HESS can integrate thermal and battery storage systems to manage both electrical and thermal energy needs, optimizing energy use while reducing emissions.

HESS significantly enhances the utilization rate of renewable energy and reduces reliance on fossil fuels, representing a transformative solution for modern energy systems.

8.1.4 Maximizing energy storage efficiency to achieve carbon neutrality

This section will focus on optimizing energy storage technologies from a technical perspective to accelerate progress toward carbon neutrality. Specific case studies from the EU's regional energy management platform, the development and application of solid-state batteries in the United States, and the efforts in Shanxi Province, China, to safely phase out high-carbon power generation provide practical insights into the contributions and pathways of energy storage in achieving carbon neutrality.

8.1.4.1 Promoting system integration and intelligent management

To improve the efficiency of energy storage, policymakers are encouraged to support the integration of energy storage technologies with smart grids and renewable energy generation systems. Through regional energy management platforms and intelligent control systems, energy storage systems can interact in real-time with the power grid, enabling dynamic adjustments in energy supply and demand. The EU, a global leader in energy transition, has made remarkable progress in regional energy management and energy storage integration, effectively addressing the challenges associated with increasing renewable energy penetration.

European regional energy management platforms utilize highly intelligent energy dispatch mechanisms to connect energy storage facilities with renewable energy generation sources like wind and solar power. These platforms, based on Distributed Energy Resource Management Systems (DERMSs), dynamically allocate energy based on real-time demand, generation conditions, and storage capacity across regions. For example:

- In Northern Germany, during periods of strong wind generation, excess electricity is transmitted to Southern Germany, where energy demand peaks [40]. Simultaneously, the surplus energy is stored in pumped hydro facilities in the south for future use. The platform also facilitates energy sharing between countries. In cross-border electricity coordination between Northern the EU and Germany, energy storage systems act as buffers to balance supply and demand discrepancies, preventing grid instability caused by imbalances.
- The Nordic Green Regional Energy Platform A prominent example of this approach is the "Nordic Green Regional Energy Platform," jointly developed by Denmark and Germany [41]. This platform integrates wind power from Nordic countries with solar power from Germany, maximizing renewable energy utilization. In the summer of 2022, when solar power generation in Southern Germany exceeded demand, the platform coordinated with Nordic pumped hydro facilities, storing approximately 30% of the surplus electricity. This efficient management reduced energy wastage and significantly lowered regional carbon emissions.

8.1.4.2 Enhancing innovation and R&D support

Increasing investments in energy storage research and development is essential to advancing next-generation storage technologies, such as solid-state batteries,

Compressed Air Energy Storage (CAES), and flow batteries. Demonstration projects and pilot programs can accelerate the application of these innovations. Among these technologies, solid-state batteries stand out for their high energy density, long lifespan, and superior safety, making them a focus of global research efforts. The United States leads in the R&D and commercialization of solid-state batteries, driven by strong government support, industrial innovation, and pilot initiatives.

- The U.S. Department of Energy (DOE) has played a pivotal role in advancing solid-state battery technology through its "Energy Storage Grand Challenge." This initiative supports the entire lifecycle of solid-state battery development, from foundational material research to commercialization. The National Renewable Energy Laboratory (NREL), funded by DOE, has developed new solid electrolyte materials that enhance battery stability and energy density [42]. Some projects have achieved an energy density of over 400 Wh/kg while significantly reducing production costs.
- Companies like Tesla and QuantumScape are driving the commercialization of solid-state batteries. QuantumScape's lithium-metal solid-state batteries boast higher safety and longer cycle life [43]. Test results show these batteries can exceed 1000 charge cycles while offering a 30% higher energy density than traditional lithium-ion batteries. This advancement positions solid-state batteries as a transformative technology for energy storage.
- In California, a distributed solid-state battery system has been installed in remote areas to provide stable electricity [44]. These systems store solar energy generated during the day for nighttime use, significantly reducing reliance on diesel generators.
- In the electric vehicle sector, companies like Ford are developing solid-state battery-powered vehicles [45]. These batteries promise reduced charging times to under 10 minutes and an extended driving range of over 600 kilometers, supporting the large-scale adoption of electric vehicles.

8.1.4.3 Coordinating energy storage with the phase-out of high-carbon technologies

The transition from high-carbon power generation technologies, such as coal-fired power plants, relies heavily on the support of energy storage technologies. Policy measures can promote the coordinated operation of energy storage systems and renewable energy to gradually replace coal power. Shanxi Province, China, a traditional coal hub, offers a compelling example of how to achieve this transition while maintaining economic stability and energy security.

- In China, Shanxi Province has adopted a "gradual phase-out and coordinated substitution" strategy, supported by policy guidance, technological innovation, and industrial restructuring [46]. Key measures include:
- Decommissioning High-Pollution Coal Power Units. Priority is given to shutting down small-scale, inefficient, and high-emission coal power plants.

As of 2023, Shanxi has decommissioned over 200 small coal-fired power plants, with a total installed capacity exceeding 10 GW.
- Scaling Up Renewable Energy Development. Shanxi has accelerated the construction of wind and solar power projects. In Datong City, for instance, large-scale solar power bases with a combined capacity of 6 GW have been established. Efforts also focus on integrating wind, solar, and storage systems to ensure efficient utilization of renewable energy.
- Leveraging Energy Storage for Grid Support. Energy storage systems play a critical role in stabilizing the grid as coal power is phased out. Shanxi has deployed large-scale lithium battery storage and pumped hydro systems to address the intermittency of wind and solar power. By the end of 2022, the province had installed over 2 GW of energy storage capacity, providing crucial peak-shaving and backup support for the grid. In Datong City, a solar farm spanning 10 square kilometers generates approximately 5 million kWh of electricity daily, with 30% of the output stored in energy storage facilities for nighttime use. This model not only bridges the energy gap left by coal plant closures but also significantly reduces carbon emissions.
- Extending Coal Plant Lifespan Through Storage Integration. Shanxi is also exploring ways to extend the operational lifespan of existing coal plants by integrating energy storage technologies. For instance, high-efficiency coal plants are adopting "storage + coal power" joint dispatch systems, storing surplus electricity during off-peak periods and releasing it during peak demand. This approach reduces coal consumption while maintaining grid reliability.

Additionally, policies favoring renewable energy, such as "renewables-first" dispatch mechanisms, have encouraged private investments in energy storage projects. The province has also provided job training and reemployment programs for coal workers, mitigating the economic and social impacts of coal plant closures.

The examples from the EU's regional energy management platform, the U.S.'s advancements in solid-state batteries, and China's practical experience in Shanxi Province collectively illustrate the critical role of energy storage in driving the energy transition and achieving carbon neutrality. These cases highlight that energy storage technologies are not just technical solutions but also essential enablers of systemic changes in the global energy landscape. By tailoring policies to local conditions and fostering innovation, energy storage can maximize the efficiency and reliability of renewable energy systems while facilitating the gradual phase-out of high-carbon technologies. As the world moves toward a low-carbon future, energy storage will undoubtedly play an increasingly pivotal role as a bridge between clean energy generation and sustainable energy consumption.

References

[1] P. Friedlingstein, M. O'Sullivan, M.W. Jones, et al., Global carbon budget 2024, Earth System Science Data 17 (1) (2025) 965–1038, https://doi.org/10.5194/essd-17-965-2025 [Online]. Available: https://essd.copernicus.org/articles/17/965/2025/.

[2] H. Ritchie, P. Rosado, M. Roser, Breakdown of carbon dioxide, methane and nitrous oxide emissions by sector, Our World in Data, 2020 [Online]. Available: https://ourworldindata.org/emissions-by-sector.

[3] United Nations Framework Convention on Climate Change, The Paris Agreement, 2015 [Online]. Available: https://unfccc.int/process-and-meetings/the-paris-agreement/the-paris-agreement. (Accessed 10 January 2025).

[4] Government of China, China's achievements and measures for carbon neutrality [Online]. Available: http://www.gov.cn, 2020. (Accessed 10 January 2025).

[5] United States Department of Energy, Net-zero economy plan [Online]. Available: https://www.energy.gov, 2021. (Accessed 10 January 2025).

[6] European Commission, The European green deal [Online]. Available: https://ec.europa.eu/clima/eu-action/european-green-deal_en, 2019. (Accessed 10 January 2025).

[7] Government of Japan, Carbon neutrality by 2050 [Online]. Available: https://www.japan.go.jp, 2020. (Accessed 10 January 2025).

[8] Bureau of Energy Efficiency, Government of India, Carbon market and net-zero goals [Online]. Available: https://beeindia.gov.in, 2021. (Accessed 10 January 2025).

[9] J.A. Smith, L.M. Brown, Drivers of global energy consumption: the role of economic growth, industrialization, and urbanization, Energy Economics 95 (2025) 105678, Elsevier.

[10] IEA, The Role of Critical Minerals in Clean Energy Transitions, International Energy Agency, 2021 [Online]. Available: https://www.iea.org/reports/the-role-of-critical-minerals-in-clean-energy-transitions.

[11] IEA PVPS, National Survey Report of PV Power Applications in China 2022, IEA International Energy Agency, 2023 [Online]. Available: https://iea-pvps.org/wp-content/uploads/2023/10/National-Survey-Report-of-PV-Power-Applications-in-China-2022.pdf.

[12] IRENA, IEA PVPS, End-of-Life Management: Solar Photovoltaic Panels, International Renewable Energy Agency, 2016 [Online]. Available: https://www.irena.org/-/media/Files/IRENA/Agency/Publication/2016/IRENA_IEAPVPS_End-of-Life_Solar_PV_Panels_2016.pdf.

[13] IEA, Net Zero by 2050: A Roadmap for the Global Energy Sector, International Energy Agency, 2021 [Online]. Available: https://www.iea.org/reports/net-zero-by-2050.

[14] World Bank, State and Trends of Carbon Pricing 2022, World Bank, 2022 [Online]. Available: https://www.worldbank.org/en/programs/pricing-carbon.

[15] H. Wang, Y. Zhao, The share of renewable energy in global electricity consumption: current status and challenges, Renewable Energy 187 (2022) 112345, Elsevier.

[16] S.H. Lee, J.W. Kim, The impact of land use changes on carbon sinks: a comprehensive assessment of deforestation and wetland disruption, Global Environmental Change 68 (2021) 102987, Elsevier.

[17] China Energy Storage Network News Center, Global Energy Storage Status and Market Development Trends Analysis, Energy Storage Tribe, Jun. 26, 2023.

[18] Carbon Exploration Energy Storage Network, Global Energy Storage Market Experiences Explosive Growth, Carbon Exploration Energy Storage Network, Dec. 4, 2023 [Online]. Available: https://www.carbonexploration.com/global-energy-storage-market-explosive-growth.

[19] Nelson Nsitem, Yayoi Sekine, Andy Leach, Headwinds in Largest Energy Storage Markets Won't Deter Growth, BloombergNEF, 2024 [Online]. Available: https://about.bnef.com/insights/clean-energy/headwinds-in-largest-energy-storage-markets-wont-deter-growth/.

[20] Sohu, Special report: energy storage industry, U.S. energy storage development leading the world in 2024, Sohu 16 (Feb. 2, 2024) 35 [Online]. Available: https://www.sohu.com/a/755720477_711768.

[21] VZKOO, Global Energy Storage Market Surges: Sungrow's Breakthrough in the U.S. Market, Nov. 29, 2024 [Online]. Available: https://m.vzkoo.com/read/202411290ccbba61238efd0beac523bd.html.
[22] IEA, Renewables 2024, IEA, Paris, 2024 [Online]. Available: https://www.iea.org/reports/renewables-2024, Licence: CC BY 4.0.
[23] M. Intelligence, Germany energy storage market size and share analysis – growth trends and forecast (2024–2029), Mordor Intelligence [Online]. Available: https://www.mordorintelligence.com/zh-CN/industry-reports/germany-energy-storage-systems-market-industry.
[24] Reuters, German Power Grid Battery Capacity up Nearly a Third in 2024, Data Shows, Reuters, 2024 [Online]. Available: https://www.reuters.com/business/energy/german-power-grid-battery-capacity-up-nearly-third-2024-data-shows/.
[25] B. Gjorgiev, J.B. Garrison, X. Han, F. Landis, R. van Nieuwkoop, E. Raycheva, M. Schwarz, X. Yan, T. Demiray, G. Hug, G. Sansavini, C. Schaffner, Nexus-e: a platform of interfaced high-resolution models for energy-economic assessments of future electricity systems, Applied Energy 307 (2022) 118193 [Online]. Available: https://www.sciencedirect.com/science/article/pii/S030626192101463X.
[26] Energy Science Center, ETH Zurich, Nexus-e: interconnected energy systems modeling platform [Online]. Available: https://nexus-e.org/.
[27] M. Intelligence, Energy storage market size and share analysis – growth trends and forecast (2024–2029), Mordor Intelligence [Online]. Available: https://www.mordorintelligence.com/zh-CN/industry-reports/energy-storage-market.
[28] IRENA, World Energy Transitions Outlook 2023: 1.5 °C Pathway, International Renewable Energy Agency, 2025.
[29] Fortune Business Insights, Energy storage system market size, share, and global trends by type (mechanical storage system, electrochemical storage system, fuel cell storage system, thermal storage system), application (residential, commercial, industrial), and geography forecast to 2023–2030, Fortune Business Insights [Online]. Available: https://www.fortunebusinessinsights.com/zh/industry-reports/energy-storage-system-market-100234.
[30] Haisheng Chen, Zhenhua Yu, Wei Liu, Energy Storage Industry White Paper 2024 (Summary Version), China Energy Storage Alliance (CNESA), 2024 [Online]. Available: https://en.cnesa.org/s/CNESA-Energy-Storage-Industry-WhitePaper-2024Summary-Version.pdf.
[31] National Development and Reform Commission and National Energy Administration, Guiding Opinions on Accelerating the Development of New Energy Storage.
[32] Reuters, China's Battery Storage Capacity Growth Likely to Slow in 2024, Reuters [Online]. Available: https://www.reuters.com/technology/chinas-battery-storage-capacity-growth-likely-slow-2024-2024-04-10.
[33] C. Howe, China's Battery Storage Capacity Growth Likely to Slow in 2024, Reuters, Apr. 10, 2024.
[34] S. Li, W. Li, L. Chen, H. Jiang, J. Zhang, D.W. Gao, Real-time robust state estimation for large-scale low-observability power-transportation system based on meta physics-informed graph timesnet, IEEE Transactions on Smart Grid 17 (6) (2024) 5500–5513.
[35] Y. Zhang, Y. Ding, X. Zhang, The role of pumped hydro storage in supporting modern power systems: a review of the practices in China, in: Proceedings of the 2019 IEEE Sustainable Power and Energy Conference (iSPEC), Beijing, China, Nov. 2019, pp. 2608–2613 [Online]. Available: https://ieeexplore.ieee.org/document/8881156.
[36] S.A.A. Shah, M.A.S. Masoum, S. Islam, Battery energy storage technology in renewable energy integration: a review, in: Proceedings of the 2022 Australasian Universities Power Engineering Conference (AUPEC), Melbourne, Australia, Nov. 2022, pp. 1–6 [Online]. Available: https://ieeexplore.ieee.org/document/9918504.
[37] M.A. Hannan, M.M. Hoque, R.A. Begum, A. Mohamed, M.N. Uddin, Hybrid energy storage systems for renewable energy integration: an overview, in: Proceedings of the 2022 IEEE Industry Applications Society Annual Meeting (IAS), Detroit, MI, USA, Oct. 2022, pp. 1–8 [Online]. Available: https://ieeexplore.ieee.org/document/10741713.

[38] A.S.A. Awadallah, M.M.A. Salama, A review of modeling and applications of energy storage systems in power grids, IEEE Transactions on Sustainable Energy 13 (1) (Jan. 2022) 309–319 [Online]. Available: https://ieeexplore.ieee.org/document/9743285.

[39] A.M. Adeyinka, O.C. Esan, A.O. Ijaola, P.K. Farayibi, Advancements in hybrid energy storage systems for enhancing renewable energy-to-grid integration, Sustainable Energy Research 11 (2024) 26.

[40] Andreas Moser, Pumped Storage Expansion to Integrate Renewable Energy in Germany, Institute of Energy Economics at the University of Cologne, 2014 [Online]. Available: https://www.econstor.eu/bitstream/10419/103278/1/wp1107.pdf.

[41] IEA Wind, Denmark 2023 Annual Report, International Energy Agency Wind Technology Collaboration Programme, 2025 [Online]. Available: https://www.iea-wind.org/wp-content/uploads/2024/11/Denmark_2023.pdf.

[42] U.S. Department of Energy, Energy storage grand challenge [Online]. Available: https://www.energy.gov/energy-storage-grand-challenge/energy-storage-grand-challenge, 2020.

[43] Bestmag, QuantumScape Solid-State Cell Completes over 1,000 Charging Cycles with Barely Any Loss of Charge, Bestmag, 2024 [Online]. Available: https://www.bestmag.co.uk/quantumscape-solid-state-cell-completes-over-1000-charging-cycles-with-barely-any-loss-of-charge/.

[44] U.S. Energy Information Administration, California residents combine battery storage with solar panels to power their homes [Online]. Available: https://www.eia.gov/todayinenergy/detail.php?id=62524, 2024.

[45] Ford Authority, Ford CEO Jim Farley Endorses Solid-State Batteries for EVs, Ford Authority, 2024 [Online]. Available: https://fordauthority.com/2024/05/ford-ceo-jim-farley-endorses-solid-state-batteries-for-evs/.

[46] China Daily, Shanxi Accelerates Green Transition with Clean Coal Use, Renewable Energy Expansion, China Daily, 2024 [Online]. Available: https://global.chinadaily.com.cn/a/202409/12/WS66e24f04a3103711928a794d.html.

[47] John B. Goodenough, Youngsik Kim, Challenges for rechargeable Li batteries, Chemistry of Materials 22 (2010) 587–603, https://doi.org/10.1021/cm901452z.

[48] Paulo Lopes, Milan Stamenkovic, Lead–Acid Batteries: Fundamentals, Technologies, and Applications, Elsevier, 2020, ISBN 978-0-444-64123-5.

[49] Yan Shi, Xiong Zhang, Supercapacitor Energy Storage Systems: Modeling, Control, and Grid Integration, Wiley, 2022, ISBN 978-1-119-79065-7.

[50] Hao Ji, Jiayu Zhao, Qiang Wu, Huaguang Zhang, Advances in flywheel energy storage systems for grid applications, Energy Storage Materials 72 (2024) 103168.

[51] Mohammad Rahman, S. Islam, A. Choudhury, Assessment of energy storage technologies: a review, Energy Conversion and Management 234 (2021) 113920.

[52] Oyeniyi C. Esan, Zhen Sun, Xiong Zhang, Flow batteries: current status, challenges, and future directions, Renewable & Sustainable Energy Reviews 134 (2020) 110252.

[53] Abdul Ghani Olabi, Tabbi Wilberforce, M. Ramadan, M.A. Abdelkareem, A.H. Alami, Compressed air energy storage systems: components and operating parameters—a review, Journal of Energy Storage 34 (2021) 102000.

[54] Xing Zhang, Yang Li, Zhenguo Gao, Siyuan Chen, Yanlin Xu, Hong Chen, Overview of dynamic operation strategies for advanced compressed air energy storage, Journal of Energy Storage 66 (2024) 107408.

[55] Luisa F. Cabeza, A. de Gracia, A. Castell, Integration of renewable technologies in historical buildings: a review, Renewable & Sustainable Energy Reviews 42 (2015) 153–165.

[56] Elisa Guelpa, Vincenzo Verda, Thermal energy storage in district heating and cooling systems: a review, Applied Energy 252 (2019) 113474.

[57] Qusay Hassan, A. Jafar, M. Zulfiqar, Hydrogen energy storage: a review of technologies, applications, and prospects, International Journal of Hydrogen Energy 48 (2023) 28452–28473.

[58] A.M. Hassan, M.A. Kibria, M. Hossain, Green hydrogen production and storage: challenges and opportunities, Renewable & Sustainable Energy Reviews 188 (2023) 113780.

[59] Tobias Terlouw, Christian Bauer, Marco Mazzotti, Hydrogen storage technologies: current status and future prospects, Applied Energy 318 (2022) 119186.
[60] Bukola B. Adetokun, Oghenewvogaga Oghorada, Sufyan J. Abubakar, Superconducting magnetic energy storage systems: prospects and challenges for renewable energy applications, Journal of Energy Storage 52 (2022) 105123.
[61] Justus P. Hoffstaedt, Daan P.K. Truijen, Jonathan Fahlbeck, L.H.A. Gans, M. Qudaih, A.J. Laguna, J.D.M. De Kooning, Low-head pumped hydro storage: a review of applicable technologies for design, grid integration, control and modelling, Renewable & Sustainable Energy Reviews 158 (2022) 112119.
[62] Shafiqur Rehman, Luai M. Al-Hadhrami, M.M. Alam, Pumped hydro energy storage system: a technological review, Renewable & Sustainable Energy Reviews 44 (2015) 586–598.

Chapter 9

Applications of energy storage systems

Xin Lei[a], Jiesheng Huang[a], Zhen Mei[a], Hongfei Yang[a], Yiwen Zhang[a], Huaiguang Jiang[a], and Yingjun Lei[b]
[a]*School of Future Technology, South China University of Technology, Guangzhou, China,* [b]*China Academy of Space Technology, Beijing, China*

Contents

9.1 Energy storage system in aerospace

Aerospace plays an increasingly important role in modern life. Conventional fuels have adverse effects on environmental sustainability, with the continuous rise in greenhouse gas emissions driving the development of next-generation aerospace technologies [1]. Data indicate that air traffic volume doubles approximately every 15 years [2], which means carbon dioxide emissions will also increase at the same rate. Moreover, the propulsion systems of existing aerospace equipment generally produce high noise levels [3]. In contrast, electric-powered systems have garnered growing attention due to their significant advantages, including low noise, low vibration, low emissions, and high efficiency. With ongoing technological advancements, particularly in Energy Storage System (ESS), electrification not only promises to greatly

Modern Energy Storage. https://doi.org/10.1016/B978-0-32-395419-8.00015-5

enhance spacecraft performance and efficiency but also offers the potential to significantly reduce operational costs [4]. Therefore the widespread adoption and application of electrification in the aerospace sector has become an inevitable trend.

9.1.1 Facilitating the lightweighting of spacecraft

Lightweight technology is a crucial approach to enhancing aircraft performance, reducing energy consumption, and increasing payload capacity. It plays a vital role in advancing the aerospace economy and facilitating large-scale space access. Studies indicate that reducing a spacecraft's weight by just one kilogram can lower launch costs by approximately 50,000 to 500,000 RMB, depending on the orbital altitude, ranging from low Earth orbit to lunar orbit. In civil aviation, lightweighting provides substantial economic and environmental benefits. For example, if a major airline reduces the weight of each aircraft by 100 kilograms, it can save nearly 5000 tons of fuel annually, cut carbon dioxide emissions by approximately 15,000 tons, and generate over 100 million RMB in additional net income [4]. Furthermore, advancements in high-energy-density battery technology allow spacecraft to significantly decrease overall weight while maintaining power capacity, thereby reducing fuel consumption and improving operational efficiency.

The power supply and distribution system of the Chang'e-5 orbiter employs $LiCoO_2$ lithium-ion battery cells with an energy density of 195 Wh/kg. Compared to conventional lithium-ion batteries with an energy density of 150 Wh/kg, these high-specific-energy batteries enable significant weight reduction for spacecraft [5]. Zinc-air batteries have also emerged as a promising option for aerospace power systems due to their low cost, high safety, environmental friendliness, and high energy density. Their theoretical energy density can reach 1084 Wh/kg, with a practical energy density exceeding 400 Wh/kg, far surpassing the 260 Wh/kg energy density of mainstream lithium-ion batteries currently used in spacecraft [6]. Additionally, the Institute of Physics at the Chinese Academy of Sciences has developed a next-generation lithium secondary battery with an energy density of up to 700 Wh/kg. This battery achieves a mass energy density of 711.30 Wh/kg during its first discharge and a volumetric energy density of 1653.65 Wh/L. This technological advancement addresses the urgent demand for high-performance power sources in specialized applications, such as high-altitude and deep-space missions, and holds significant promise for the future of electric aviation [7].

9.1.2 Extraterrestrial energy harvesting and utilization

As human exploration of the Moon and other celestial bodies advances, establishing long-term sustainable habitats in extreme environments has become a significant challenge in space technology. Solar energy, as the most readily available power source, offers a feasible solution for these missions. However, its acquisition is not always stable or reliable under all conditions. For example, the lunar night, which lasts approximately 14 Earth days, presents substantial challenges to solar energy storage. To ensure a continuous power supply in such extreme environments, the development and application of ESS and management technologies are essential. ESS mitigates solar energy fluctuations, ensuring the stable operation of spacecraft and base equipment, while energy management technologies optimize the use of limited energy.

The absence of an atmosphere and clouds on the Moon ensures that solar radiation is nearly unaffected, resulting in a higher solar energy density than anywhere on Earth [8]. Calculations indicate that the annual solar radiation energy reaching the Moon is approximately 12 trillion kilowatts, equivalent to 25,000 times the total energy consumed each year [10]. The lunar day, lasting about 14 Earth days, provides an extended period for solar energy harvesting. Moreover, remote sensing and sample-return analysis have shown that lunar regolith contains 40% oxygen and 20% silicon. Since manufacturing one square meter of solar panel requires only 400–500 g of silicon, the silicon content in the lunar regolith is more than sufficient to support solar panel production.

Solar panels can capture solar energy during the lunar day to provide power. However, solar energy cannot meet the energy demands during the lunar night. The Moon's day-night cycle is exceptionally long, lasting approximately 29.5 Earth days, with nights spanning 14 Earth days. The temperature difference between day and night is extreme due to the absence of an atmosphere. During the day, solar radiation causes surface temperatures to rise to 127 °C, while at night, temperatures plummet to –173 °C. This substantial temperature variation presents an opportunity for thermoelectric power generation systems, which can convert the temperature differences on the Moon's surface into thermal energy for power generation. However, the complexity and low reliability of thermoelectric systems pose significant challenges for their practical application on the Moon. Consequently, even with abundant solar energy, achieving efficient and stable solar energy capture and storage remains a major technological challenge.

With the continuous development of technology, small nuclear reactors have emerged as another solution and are gradually gaining attention. They can provide a high-power, compact, and stable energy supply on the Moon. Both Russia and the United States have considered establishing nuclear power stations on the Moon to support long-term stays and research activities. In addition, the Moon's surface is rich in helium-3, an ideal fuel for nuclear fusion. When helium-3 undergoes nuclear fusion with deuterium, it releases a huge amount of energy without producing radioactive waste [9]. If it can be successfully extracted and used for fusion power generation, it will provide a long-term, sustainable solution for the Moon's energy supply. Thus, despite the Moon's abundant resources, how to efficiently and stably capture and store energy remains a key technological challenge.

From the perspectives of technology maturity and launch economic costs, solar power generation will be the primary source of energy supply for the lunar research station over the long term, with ESS integrated to ensure the normal energy demands of the lunar base. Considering the basic environment of the Moon, the ESS must meet several fundamental requirements: large capacity, high efficiency, high stability, and low-temperature adaptability.

Current energy storage technologies include batteries, superconducting energy storage, thermal energy storage, and hydrogen energy storage. Among these, batteries—particularly lithium-ion batteries—are the most widely used due to their high energy density and long service life, especially in spacecraft applications. However, lithium-ion batteries experience reduced efficiency and capacity degradation in extremely low-temperature environments, necessitating improvements in their adaptability. In contrast, superconducting ESS utilizes the properties of superconducting materials at low temperatures to store large amounts of electrical energy with exceptionally high efficiency, making them suitable for a stable, long-term energy supply. Nevertheless, these technologies remain in the exploratory phase.

Thermal energy storage, which uses specialized materials to absorb and store heat, is well-suited for energy supply during lunar nights or in the absence of sunlight. Technologies such as Phase Change Materials (PCMs) enable these systems to absorb solar heat during the day and release it at night, maintaining the base's temperature and energy supply. However, the 300 °C temperature difference between lunar day and night poses significant challenges for the design and stability of these systems. Hydrogen energy storage, a highly promising technology, offers solutions for long-term energy storage and transportation by producing hydrogen through water electrolysis. This hydrogen can later be converted into electrical energy via fuel cells, providing reliable energy support when needed.

In summary, ESS plays a crucial role in the lunar energy supply system by addressing the challenges of extreme environmental conditions, such as long lunar nights and significant temperature fluctuations. Advanced technologies like lithium-ion batteries, superconducting energy storage, thermal energy storage, and hydrogen energy storage provide essential solutions for a stable and reliable power supply, enabling the continuous operation of lunar bases and supporting long-term space exploration.

9.1.3 Intelligent energy management technology

To ensure a reliable energy supply for lunar and Martian bases, both ESS and autonomous energy cycling systems are essential. These systems enable automatic regulation and optimization of energy distribution and usage through intelligent control, minimizing waste and enhancing efficiency. As illustrated in Fig. 9.1, researchers have developed an integrated energy system combining solar power and fuel cells, which leverages solar energy and closed-loop recycling technologies to maintain a stable energy supply [8].

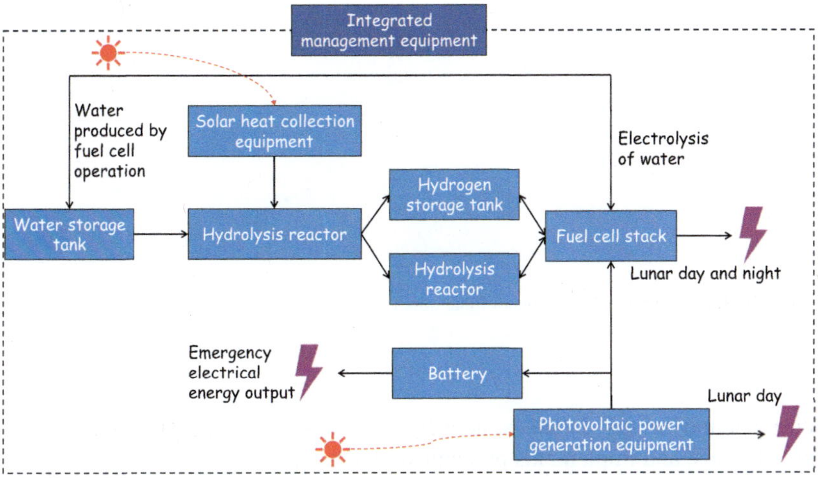

FIGURE 9.1 Lunar energy structure.

During the lunar day, the system captures solar energy through two primary methods: photovoltaic power generation, which directly converts sunlight into electricity for daily use, and solar thermal devices, which transform sunlight into thermal energy to support subsequent water electrolysis. Water serves as a crucial "energy carrier" within this system, undergoing a closed-loop recycling process. Initially, water is stored in a storage tank and transported through pipes to an electrolyzer, where solar-generated heat facilitates its decomposition into hydrogen and oxygen. This process effectively converts solar energy into storable chemical energy in the form of hydrogen and oxygen. During the lunar night, when sunlight is unavailable, the system depends on fuel cells to supply energy. These fuel cells combine the stored hydrogen and oxygen from the day to generate electricity and thermal energy [9].

By coupling solar energy, thermal conversion, and fuel cell technologies, this system offers a reliable and sustainable energy supply solution for the Moon. It efficiently utilizes the abundant solar resources during the lunar day while ensuring a continuous energy supply throughout the long lunar night. Such technology will provide robust energy support for future long-term human presence and research activities on the Moon.

Once the energy system structure is established, Artificial Intelligence (AI) technology becomes a crucial component for constructing a lunar microgrid. As shown in Fig. 9.2, the intelligence of the lunar microgrid is achieved through the following approaches:

1. High-Precision Forecasting and Early Warning Capabilities. Utilizing AI and big data analysis, the system precisely forecasts lunar environmental

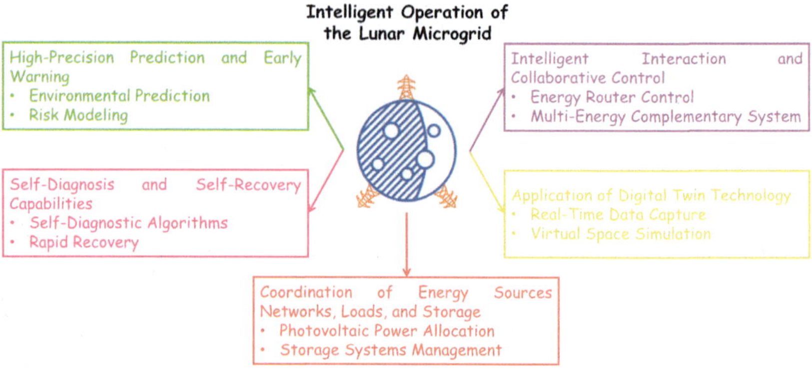

FIGURE 9.2 Technical key points of lunar smart microgrid.

changes and energy demands, actively identifies potential risks, conducts risk simulations, and implements preemptive defense measures.

2. Intelligent Interaction and Collaborative Control. By leveraging advanced information and communication technologies, the system enables intelligent coordination among energy generation, storage, and consumption within the base. Energy routers serve as the core to flexibly control voltage and power, optimizing the operation of multienergy complementary systems.

3. Application of Digital Twin Technology. With high-precision sensors collecting real-time data from the lunar microgrid, a digital twin system is established on Earth. This enables comprehensive, multidimensional real-time simulation and prediction of the energy base in the digital space, improving decision-making efficiency and operational optimization.

4. Self-Diagnosis and Self-Healing Capabilities. Utilizing intelligent algorithms, the system can automatically diagnose faults within the energy base and quickly restore operation, effectively responding to disruptions and challenges posed by the extreme environment.

5. Coordinated Optimization of Energy Generation, Storage, and Consumption. Through flexible configuration of energy supply (e.g., photovoltaic power generation), storage systems (electricity, hydrogen, oxygen), and energy-consuming devices, the lunar research station achieves dynamic energy balance and autonomous operation.

The intelligence of the lunar microgrid's integrated energy base lies in the seamless integration of advanced digital technologies with efficient operational control and scheduling methods. By leveraging high-precision sensing, big data, AI, and digital twin technologies, the system achieves exceptional self-healing capabilities, reliability, and stability. This intelligent framework actively adapts to the Moon's extreme environmental conditions, ensuring robust support for the survival and long-term stable operation of the research station.

9.2 Frontiers in carbon fixation and fuel cells exploration

9.2.1 Recent advances in electricity-driven carbon fixation and metabolic pathway engineering

The development of efficient technologies for the resource utilization of carbon dioxide (CO_2) can simultaneously alleviate urgent environmental and energy pressures, serving as a crucial pathway to achieve the "dual carbon" goals. Microbial Electrosynthesis (MES) is a sustainable bioelectrochemical process that utilizes electricity to provide reducing power and energy to microorganisms, enabling the fixation of CO_2 and the production of valuable biofuels under ambient temperature and pressure conditions [11]. In MES, the process of electron transfer from the anode to the cathode is critical, and there are three primary transfer mechanisms: direct Electron Transfer (DET), Mediated Electron Transfer (MET), and Indirect Electron Transfer (IET). DET refers to the direct contact between microorganisms and the electrode, allowing the microorganisms to utilize electrons directly through electron transfer proteins on their cell surfaces. Known mediators include cytochrome complexes, surface cytochromes, conductive nanowires, and other redox proteins such as copper and iron-sulfur proteins [12,13]. MET relies on electroactive molecules to mediate the transfer of electrons between microorganisms and electrodes. Commonly used electroactive molecules include Methyl Viologen (MV), AnthraQuinone-2,6-DiSulfonate (AQDS), and the organic dye Neutral Red (NR), and this mechanism can also promote the regeneration of NAD(P)H [14]. IET typically depends on the transfer of H_2. When the cathode's negative potential is sufficiently low, electrons and H^+ can generate H_2, which certain microorganisms, such as methanogens, can directly utilize [15].

Cyanobacteria are important photosynthetic carbon-fixing microorganisms and serve as a major source of primary productivity in the biosphere, making them a significant model system for studying photosynthesis. In recent years, microalgae have also been recognized as a highly promising novel microbial photosynthetic platform, possessing the potential to directly convert solar energy and CO_2 into various bio-based products. Building on this foundation, the integration of electro-driven carbon fixation with metabolic pathway engineering allows for the utilization of electrical energy to convert CO_2 into useful compounds, combined with the optimization of metabolic pathways to enhance the efficiency and sustainability of biological production. Through electrochemical reactions, CO_2 can be reduced to organic substances (such as glucose, alcohols, or other compounds). This process typically involves electrodes and catalysts that can significantly accelerate the rate of CO_2 reduction reactions using electric current, thereby effectively reducing greenhouse gas emissions and advancing the achievement of carbon neutrality goals. Additionally, metabolic pathway engineering involves optimizing the metabolic networks of microorganisms or plants through genetic modification or metabolic regulation to improve the yield and efficiency of target products.

Techniques such as CRISPR can be employed to modify specific genes, thereby enhancing or suppressing particular metabolic pathways.

TABLE 9.1 Reduction potentials of different products [16].

Reaction	Reduction Potential E° (V)
$CO_2 + 2\ H^+ + 2\ e^- \rightarrow HCOOH$	–0.61
$CO_2 + 2\ H^+ + 2\ e^- \rightarrow CO + H_2O$	–0.53
$CO_2 + 8\ H^+ + 8\ e^- \rightarrow CH_4 + 2\ H_2O$	–0.24
$2\ CO_2 + 12\ H^+ + 12\ e^- \rightarrow C_2H_4 + 4\ H_2O$	–0.349
$2\ CO_2 + 12\ H^+ + 12\ e^- \rightarrow C_2H_5OH + 3\ H_2O$	–0.329

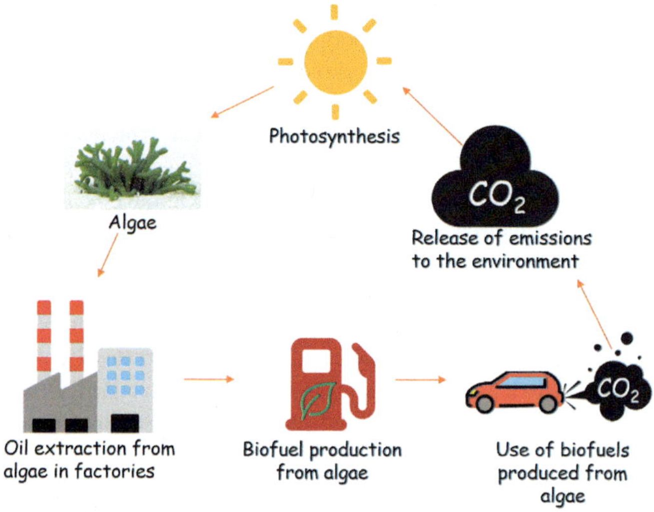

FIGURE 9.3 "Artificial PSII" electro-driven high-efficiency carbon fixation platform.

However, the carbon in CO_2 is in its highest oxidation state, and its activation into organic matter requires a substantial energy input. The reduction potentials of common products are shown in Table 9.1 [16]. Relying solely on natural light-driven photosynthesis results in low efficiency, necessitating additional electrical energy input to enhance the effectiveness of photosynthesis. To address the issue of low natural photosynthetic efficiency, researchers at the U.S. National Renewable Energy Laboratory [17] proposed the use of externally supplied electricity to drive the PhotoSystem I (PSI) of cyanobacteria, as illustrated in Fig. 9.3. By supplying current from an external source to the photosynthetic electron transport chain, CO_2 can be efficiently reduced. Experimental results indicate that the conversion efficiency of CO_2 is significantly improved when external power is applied; the current enhances the electron transfer process, thereby increasing the yield of photosynthetic products. This finding offers new insights for the future utilization of microorganisms in CO_2 valorization, with significant application potential, especially in the fields of renewable energy and environmental protection. The study also proposes further optimization of the integration of current and photosynthesis to achieve higher energy conversion efficiency, with an average external electron utilization rate of 74.9%. In [18] the authors discuss the ultrafast electron transfer mechanisms in a hybrid system combining gold nanoparticles and cyanobacteria, along with its applications in converting solar energy into chemical energy. The research demonstrates that Au nanoparticles significantly accelerate the electron transfer process in cyanobacteria, showcasing excellent electron dynamics that effectively facilitate CO_2 reduction and organic matter synthesis. The high solar energy conversion capability of this hybrid system provides new directions for the development of novel photocatalytic materials and the enhancement of photosynthetic efficiency, holding broad application potential. This study lays an important foundation for the advancement of future renewable energy technologies. Work [19] explores how the integration of microalgae with material hybrid systems can enhance energy conversion during photosynthesis, providing feasible solutions for achieving carbon neutrality. The research shows that this hybrid system significantly improves photosynthetic efficiency, promoting electron transfer and the conversion of CO_2 into organic matter. The paper emphasizes the potential of microalgae-material hybrid systems in practical applications such as wastewater treatment and biofuel production, highlighting their importance in achieving sustainable carbon neutrality goals. This research provides a vital theoretical foundation and practical guidance for the development of new carbon neutrality technologies.

The development of efficient CO_2 utilization technologies is not only a key pathway to achieving the "dual carbon" goals but also an effective means to alleviate environmental and energy pressures. As important photosynthetic microorganisms, cyanobacteria and microalgae demonstrate significant potential for converting solar energy and CO_2 into bio-based products. By integrating electro-driven carbon fixation with metabolic pathway engineering, the conversion efficiency of CO_2 and the yield of target products can be significantly enhanced, thereby advancing renewable energy technologies. Future research should focus on further optimizing the integration of current and photosynthesis, exploring new catalytic materials, and applying microalgae-material hybrid systems to achieve higher energy conversion efficiency and sustainability. These efforts will provide practical solutions for achieving carbon neutrality goals and promoting environmental protection.

9.2.2 Advances in biofuel cells and their industrial applications

Plants and microalgae convert light energy into chemical energy through photosynthesis, while also delivering additional electrons to the Calvin cycle

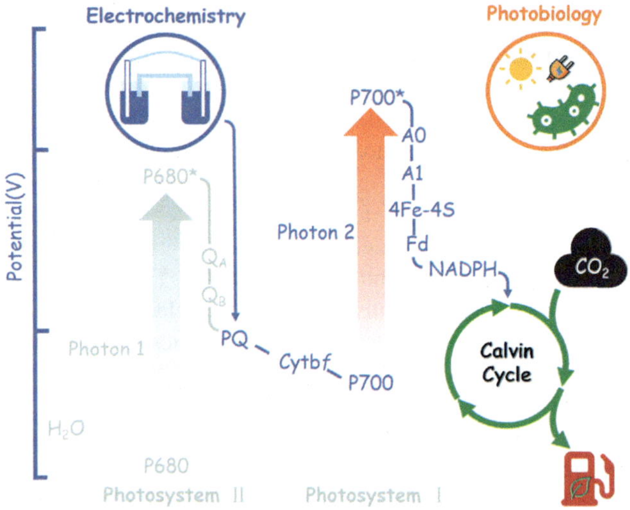

FIGURE 9.4 Biofuel production process.

via electrochemical reactions, ultimately synthesizing organic compounds (such as sugars). These organic compounds serve as the foundation for biofuels, and certain microorganisms convert sugars and other organic materials into biofuels (such as ethanol and biogas) through fermentation processes under anaerobic conditions. This process of utilizing cellular metabolic pathways to obtain energy and produce fuel is known as Microbial Fuel Cells (MFCs). The entire cyclical process from photosynthesis to biofuel production is illustrated in Fig. 9.4. Biofuels have garnered widespread attention due to their relatively low cost, clean combustion products, and renewability.

Based on this, numerous scholars have conducted extensive research on biofuels and their industrial applications. In [20] the authors investigated the process of producing microbial diesel using Chlorella protothecoides in bioreactors. By optimizing fermentation conditions such as temperature, light, pH, and nutrient concentration, the study found that high-density fermentation significantly enhanced microalgal growth and lipid accumulation, thereby improving diesel production efficiency. The paper discusses the advantages of high-density fermentation, including resource utilization efficiency and cost reduction, providing a new perspective on the application of microalgae as a feedstock for biodiesel production and laying a foundation for the promotion of renewable energy. Future research could further explore combinations of different microalgal species and fermentation conditions to achieve more efficient energy conversion. [21] explored how a dual CO_2 fixation mechanism can enhance algal lipid synthesis within a photosynthesis-fermentation model to increase biodiesel production efficiency. The research indicated that under conditions of ample light and CO_2 supply, dual CO_2 fixation significantly boosted algal lipid yields, offering a new strategy for potential biodiesel production. The paper emphasizes the advantages of this method in enhancing algal productivity and reducing costs, laying the groundwork for the development of more efficient biodiesel production processes while demonstrating significant application potential in renewable energy. [22] investigated the characteristics of Ethylene-Forming Enzyme (EFE) and its application in biological ethylene production, highlighting the feasibility and sustainability of biological ethylene production. The study showed that EFE can effectively catalyze the conversion of S-AdenosylMethionine (SAM) into ethylene, and optimizing the cultivation conditions of microbial or plant cells can significantly enhance ethylene yield. The paper discusses the economic and industrial application potential of biological ethylene, pointing out its environmental friendliness compared to traditional petrochemical production methods, thereby providing new ideas and foundations for the development of renewable chemicals. Work [23] examined how to improve the yield and quality of HydroThermal Liquefaction (HTL) biofuels by adjusting the composition of cyanobacterial biomass. The research indicated that increasing the lipid content in cyanobacteria can significantly enhance the oil yield and fuel quality of HTL products, offering a new strategy for cyanobacteria as a renewable biofuel source. By altering cultivation conditions such as light intensity and nutrient concentration, the biomass composition of cyanobacteria can be optimized, thereby enhancing the commercialization potential of HTL biofuels. This study lays a theoretical foundation for the industrial application of cyanobacteria and provides important guidance for the development of renewable energy. In [24] the authors explored a system-level optimization approach that combines genetic engineering with hydrothermal liquefaction to enhance biofuel production potential. The research demonstrated that optimizing gene expression can significantly increase the ratio of lipids and carbohydrates in biomass, thus improving the yield and quality of HTL products. Specific combinations of genetic modifications showed higher liquid fuel conversion efficiency, emphasizing the synergistic effects of combining genetic engineering with HTL. This study provides new insights for achieving efficient biofuel production and lays the groundwork for further development of renewable energy technologies.

Biofuels hold significant appeal for power generation applications that require low power but may find battery replacement impractical, such as in wireless sensor networks [25]. The Zoological Society of London (ZSL) has successfully implemented the world's first biological selfie system, a technology poised to revolutionize conservation efforts in the field (Fig. 9.5). In this project, a fern named Pete is equipped with a microbial fuel cell that leverages the biomass naturally deposited during the plant's growth to nourish bacteria in the soil. The energy produced is collected by the fuel cell and used to power a camera. This technology not only demonstrates the potential of bioenergy but also highlights its nearly limitless possibilities, as the plant can thrive in shaded areas and naturally move to maximize sunlight absorption. Pete is outfitted with an ultra-low-power camera manufactured by the American AI company

FIGURE 9.5 Micro low-power camera powered by biofuels [25].

Xnor.ai, which consumes such minimal energy that it can operate around the clock, entirely powered by a small plant. This technology can be utilized in remote and harsh environments that are difficult for humans to access, where battery replacement is often necessary. Through this technology, critical data on temperature, humidity, and plant growth can be recorded, aiding experts in better understanding threats such as climate change and habitat loss. ZSL plans to conduct a second field trial to further refine this technology. If successful, it will provide a new method for monitoring remote rainforest areas, contributing to conservation efforts and scientific research. Pete's successful selfies not only represent a technological breakthrough but also an innovation in the field of conservation science, showcasing how natural forces can be harnessed to support and enhance our monitoring and protection of the natural environment.

Electrochemical reactions, photosynthesis, and fermentation processes play crucial roles in the production of biofuels, with microalgae and plants effectively converting light energy and organic matter into renewable fuels through these mechanisms. Research indicates that optimizing cultivation conditions, utilizing dual CO_2 fixation mechanisms, and employing genetic engineering techniques can significantly enhance the yield and quality of biodiesel and other biofuels, offering new perspectives and strategies for the promotion of renewable energy. Future research may further explore combinations of various microalgae and their biomass components, as well as optimize fermentation and hydrothermal liquefaction processes to achieve more efficient energy conversion. Additionally, developing novel microbial and material hybrid systems, along with advancing the application of genetic engineering technologies in biofuel production, will provide a more solid foundation for achieving sustainable development and carbon neutrality goals.

9.3 Application of ESS in big data centers

The rapid growth in data processing and storage demands has driven the widespread development of large-scale data centers worldwide. As critical infrastructure, these data centers support a diverse range of applications, including cloud computing, AI, and big data analytics. However, their expanding scale and complexity have led to significantly increased energy demands, posing substantial challenges to power supply systems and environmental sustainability. To address these challenges, ESS has emerged as an essential solution, enabling the reliable and efficient operation of data centers while facilitating the integration of renewable energy sources.

ESS has proven to be a versatile and indispensable component of modern data center energy management strategies. By addressing the unique and dynamic energy consumption patterns of data centers, ESS can mitigate the impact of power demand fluctuations, enhance power reliability, and lower energy costs. Furthermore, ESS provides a critical pathway for data centers to transition to low-carbon energy systems by increasing the utilization of renewable energy sources. This section examines the application of ESS in data centers, emphasizing its role in addressing energy demand challenges, enhancing system reliability, and promoting sustainable operations.

To better illustrate the pivotal role of ESS, Fig. 9.6 presents a typical ESS configuration in large-scale data centers. The diagram demonstrates the integration of ESS with various components of the data center, such as renewable energy sources, grid connections, and backup power systems. As depicted, ESS serves as a central hub for energy flow management, ensuring stable power delivery and optimizing energy utilization across different operational scenarios. This architecture enables data centers to maintain an uninterrupted energy supply, even during grid fluctuations or renewable energy intermittency, ultimately enhancing their resilience, efficiency, and sustainability.

9.3.1 Energy storage requirements for big data centers

In recent years, the development and rise of new infrastructure, such as big data centers and 5G base stations, have driven the comprehensive upgrade of an infrastructure system based on information networks. This system aims to meet the demands of high-quality development by enabling services such as digital transformation and integrated innovation. However, these advancements have also significantly increased energy supply demands. For example, in 2024, Huawei Cloud's Shangtun Data Center expanded its operations, but the existing No. 1 and No. 2 main transformers were unable to meet the growing electricity

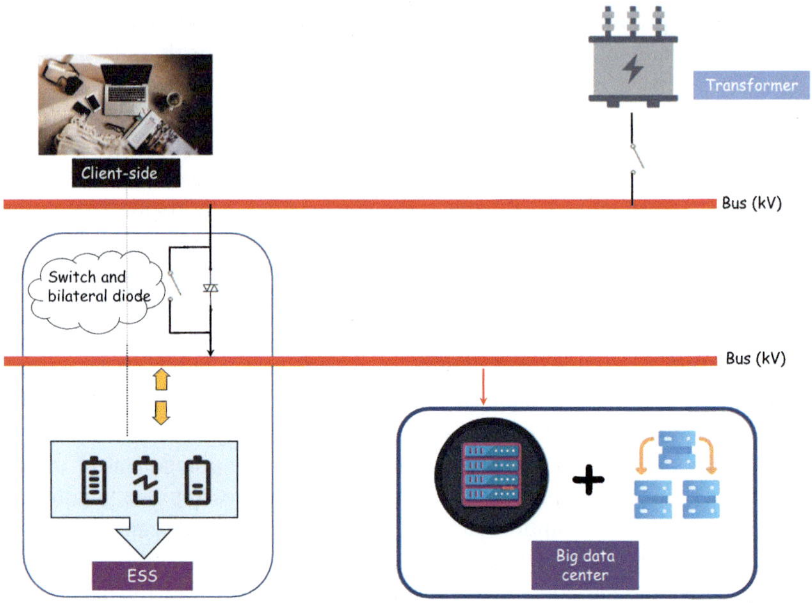

FIGURE 9.6 ESS configuration in large-scale data centers.

requirements. Consequently, on-site teams enhanced power supply stability by upgrading the capacity of the main transformers [26]. In the first seven months of 2024, the electricity consumption of the Guian New Area Data Center cluster increased by 43.96% [26]. Furthermore, data from the China Electricity Council indicates that, from 2018 to 2023, the annual electricity consumption of internet data services—including big data processing, cloud storage, and cloud computing—grew at an average rate of approximately 28%, with a year-on-year increase of 33% in the first half of 2024 [26].

To address the issue of high energy consumption, various regions have introduced a series of energy-saving policies in recent years. For instance, in 2019, Shenzhen mandated that newly built data centers must achieve a Power Usage Effectiveness (PUE) below 1.25 to qualify for a policy granting "support for more than 40% of the actual added energy consumption replacement" [27]. Similarly, in April 2021, the Guangdong Provincial Energy Administration issued a notice requiring that, during the "14th Five-Year Plan" period, the PUE of outdated data centers must be reduced to below 1.3 [27]. While these policies have alleviated some of the pressure from the rapidly increasing energy demands and improved energy efficiency to a certain extent, they fall short of providing direct technical solutions to the fundamental issue of high energy consumption in data centers. This challenge therefore warrants further research.

1. The Necessity of ESS Implementation. Large-scale data centers, with their continuous operation and high-intensity computational demands, face significant challenges related to electricity consumption, alongside stringent requirements for power stability, reliability, and resilience. Any disruption in power supply can result in business interruptions, data loss, and substantial economic losses. As energy consumption in data centers becomes increasingly critical, ESSs have emerged as a pivotal technological solution. ESS ensures seamless data center operations, enhances power reliability, optimizes energy utilization, promotes renewable energy integration, supports resource-intensive management, and enables intelligent energy regulation.

Data centers are highly dependent on electricity, especially during periods of high computational load, where power supply stability is paramount. ESS can provide instantaneous backup power during grid fluctuations or sudden outages, preventing power instability from impacting equipment performance. By responding rapidly to power demands during grid disruptions, ESS plays a vital role in mitigating the risks of system crashes or equipment damage caused by power interruptions.

Typically located in areas with concentrated electricity demand, large-scale data centers are often subject to power supply fluctuations. These challenges are particularly pronounced during peak electricity pricing periods, where costs can vary significantly. ESS can store electricity during off-peak periods, when prices are lower, and release it during peak periods, effectively reducing procurement costs and enabling data centers to better manage energy expenditures. Additionally, ESS addresses power supply instabilities, ensuring uninterrupted, 24/7 operation of data centers. During peak computational loads, when power demand surges, ESS can store energy during low-demand periods and release it during high-demand periods, balancing power requirements and avoiding overloading the power supply system. In this way, ESS not only improves the efficiency of energy utilization in data centers but also prevents potential service disruptions caused by power shortages during peak periods.

For instance, Google integrates renewable energy sources, such as solar power, with large-scale battery storage systems in its data centers to balance grid loads. During off-peak hours, Google stores solar energy or other low-cost electricity in batteries and releases it during peak pricing periods, significantly reducing electricity procurement costs. Similarly, Microsoft has partnered with energy storage companies to implement Liquid Air Energy (LAE) storage technology in some of its data centers. This technology converts electricity into low-temperature liquid air through compression during periods of low electricity prices and later reconverts it into electricity during high-price periods through an expansion process [30].

The application of ESS in data centers not only helps internet companies reduce electricity costs but also enhances energy management efficiency, supports renewable energy integration, and strengthens the sustainability and reliability of data center operations. With ongoing technological advancements and growing market demand, ESS is poised to play an increasingly critical role in the operation of data centers managed by internet enterprises.

2. Role of ESS in Enhancing Reliability, Efficiency, and Sustainability of Data Centers. ESS plays a critical role in ensuring the uninterrupted operation of data centers in the face of unexpected events such as grid failures or natural disasters. Even during power system outages, ESS can provide emergency power for durations ranging from several minutes to several hours, allowing essential data center equipment to remain operational, preventing system crashes, and buying time for fault recovery. Additionally, ESS can work seamlessly with backup generator sets, enhancing the resilience of data centers, accelerating fault recovery, and minimizing the risk of operational disruptions.

According to the Data Center Design Specifications (GB 50147), Tier A data centers are required to have dual power supplies and independent backup power systems. The backup power must operate separately from the primary power supply, with diesel generator sets or dedicated feeder lines being recommended. For Tier B data centers, dual power configurations are prioritized; however, if only a single power source is available, diesel generator sets must be installed as backup power. Furthermore, the specifications encourage the adoption of distributed energy systems in data centers. Based on these requirements, ESS applications in data centers primarily include the following: smoothing the output of renewable energy to ensure a stable supply; serving as a reliable backup power source in emergencies; replacing traditional Uninterruptible Power Supply (UPS) systems with specially designed ESS to enhance energy efficiency and response speed; and supporting multistation integration to achieve efficient energy utilization and resource scheduling [28].

As global concerns about energy efficiency and environmental impact intensify, data centers are increasingly expected to integrate renewable energy sources and reduce carbon emissions. ESS can be effectively combined with renewable energy sources such as solar and wind power, storing surplus energy during periods of excess generation and releasing it during periods of high demand. This reduces dependency on traditional fossil fuels and lowers carbon footprints. Moreover, ESS improves the utilization rate of renewable energy, supporting the green operation of large-scale data centers and aligning with global sustainability goals. By providing a stable power supply, ESS helps mitigate the impact of power fluctuations on data center equipment. Power instability or sudden voltage surges can damage servers, networking devices, and storage systems, significantly shortening their lifespans. ESS minimizes these risks by delivering a consistent energy input, reducing equipment failure rates, and enhancing overall system efficiency and reliability.

ESS can also be integrated with a data center's Energy Management System (EMS) to further optimize energy usage through intelligent scheduling and real-time monitoring. By analyzing variables such as current load, electricity prices, and energy demand, ESS can automatically adjust its charging and discharging strategies to optimize energy costs and operational efficiency. Furthermore, ESS provides real-time monitoring of battery status, offering detailed energy usage data and reports. These insights enable data center operators to make

FIGURE 9.7 Approaches and processes for achieving sustainable green computing.

more informed decisions, improving the transparency and intelligence of energy management.

For large-scale data centers, maintaining continuous and stable service is a top priority. Any interruption in power supply can lead to service outages, negatively affecting customer experience and trust. ESS acts as a buffer, ensuring business operations can continue during power disruptions, minimizing downtime, and enhancing business continuity and customer satisfaction.

The importance of ESS in data centers cannot be overstated. It ensures a stable and reliable power supply, reduces energy costs, strengthens resilience to risks, supports the integration of renewable energy, extends equipment lifespans, and improves overall system performance. Against the backdrop of rising energy demands and increasingly volatile electricity costs, ESS provides a flexible and efficient solution for data centers, enabling them to operate in a stable, efficient, and sustainable manner. Consequently, the deployment of ESS has become a key factor in achieving the long-term stability and reliability of large-scale data centers.

9.3.2 Green computing

From the perspective of power system scheduling, large-scale data centers, as major electricity consumers, present significant challenges to grid stability, scheduling flexibility, and energy demand due to their rapid growth and large-scale operations [29]. In particular, against the backdrop of an accelerated transition toward cleaner and low-carbon energy systems, data centers have become key factors influencing grid load peak-valley characteristics, renewable energy integration capacity, and overall carbon emissions. As a result, optimizing energy efficiency management in data centers and enhancing the utilization of renewable energy through advanced technological solutions has become an urgent issue that demands attention.

Fig. 9.7 illustrates the primary approaches and processes for achieving sustainable green computing in data centers. The diagram emphasizes three core strategies: the integration of ESS, the optimization of computing power allocation, and the implementation of advanced energy management technologies. By utilizing ESS, data centers can store excess energy during periods of low demand and discharge it during peak demand, effectively mitigating grid load fluctuations. Simultaneously, optimized computing power allocation ensures that energy-intensive tasks are scheduled when renewable energy output is high or when electricity prices are low, thereby reducing dependence on traditional fossil fuels. Furthermore, advanced energy management technologies—such as AI-driven predictive analytics and real-time monitoring—facilitate precise load balancing and energy optimization, further enhancing the efficiency, reliability, and sustainability of data center operations.

By strategically deploying ESS and developing efficient computing power allocation strategies, it is possible to achieve seamless coordination between electricity consumption and supply. This not only mitigates the pressure of grid load fluctuations but also supports a high penetration of renewable energy while advancing the goals of green and sustainable computing. The following sections delve into how these strategies can be effectively implemented to achieve green computing objectives.

1. Configuration of ESS. ESS, often referred to as "buffers" in power systems, play a vital role in peak shaving and valley filling. By deploying ESS with sufficient capacity, stored energy can be discharged during peak grid load periods to alleviate instantaneous stress on the grid, while charging during off-peak periods ensures readiness for future load demands. This dynamic regulation mechanism helps mitigate the impact of high data center loads on grid stability.

The intermittent and fluctuating nature of renewable energy sources, such as wind and solar power, poses challenges to grid stability when directly integrated. ESS addresses this issue by storing surplus energy during periods of high renewable generation and discharging it during periods of low output. This not only enhances the utilization of renewable energy but also improves the flexibility and resilience of grid operations.

Optimizing ESS configuration involves considering multiple factors, including grid operating costs, storage capacity, charging and discharging efficiency, and lifecycle performance. By incorporating real-time electricity prices from power markets and dynamically optimizing ESS charging and discharging strategies through economic dispatch models, operational costs can be significantly reduced while simultaneously minimizing carbon emissions.

2. Optimization of Computing Power Allocation Strategies. The computing power demand of large-scale data centers is highly dynamic, and optimizing computing power allocation strategies can enable efficient coupling between computational tasks and power demand. Specifically, computational tasks can be dynamically aligned with the grid's power supply capacity and real-time electricity prices. High-energy-consuming tasks can be prioritized during periods of low electricity prices or high renewable energy generation, while low-priority tasks can be postponed or transferred to data centers in other regions when the grid experiences high stress. This approach effectively leverages regional electricity market price differences and low-carbon energy resources.

Furthermore, cross-regional scheduling allows computational tasks to be migrated to areas with higher shares of renewable energy or more abundant power resources, achieving global optimization of resource allocation. By integrating demand response strategies, large-scale data centers can further flexibly adjust their energy consumption based on grid load fluctuations. For instance, during peak load periods, data centers can reduce task execution frequency or temporarily shut down noncritical servers, while accelerating task execution during off-peak periods. This enables data centers to achieve deep coordination and efficient integration with grid scheduling, promoting optimal operation and resource utilization.

3. The Enhancement of Renewable Energy Penetration. Enhancing the penetration of renewable energy is a vital objective in achieving green computing, a goal that hinges on the support of advanced technologies. Large-scale data centers can function as key components of Virtual Power Plants (VPPs), working in coordination with renewable energy generation facilities and ESS. By participating in ancillary grid services such as frequency and voltage regulation, data centers can improve their adaptability to the intermittency of renewable energy while boosting the overall utilization of renewable energy in the power system.

The integration of edge computing technologies enables real-time collection and processing of electricity load data from data centers. When combined with AI algorithms, this facilitates accurate load forecasting, providing the grid with reliable electricity demand information and optimizing the scheduling of renewable energy resources. Additionally, precise load forecasting enhances the internal energy management efficiency of data centers, reducing their dependence on traditional energy sources. By incorporating distributed energy sources, such as solar and wind power, with flexible load-scheduling technologies, data centers can establish "microgrids" that achieve localized energy self-balancing and efficient utilization, further advancing the penetration of renewable energy. For example, Google has achieved 100% renewable energy operation across its data centers and has invested in wind and solar energy projects to further reduce reliance on traditional fossil fuels, driving green operations. Similarly, Microsoft has launched "zero-carbon emission" initiatives in several data centers, leveraging wind and solar energy to significantly reduce its electricity-related carbon footprint. Facebook has committed to powering all of its data centers with 100% renewable energy and has supported this goal by building its own wind and solar power facilities.

Green computing not only advances sustainability but also delivers significant and multifaceted benefits. Through the strategic deployment of ESS and optimized computing power allocation strategies, data centers can significantly improve their overall energy efficiency. This shift reduces reliance on fossil fuels and accelerates the transition to a power supply structure dominated by renewable energy. Moreover, effective scheduling strategies can dramatically decrease fossil fuel consumption, lower carbon emissions, and shrink the carbon footprint of data centers. By maximizing the utilization of renewable energy, these strategies also contribute to achieving carbon neutrality. Furthermore, flexible computing power scheduling and optimized ESS charging and discharging strategies not only alleviate grid pressure but also reduce grid operating costs.

Additionally, by participating in ancillary grid services, data centers can generate extra revenue, achieving a dual benefit of economic and environmental gains.

In the context of the deep integration of power systems and information technology, the green transformation of data centers—as major energy-intensive entities—requires the collaborative support of multidisciplinary technologies. Looking to the future, as AI, blockchain, and big data technologies continue to advance, data-driven intelligent scheduling systems will further optimize the interaction between data centers and power systems. These innovations will provide robust technological support for building a low-carbon, efficient, and sustainable energy ecosystem.

Through the scientific deployment of ESS, optimized computing power allocation strategies, and the coordinated application of advanced technologies, data centers can not only meet their own electricity demands effectively but also significantly enhance the power system's ability to integrate renewable energy and improve operational efficiency. This transformation will accelerate the transition of the energy system toward greater sustainability, cleanliness, and intelligence, laying a solid foundation for a greener energy future.

9.4 ESS in extreme environments

9.4.1 ESS in deep-sea

The ocean, covering approximately 70% of the Earth's surface, constitutes the planet's largest ecosystem, harboring abundant natural resources and significant scientific research value. The exploration of the deep sea is driven not only by human curiosity about the unknown but also by the pressing needs of modern society in areas such as resource development, environmental protection, and scientific research. With technological advancements, deep-sea exploration and operational activities have become increasingly frequent. These activities range from the extraction of seabed mineral resources to the study of marine biodiversity, the construction of underwater communication infrastructure, and marine environmental monitoring. All these endeavors require reliable energy supplies to support the long-term operation of equipment.

The deep-sea environment is extremely harsh, presenting significant challenges to deep-sea equipment. At the seabed, pressure increases with depth, with an additional 10 meters of depth corresponding to an increase of approximately one atmosphere (about 101.3 kPa) [31]. Equipment must not only withstand the immense pressure of the deep sea but also operate stably under varying pressure conditions. The salinity and other chemical substances in seawater are highly corrosive, and prolonged exposure to such an environment can severely impair the smooth operation of equipment. Given the extended operational periods and high recovery costs associated with deep-sea operations, equipment must be capable of stable operation over long durations, necessitating exceptional sealing and stability.

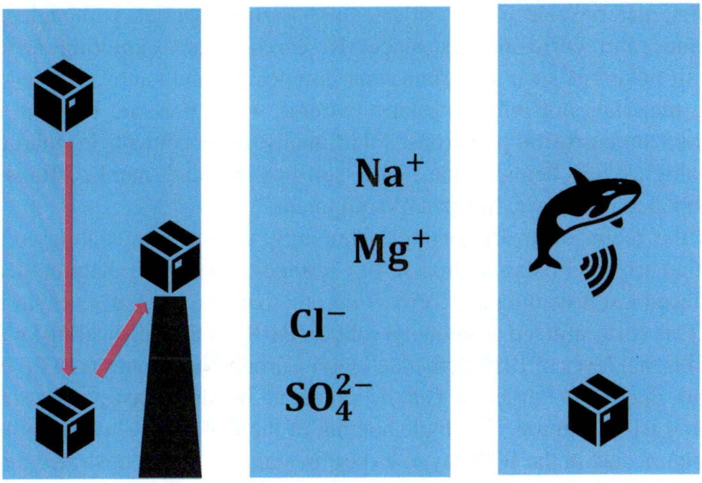

FIGURE 9.8 Challenges of ESS in deep-sea environments.

It is particularly important to emphasize that a reliable energy supply is crucial for the normal operation of deep-sea equipment. Deep-sea clean energy sources, including tidal energy, tidal current energy, wave energy, thermal gradient energy, salinity gradient energy, and wind energy [32], not only provide a sustainable energy supply but also minimize environmental impact.

The potential and significance of developing these energy sources cannot be overlooked. For instance, the global ocean energy reserve is estimated at approximately 7.4 TW (terawatts), with the installed capacity for power generation currently only at 0.02 TW due to the nascent stage of ocean energy development [33]. In China, the total carbon-free ocean energy reserves amount to 680 million kilowatts, with a theoretical annual power generation capacity of 2.2 trillion kilowatt-hours [34].

At the same time, the operation of equipment must not adversely affect the marine ecosystem. It is essential to ensure that equipment operates smoothly without impacting the marine environment and that, in the event of a malfunction, it can be promptly and completely recovered to prevent the leakage of chemical substances from ESS into the ocean.

The use of cables and other wired devices to provide energy in deep-sea environments presents numerous limitations. The nearly vertical cables extending from the surface to the seabed are primarily constrained by the oceanic topography and may also impact the marine ecosystem. Moreover, manufacturing cables that can withstand seawater corrosion and high pressure is highly costly. Consequently, ESS has emerged as a crucial solution to address the energy supply issue. To tackle these challenges in Fig. 9.8, current research trends in deep-sea ESS focus on several key areas.

First, material selection is critical, emphasizing materials with high pressure resistance and corrosion resistance. Researchers are exploring new high-strength materials such as carbon fiber composites and nano-enhanced alloys. These materials not only withstand extreme water pressure but also exhibit excellent anticorrosion properties. Additionally, advancements in smart coating technologies are being made, with self-healing and nano-coatings being developed to extend the lifespan of equipment.

In the realm of smart coating technology, self-healing coatings represent a particularly promising approach. For instance, a study by Kim et al. (2023) introduced a self-healing anticorrosion protective film for marine environments [35]. This study utilized bromobutyl rubber (BIIR) as the self-healing layer, with an additional layer of BIIR embedded with Carbon NanoTubes (CNTs) beneath it. This dual-layer structure can be prepared by drop casting, brushing, or spraying BIIR solution. The Joule heating of the CNT/BIIR layer rapidly heats and repairs cuts in the BIIR layer. Experimental results demonstrated that BIIR could successfully restore razor-cut cracks within a few hours, even underwater. Furthermore, the CNT/BIIR dual-layer coating exhibited good durability in saltwater corrosion tests and remained effective under wave action generated by impellers. Therefore this method is suitable for anticorrosion protection of large marine structures.

Second, harnessing seabed renewable energy to power ESS is a significant development direction. For instance, Ocean Thermal Energy Conversion (OTEC) technology, which generates electricity by exploiting the temperature difference between surface seawater and deep seawater, represents a highly promising form of renewable energy [36]. By deploying OTEC systems in deep-sea environments, it is possible to provide a continuous and stable power supply to ESS, thereby reducing dependence on external energy sources (see Fig. 9.9).

Finally, the introduction of intelligent management and monitoring systems is also an important future direction. By integrating sensors and Internet of Things (IoT) technologies, the status of ESS can be monitored in real time, enabling remote control and fault diagnosis. This not only enhances system reliability but also reduces maintenance costs by allowing the retrieval of equipment with operational risks, thus preventing leakage of contents into the deep sea due to equipment failure, which could lead to environmental pollution.

Future deep-sea ESS are poised to achieve breakthroughs in multiple aspects. These innovations will further enhance the performance and reliability of deep-sea ESS, providing robust support for deep-sea scientific exploration, resource development, and environmental protection.

9.4.2 Innovations for polar and high-altitude regions

Polar and high-altitude regions represent some of the most extreme and unique environments on Earth. These areas are characterized not only by harsh climates but also by complex geographical conditions, remoteness from power grids,

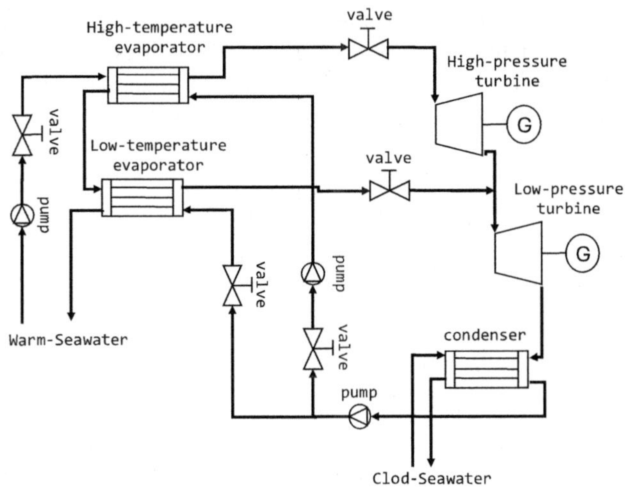

FIGURE 9.9 Overall framework of ocean thermal energy conversion (OTEC) technology, illustrating the process of generating electricity by exploiting the temperature difference between surface seawater and deep seawater. Adapted from [36].

and logistical challenges. In such environments, conducting scientific research, resource extraction, and other critical tasks necessitates the provision of stable and reliable power through ESS.

In these contexts, ESS generally serves two primary applications: providing energy support for scientific expeditions and ensuring the stable and reliable operation of equipment.

ESS in these unique environments faces multiple challenges. The first consideration is the operating temperature of the system. In polar regions, extreme low temperatures are a concern, while in high-altitude areas, temperature variations can be significant due to altitude differences and climatic influences. ESS must adapt to these extreme climate conditions. Additionally, many polar and high-altitude regions are far from traditional power grids and cannot access stable electrical networks, necessitating localized energy solutions. Logistics and assembly also pose significant challenges; due to the extreme remoteness of these areas, transportation costs are high, and lead times are long. Therefore energy storage equipment must have a long lifespan and low assembly and maintenance requirements to minimize the inconvenience of frequent replacements and repairs—challenges common to both polar and high-altitude regions.

To meet the ESS needs under extreme temperatures, wide-temperature batteries represent a highly promising research direction. Sodium-ion batteries (NIBs) have garnered attention for their stable performance in extreme temperatures. Compared to lithium-ion batteries, sodium-ion batteries exhibit superior discharge performance in low-temperature environments. This is primarily due to the diffusion kinetics of sodium ions and the stability of the electrolyte. Furthermore, the abundance of sodium in the Earth's crust is much higher than that of lithium, making sodium-ion batteries more cost-effective. Studies have shown that sodium-ion batteries can maintain a high discharge capacity even at temperatures below –20 °C [37] (see Table 9.2).

TABLE 9.2 Desolvation energies of Li and Na to common organic electrolyte solvents (kJ/mol) [37].

Solvent	Li	Na
EC	211.3	151.9
PC	218	157.3
FEC	191	135.4
DEC	207.7	147.5
EMC	201.3	142.7
THF	183.4	124.5
ES	194.1	135.4

Utilizing renewable energy storage facilities provides a localized energy solution that does not rely on traditional power grids. Solar and wind energy are two widely used renewable resources in polar and high-altitude regions. At Arctic research stations, solar panels are extensively installed, and efficient ESS can provide a continuous power supply even during the prolonged polar night. The Princess Elisabeth Antarctica Station established by Belgium is the world's first Antarctic research station powered solely by solar and wind energy [38]. The station's ESS forms an independent microgrid, ensuring that electrical energy from photovoltaic modules or wind turbines is stored in batteries for backup.

Utilizing renewable energy storage facilities provides a localized energy solution that does not rely on traditional power grids. Solar and wind energy are two widely used renewable resources in polar and high-altitude regions. At Arctic research stations, solar panels are extensively installed, and efficient ESS can provide a continuous power supply even during the prolonged polar night. The Princess Elisabeth Antarctica Station established by Belgium is the world's first Antarctic research station powered solely by solar and wind energy [38]. The station's ESS forms an independent microgrid, ensuring that electrical energy from photovoltaic modules or wind turbines is stored in batteries for backup (see Fig. 9.10).

To address logistical challenges and ensure that energy storage equipment has a long lifespan and low maintenance requirements, modular design and the construction of intelligent detection systems for ESS can be employed. Developing modular, easy-to-install, and maintain energy storage units not only facilitates transportation and installation but also allows for flexible expansion

FIGURE 9.10 The Princess Elisabeth Antarctica Station, the world's first antarctic research station powered solely by solar and wind energy [38].

based on actual needs. Each module can operate independently, and in the event of a module failure, it can be quickly replaced without affecting the overall system operation. Constructing intelligent detection systems for energy storage devices enables real-time monitoring of device status and early warning of potential issues, thereby extending the lifespan of the equipment and reducing maintenance costs.

The unique environments of polar and high-altitude regions impose stringent requirements on ESS. The combination of extreme temperatures, remoteness from power grids, and logistical difficulties makes a stable and reliable energy supply a complex and arduous task. However, continuous technological innovation and scientific research have led to the emergence of many effective solutions. The application of these technologies not only supports scientific activities and resource extraction but also ensures the safety and quality of life for personnel involved. With advancements in new material sciences and technology development, more efficient and environmentally friendly energy storage solutions are expected to emerge, further promoting sustainable development in these extreme environments.

References

[1] M. Ayar, S. Ekici, T.H. Karakoc, Energy storage technologies in aircraft hybrid-electric propulsion systems, in: International Symposium on Unmanned Systems and The Defense Industry, Springer, 2022, pp. 1–7.

[2] R. De Neufville, Airport systems planning, design, and management, in: Air Transport Management, Routledge, 2020, pp. 79–96.

[3] D.L. Huff, Noise reduction technologies for turbofan engines, in: 35th International Congress and Exposition on Noise Control Engineering (INTER-NOISE 2006), 2007, no. E-15787.

[4] W. Zhang, T. Changhong, Lightweighting of aerospace and aero-nautical equipment: Challenges and perspectives, Acta Aeronautica et Astronautica Sinica 45 (5) (2024).
[5] L. Zhiqiang, Z. Chen, C. Yan, C. Jianyue, Y. Min, L. Tianyi, High specific energy design of power supply and distribution system for chang'e-5 orbiter, Journal of Deep Space Exploration 8 (3) (2021) 237–243.
[6] X. Guo, T. Zheng, G. Ji, N. Hu, C. Xu, Y. Zhang, Core/shell design of efficient electrocatalysts based on $NiCo_2O_4$ nanowires and nimnldh nanosheets for rechargeable zinc–air batteries, Journal of Materials Chemistry A 6 (22) (2018) 10 243–10 252.
[7] Q. Li, Y. Yang, X. Yu, H. Li, A 700 Wh/kg rechargeable pouch type lithium battery, Chinese Physics Letters 40 (4) (2023) 048201.
[8] R. Depeng, L. Qing, X. Yingqiao, Preliminary research on the lunar base energy system, Journal of Deep Space Exploration 5 (6) (2018) 561–568.
[9] Y. Jiang, Y. Zhang, J. Wan, Lunar microgrid——integrated energy base for deep space exploration, New power system 2 (02) (2024) 190–200.
[10] J.Z. Liu, L. Xiongyao, Z. Kai, Key science and technology issues of lunar in situ resource utilization, Bulletin of National Natural Science Foundation of China 36 (6) (2022) 907–918.
[11] Y. Liu, M. Zhou, M. Ma, K. Wang, T. Tan, Research progress on biocatalytic fixation of CO2 driven by renewable energy sources, Chemical Industry Progress 42 (1) (2023) 1–15.
[12] Y. Chen, P. Li, J. Zhou, C.T. Buru, L. orevic, P. Li, X. Zhang, M.M. Cetin, J.F. Stoddart, S.I. Stupp, et al., Integration of enzymes and photosensitizers in a hierarchical mesoporous metal–organic framework for light-driven co2 reduction, Journal of the American Chemical Society 142 (4) (2020) 1768–1773.
[13] P. Dess'ı, L. Rovira-Alsina, C. Sánchez, G.K. Dinesh, W. Tong, P. Chat-terjee, M. Tedesco, P. Farr'as, H.M. Hamelers, S. Puig, Microbial electrosynthesis: towards sustainable biorefineries for production of green chemicals from co2 emissions, Biotechnology Advances 46 (2021) 107675.
[14] H. Chen, F. Dong, S.D. Minteer, The progress and outlook of bioelectrocatalysis for the production of chemicals, fuels and materials, Nature Catalysis 3 (3) (2020) 225–244.
[15] F. Mayer, F. Enzmann, A.M. Lopez, D. Holtmann, Performance of different methanogenic species for the microbial electrosynthesis of methane from carbon dioxide, Bioresource technology 289 (2019) 121706.
[16] Z. Sun, T. Ma, H. Tao, Q. Fan, B. Han, Fundamentals and challenges of electrochemical co2 reduction using two-dimensional materials, Chem 3 (4) (2017) 560–587.
[17] Z. Li, C. Wu, X. Gao, B. Addison, S. Shinde, X. Wang, X. Chen, J. Yu, D. Svedruzic, J.L. Blackburn, et al., Exogenous electricity flowing through cyanobacterial photosystem i drives co 2 valorization with high energy efficiency, Energy & Environmental Science 14 (10) (2021) 5480–5490.
[18] Q. Hu, H. Hu, L. Cui, Z. Li, D. Svedruzic, J.L. Blackburn, M.C. Beard, J. Ni, W. Xiong, X. Gao, et al., Ultrafast electron transfer in au–cyanobacteria hybrid for solar to chemical production, ACS Energy Letters 8 (1) (2022) 677–684.
[19] W. Xiong, Y. Peng, W. Ma, X. Xu, Y. Zhao, J. Wu, R. Tang, Microalgae–material hybrid for enhanced photosynthetic energy con- version: a promising path towards carbon neutrality, National Science Review 10 (10) (2023) nwad200.
[20] W. Xiong, X. Li, J. Xiang, Q. Wu, High-density fermentation of microalga chlorella protothecoides in bioreactor for microbio-diesel production, Applied microbiology and biotechnology 78 (2008) 29–36.
[21] W. Xiong, C. Gao, D. Yan, C. Wu, Q. Wu, Double co2 fixation in photosynthesis–fermentation model enhances algal lipid synthesis for biodiesel production, Bioresource technology 101 (7) (2010) 2287–2293.
[22] C. Eckert, W. Xu, W. Xiong, S. Lynch, J. Ungerer, L. Tao, R. Gill, P.-C. Maness, J. Yu, Ethylene-forming enzyme and bioethylene production, Biotechnology for biofuels 7 (2014) 1–11.

[23] J. Yu, W. Xiong, T. Dong, N. Sweeney, P.T. Pienkos, Manipulation of cyanobacterial biomass composition for enhanced htl biofuel yield and quality, in: 2017 SIMB Annual Meeting and Exhibition, SIMB, 2017.

[24] T. Dong, B. Wang, W. Xiong, N. Sweeney, P.T. Pienkos, J. Yu, System-level optimization to improve biofuel potential via genetic engineering and hydrothermal liquefaction, ACS Sustainable Chemistry & Engineering 8 (7) (2020) 2753–2762.

[25] CNNWorld, Plant takes 'selfies,' showing possibility of plant-powered devices, https://edition.cnn.com/2019/10/16/europe/plant-selfie-london-scli-intl-gbr-scn/index.html.

[26] Y. Ting, Viewing China's Economic Growth through Electricity Consumption, People's Daily (Aug. 23, 2024) 006, https://doi.org/10.28655/n.cnki.nrmrb.2024.009118.

[27] H. Dou, Q. Chen, Y. Yang, et al., Summary of Energy-Saving Operation Optimization Practices in a Data Center in Shenzhen, HVAC 53 (1) (2023) 112–116, https://doi.org/10.19991/j.hvac1971.2023.01.16.

[28] Y. Zhou, W. Hao, Research on the Application of Energy Storage Systems in Data Centers, Southern Energy Construction 8 (3) (2021) 58–62, https://doi.org/10.16516/j.gedi.issn2095-8676.2021.03.008.

[29] S. Li, W. Li, L. Chen, H. Jiang, J. Zhang, D.W. Gao, Real-time robust state estimation for large-scale low-observability power-transportation system based on meta physics-informed graph timesnet, IEEE Transactions on Smart Grid 15 (6) (2024) 5500–5513.

[30] Microsoft, Emissions Impact Dashboard, Available: https://www.microsoft.com/en-us/sustainability/emissions-impact-dashboard.

[31] T. Garrison, Oceanography: An Invitation to Marine Science, Cengage Learning, 2016.

[32] D.O.B. Jones, et al., Biological responses to disturbance from simulated deep-sea polymetallic nodule mining, PLOS ONE 12 (2) (2017) e0171750.

[33] J.X. Han, Marine renewable energy in China's offshore areas [Marine Renewable Energy], China Ocean Press, ISBN 978-7-5027-8375-4, 2014.

[34] International Renewable Energy Agency, Offshore renewable energy boosts blue economic development, Marine Development and Management 35 (4) (2021) 10–15.

[35] H. Kim, A.L. Yarin, M.W. Lee, Self-healing corrosion protection film for marine environment, Composites Part B: Engineering 182 (2020) 107598.

[36] J.D. Smith, Review of Enhancement for Ocean Thermal Energy Conversion System, Journal of Ocean Energy 15 (3) (2020) 123–145.

[37] M. Li, H. Zhuo, Q. Jing, Y. Gu, Z. Liao, K. Wang, J. Hu, D. Geng, X. Sun, B. Xiao, Low-temperature performance of Na-ion batteries, Journal of Power Sources 489 (2021) 229356.

[38] S.M.A. Sunny, Solar plants: Emissions-free research station in Antarctica [Web page], Available: https://www.sma-sunny.com/zh/solar-plants-emissions-free-research-station-in-antarctica/.

Chapter 10

Parallel energy storage: new systems with new AI beyond LLMs

Jing Yang[a], Fei Lin[b], and Fei-Yue Wang[a,c]
[a]*The State Key Laboratory for Management and Control of Complex Systems, Institute of Automation, Chinese Academy of Sciences, Beijing, China,* [b]*The Department of Engineering Science, Faculty of Innovation Engineering, Macau University of Science and Technology, Taipa, Macau, China,* [c]*Macau Institute of Intelligent Science and Interdisciplinary Engineering, Faculty of Innovative Engineering, Macau University of Science and Technology, Taipa, Macau, China*

Contents

10.1 Motivation and background

The global transition to renewable energy stands as an urgent and transformative endeavor [1,2], essential to confronting the existential threats posed by climate change, the relentless depletion of finite resources, and the escalating toll of environmental degradation. Wind and solar energy emerge as twin pillars of this transformative vision due to their unparalleled abundance, environmental benevolence, and rapid technological advancements. However, their inherent variability, dependent on the whims of weather and daylight, can disrupt grid stability, create imbalances between supply and demand, and undermine the reliability of our energy systems. Energy storage systems emerge as the elegant solution, capturing surplus energy during peak generation and releasing it seamlessly when the sun sets or the winds still, ensuring a harmonious flow of power and transforming intermittency into dependability for a sustainable future.

With the rapid advancement of information and engineering technologies, the traditionally distinct facets of energy systems, such as carbon emissions, energy security, multisource energy pricing, and demand-side load regulation, are becoming increasingly interconnected and interdependent. This growing integration fosters a complex nexus of interactions among physical energy infrastructures, societal information systems, and human behavioral patterns, creating a dynamic and highly coupled ecosystem. Therefore the socio-technical nature of energy systems requires guidance from parallel intelligence theory [3] and the infrastructure of Cyber-Physical-Social Systems (CPSSs) [4,5] to design and build parallel energy storage systems that effectively reconcile technological capabilities, societal demands, and human-centric dynamics.

Recently, Large Language Models (LLMs) like GPT series [6] and DeepSeek [7] have breathed new life into parallel energy storage systems. Their exceptional language comprehension and generation capabilities enable a more accurate understanding of human needs, facilitating natural and seamless communication with users, thereby lowering the professional and knowledge barriers for human participation in energy management. Their reasoning and logical analysis abilities make it possible to create digital humans virtual assistants capable of simulating human decision-making processes, optimizing energy distribution and storage strategies in real time, and even providing multidimensional solutions in complex scenarios. Furthermore, through deep learning and big data analysis [8], these models can predict energy demand fluctuations, weather changes, and market dynamics, offering precise decision-making support for parallel energy storage systems. This integration of technologies not only enhances the intelligence level of energy systems but also democratizes energy management, allowing ordinary users to easily participate in energy optimization and sustainable development. Therefore this chapter proposes an LLM-based parallel energy storage framework, explores its key technologies, and envisions a future where energy storage systems evolve in the 6S goal: Safe in the physical world, Secure in the cyber world, Sustainable in the ecological world, Sensitive to individual needs, Services for all and Smart in all.

10.2 Development history

Energy exploration and storage technologies have continuously evolved and advanced in tandem with shifting human needs, as depicted in Fig. 10.1. From the humble beginnings of the firewood era to the widespread adoption of fossil fuels like coal and oil, and now to the dawn of renewable energy with the rise of wind and solar power, each transformative leap in energy technology has profoundly reshaped the fabric of human society, revolutionizing how we produce, consume, and live.

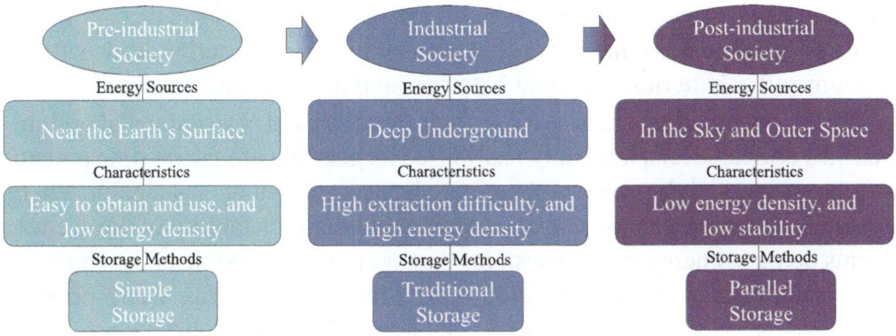

FIGURE 10.1 Development and evolution of energy storage.

Specifically, in the preindustrial era, constrained by limited productivity, humanity's demand for energy was relatively modest. The energy required is primarily sourced from surface-level resources, such as plants and shallow fossil fuels, which are stable and readily available. During this period, energy storage methods are simple and small-scale, perfectly suited to the uncomplicated needs of the period. Following the Industrial Revolution, electricity emerged as a quintessential secondary energy source, celebrated for its cleanliness, convenience, efficiency, and versatile conversion capabilities, leading to its widespread adoption. The proliferation of large-scale machinery and electrical appliances in both industrial production and daily life has significantly heightened the electricity demand and underscored the need for its reliable and stable supply. A stark illustration of the consequences of power instability occurred in 2003 [9], when a massive blackout swept through parts of the northeastern United States and eastern Canada. The outage left 10 million Canadians and 45 million residents across eight U.S. states without power, with some areas enduring outages for up to four days. Critical infrastructure, including transportation networks, urban rail systems, communication networks, healthcare facilities, urban utilities (water supply, drainage, and residential power distribution), and various industrial operations (chemical refineries, mining facilities, etc.), was severely disrupted. The economic toll exceeded tens of billions of dollars, and tragically, over 100 lives were lost. This event highlighted the critical importance of power storage in mitigating the impact of such emergencies. Unlike other resources, electricity cannot be stored directly; it must first be converted into intermediate and storable forms of energy, such as potential, kinetic, chemical, or thermal energy, before being restored for future use [10]. Representative methods include pumped storage [11], compressed air energy storage [12], flywheel energy storage [13], and superconducting magnetic energy storage [14]. However, these approaches are often constrained by geographical limitations, energy losses, response times, and operational flexibility, preventing their widespread and convenient use [15]. In 1800 Italian scientist Alessandro Volta invented the first true battery [16], the voltaic pile marking the beginning of humanity's century-long journey toward flexible, efficient, and convenient battery-based energy storage. Since then, a diverse array of batteries has emerged [17], such as lithium-ion batteries, flow batteries, sodium sulfur batteries, lead acid batteries, and magnesium batteries.

In the postindustrial age, as fossil fuels diminish and environmental pollution reaches critical levels, humanity has shifted its gaze toward the pristine potential of the skies and beyond, embracing cleaner and more sustainable energy sources. However, the low density and inherent unpredictability of these energy forms demand energy storage systems that are not only flexible and adaptive but also economically viable. To address this, Professor Fei-Yue Wang proposed the concept of parallel batteries [18] based on parallel intelligence [3,19]. By constructing several virtual, distributed artificial parallel battery systems, this approach models, reasons, analyzes, and optimizes the control of real-world parallel battery systems. It enables efficient interaction, mutual guidance, and iterative updates between the physical battery systems and their virtual counterparts. With further breakthroughs in intelligent technologies, this chapter delves into an expanded and upgraded version of parallel energy storage systems, exploring their various dimensions and potential in the new era, including their basic framework, operational process, key enabling technologies, and future trends.

10.3 Parallel energy storage systems

10.3.1 Basic framework

The basic framework of parallel energy storage systems, illustrated in Fig. 10.2, consists of three interconnected layers: a storage layer, a computing layer, and an execution layer. The storage layer serves as the data-driven foundation for energy capture and release; the computing layer harnesses LLMs and related technologies for intelligent optimization; and the execution layer translates computational insights into real-world operations. Together, these layers form a cohesive and adaptive system that enhances efficiency, reliability, and sustainability. There are three types of humans: biological humans, digital humans, and robotic humans. Biological humans can seamlessly convey their intentions and insights to digital humans through natural dialogue, while digital humans and robotic humans engage in dynamic information exchange and communication, creating a harmonious interplay between the physical and digital realms. A Human-Oriented Operating System (HOOS) serves as the center of the parallel energy storage systems, responsible for coordinating these three layers and three types of humans efficiently and orderly through an interrupt mechanism to accomplish tasks.

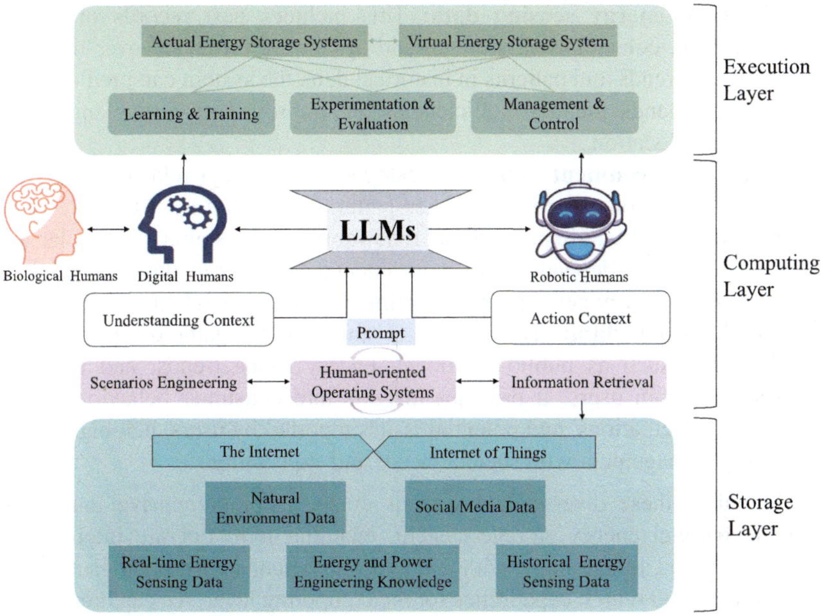

FIGURE 10.2 Basic framework of parallel energy storage systems.

10.3.1.1 Storage layer

The storage layer, as the bottom tier, utilizes the Internet of Things (IoT) and the Internet to store data encompassing the entire lifecycle of energy, from production to recycling. This layer not only captures real-time data on energy production, transmission, storage, and usage but also includes multidimensional information such as equipment status, environmental parameters, and user behavior. These data can be categorized into real-time energy sensing data, energy and power engineering knowledge, historical energy sensing data, natural environment data, and social media data. Their descriptions and functions are as follows.

- **Real-time energy sensing data:** This refers to instantaneous data collected from sensors and monitoring devices across the energy system, such as current energy production levels, grid load, battery charge/discharge rates, and equipment performance metrics. It provides a live snapshot of the system's operational status, enabling immediate adjustments and responses.
- **Energy and power engineering knowledge:** This encompasses technical expertise, theoretical models, and operational guidelines related to energy systems. It includes information on energy conversion efficiency, power grid stability, storage technologies, and best practices for system design and maintenance. This knowledge forms the backbone of intelligent decision-making and optimization.

- **Historical energy sensing data:** This includes past records of energy production, consumption, storage, and system performance over time. By analyzing trends and patterns in historical data, the system can predict future energy demands, identify inefficiencies, and improve long-term planning and resource allocation.
- **Natural environment data:** This category covers data related to environmental conditions, such as weather patterns, temperature, humidity, solar irradiance, and wind speed. Since renewable energy sources like solar and wind are highly dependent on environmental factors, this data is crucial for forecasting energy generation and optimizing storage strategies.
- **Social media data:** This refers to information gathered from social platforms, such as public sentiment, energy usage trends, and real-time feedback from users. It provides insights into consumer behavior, energy demand fluctuations, and potential issues reported by users, helping to align energy management with societal needs and expectations.

Together, these diverse data streams create a comprehensive foundation for the parallel energy storage system, enabling it to operate intelligently, adaptively, and in harmony with both technological and human factors. To streamline rapid search capabilities for the computing layer, we can break down large-scale text data into segments or chunks and encode them using advanced encoders [20] to build efficient indexes. This method not only boosts the speed and precision of data retrieval but also optimizes computational resource usage. By transforming intricate data structures into manageable, bite-sized pieces, the system can swiftly pinpoint and extract vital information, significantly speeding up analysis and enabling faster, more informed decision-making.

10.3.1.2 Computing layer

The computing layer, serving as the intermediary tier, merges the retrieved understanding contexts or action contexts from the storage layer with task-specific prompts, prompting LLMs to engage in logical reasoning and computational analysis, and generates a range of potential solutions, whose credibility is then rigorously validated through scenarios engineering. For instance, in energy dispatch, the computing layer can integrate real-time energy data, historical usage patterns, and weather forecasts to devise optimal energy distribution strategies. These strategies are subsequently tested through simulations to evaluate their effectiveness in real-world applications. The retrieved contexts adapt to different activation targets. Understanding contexts primarily emphasizes information processing, enhancing the reasoning capabilities of digital humans. In contrast, action contexts, centered on the manipulation of physical tools and action generation, are designed to enhance the operational capabilities of robotic humans. Both the "soul" of digital humans and the "brain" of robotic humans are powered by LLMs as their cognitive core, but they also require the assistance of other modal models, such as:

- **Image-to-Text Models (e.g., CLIP, Contrastive Language Image Pre-training [21]):** These models can transform satellite images of solar panels into textual descriptions, analyzing surface contamination or damage to inform maintenance decisions.
- **Text-to-Video Models (e.g., Sora [22]):** They can convert textual descriptions of energy dispatch plans into dynamic, visual videos, vividly illustrating the flow and distribution of energy, making it easier for decision-makers to grasp complex scenarios.
- **Audio-to-Text Models (e.g., Whisper [23]):** These models can translate the sounds of wind turbines into textual data, identifying abnormal noises and predicting potential failures before they occur.
- **Multimodal Fusion Models (e.g., Flamingo [24]):** Capable of integrating weather data (text), satellite cloud images (visual), and user feedback (audio), these models generate comprehensive energy prediction reports, optimizing the operational strategies of energy storage systems.

These models are guided and activated by various prompting techniques to generate high-quality outputs, such as in-context learning [25], Chain-of-Thoughts (CoT) [26], ReAct [27], and Retrieval Augmented Generation (RAG) [28]. Through collaborative efforts, these models significantly enhance the intelligence of the computational layer, enabling it to handle more complex and diverse tasks. This synergy ultimately provides more accurate and efficient solutions for parallel energy storage systems.

10.3.1.3 Execution layer

The execution layer, as the topmost tier, leverages the ACP (Artificial systems, Computational experiments, and Parallel execution) approach [19] of parallel intelligence to efficiently mobilize digital and robotic humans to execute various tasks, providing users with diversified energy storage services. A prime example of such services is Electric Vehicle (EV) charging management, where digital humans optimize charging schedules based on real-time energy demand, while robotic humans physically connect and disconnect charging cables. Specifically, task execution begins with the construction of multiple artificial counterparts that correspond to physical systems, representing realities with divergent evolutionary paths. Subsequently, computational experiments are conducted by digital humans within these artificial systems to analyze the causes of observed phenomena, potential consequences, and corresponding solutions. Finally, robotic humans implement these solutions in actual physical systems while continuously optimizing them through virtual-real interactions, ensuring adaptability to complex and dynamic operational conditions. Clearly, there are three operational modes here: learning and training, experimentation and evaluation, and management and control.

- **Learning and Training:** In artificial systems, real-world scenarios are simulated, where digital humans and robots continuously engage in trial and error to achieve knowledge accumulation and skill training. Specifically, in the simulated environments, digital humans refine operational strategies, such as charge-discharge scheduling and emergency response protocols, using reinforcement learning or transfer learning, thereby creating standardized operational guidelines; while robotic humans perform physical operation drills, including tasks like plugging and unplugging charging cables or replacing battery modules, to verify the safety and efficiency of these actions.

- **Experimentation and Evaluation:** Digital humans design computational experiments (e.g., testing new battery performance, evaluating the economic impact of charging/discharging strategies) within artificial systems, using large-scale simulations to predict the long-term effects of various solutions. Robotic humans, on the other hand, conduct small-scale physical experiments (e.g., battery cycle testing) in controlled environments, collecting real-world data and comparing it with simulation results to calibrate model accuracy, ensuring the scientific validity and feasibility of the strategies.
- **Management and Control:** The artificial systems and virtual systems are connected in real time, forming a highly coordinated intelligent network. Digital humans dynamically generate optimal scheduling instructions (e.g., adjusting the output priority of energy storage systems) based on real-time data (e.g., grid load, equipment status) and achieve closed-loop control through digital twin technology. Robotic humans, on the other hand, are responsible for executing physical operations (e.g., switching operating modes, repairing faults) while equipped with sensors to monitor environmental conditions (e.g., battery temperature, connector status), ensuring safe and efficient system operation and rapid response to emergencies. In other words, digital humans continuously optimize scheduling strategies by learning from historical data and real-time feedback, while robotic humans provide more accurate inputs through high-precision operations and real-time data collection, creating a closed loop of "perception-decision-execution-optimization".

In summary, the execution layer, through these three operational modes, enables virtual-real interaction between artificial systems and physical systems, as well as between digital humans and robotic humans. This creates a robust, flexible, and adaptive execution process that significantly enhances operational efficiency and system resilience.

10.3.2 Operational process

The deployment of energy storage systems in distributed networks enables peak shaving, load shifting, and frequency regulation, achieving supply-demand balance. The operational process of parallel energy storage systems is represented as four stages spanning from design to recycling, as illustrated in Fig. 10.3. It can make full use of the computational rationality of digital humans and the adaptive rationality of robotic humans by amplifying small data into big data and extracting it to deepen intelligence.

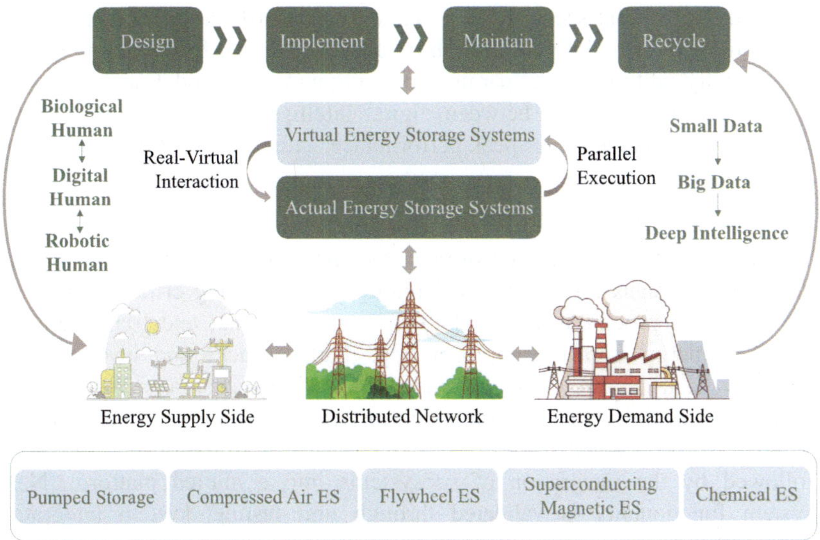

FIGURE 10.3 Operational process of parallel energy storage systems.

10.3.2.1 Design stage

The design of energy storage systems involves strategically planning, selecting, and configuring storage technologies and their components to achieve efficient, reliable, and secure energy storage and release, tailored to specific application scenarios and requirements. This process encompasses: technology selection (e.g., electrochemical, mechanical, or thermal storage), system architecture design (modular layouts, multienergy integration), hardware engineering (storage units, power electronics, thermal management systems), and software development (control algorithms, real-time monitoring platforms). Through a systematic workflow, demand analysis, technology evaluation, system modeling, detailed design, safety validation, and cost optimization, supported by advanced simulation tools and multidisciplinary expertise, the system is crafted to meet rigorous performance, economic, safety, and sustainability standards, laying the groundwork for seamless implementation, maintenance, and end-of-life recycling.

In parallel energy storage design, LLM-based digital humans harness industry insights and historical data (e.g., grid load patterns, user consumption behaviors) to define requirements. Subsequently, they select the appropriate energy storage technology according to these requirements and specific application scenarios, design the system architecture, hardware, and software of multienergy hybrid storage systems and architect their virtual models in artificial systems. Additionally, these digital humans refine design strategies using reinforcement learning and predictive analytics. Robotic humans rigorously test physical deployment scenarios (e.g., battery-array installation, heat-exchanger integration) in artificial systems. Through sensor-driven feedback loops, they dynamically calibrate design parameters, bridging virtual models and real-world constraints. This synergy between digital intelligence and robotic precision ensures the system achieves peak performance, safety, and cost-efficiency.

10.3.2.2 Implementation stage

The implementation of energy storage systems is the process of transforming design-phase plans into operational reality, encompassing critical steps such as equipment procurement, installation, commissioning, and trial operation. Initially, energy storage equipment (e.g., battery packs and inverters) and auxiliary facilities are procured and inspected in accordance with design specifications. Subsequently, equipment installation, wiring, and safety system deployment are carried out strictly in accordance with the design blueprints, followed by the integration of subsystems into a unified platform. Next, system functionality is validated through unit testing, system integration testing, and performance testing to ensure efficient and safe operation. Finally, during the trial operation phase, system performance is monitored, parameters are optimized, and user training is completed to facilitate formal system deployment.

In the implementation of parallel energy storage systems, digital humans generate dynamic deployment plans (e.g., equipment installation sequences and power electronics interface configurations) through artificial systems, guiding robotic humans to perform high-precision tasks on-site (e.g., cable laying and energy storage unit assembly) while using sensors to calibrate equipment positions in real time. The actual system and the artificial system remain synchronized via the IoT, transmitting commissioning, testing, and trial operation data to digital humans in real time. Leveraging LLMs, digital humans analyze this data along with construction logs, continuously refining deployment strategies to ensure the implementation process is both efficient and precise.

10.3.2.3 Maintenance stage

The maintenance of energy storage systems involves a series of activities, including regular inspections, upkeep, performance testing, and fault handling, to ensure the system operates efficiently and safely over time. Maintenance tasks encompass routine inspections (e.g., equipment status checks and environmental monitoring), periodic upkeep (e.g., cleaning equipment and replacing wear-prone components), performance testing (e.g., capacity testing, efficiency testing), and fault management (e.g., diagnosis, repair, and postincident analysis). By integrating preventive, predictive, and corrective maintenance approaches and utilizing monitoring equipment, data analysis tools, and management software, energy storage systems can extend equipment lifespan, optimize performance, reduce operational costs, and deliver reliable energy storage services to users.

In the maintenance of parallel energy storage systems, LLM-powered digital humans rely on artificial systems to monitor equipment status (e.g., battery health and hydrogen tank pressure), predict potential failures, and generate maintenance work orders. Meanwhile, robotic humans execute preventive tasks (e.g., cleaning solar panels and replacing electrolytes) and recover faulty components. For instance, when a digital human detects abnormal capacity degradation in a battery module, it dispatches robotic humans for on-site inspection, updates the maintenance strategy based on the results, and stores the results in the database, creating a closed loop of "perception-decision-execution."

10.3.2.4 Recycling stage

The recycling of energy storage systems refers to the process of disassembling, sorting, processing, and reusing components such as battery packs and power electronics after their lifecycle ends, aiming to maximize resource utilization, reduce environmental pollution, and promote a circular economy. The recycling process includes system decommissioning, equipment disassembly, material recovery (e.g., extracting rare metals like lithium, cobalt, and nickel through physical and chemical methods), harmless treatment (e.g., professional handling of electrolytes and heavy metals), and resource reuse (e.g., using recycled materials for new product manufacturing). Through methods such as manual disassembly, automated recycling, and collaborative recycling, combined with disassembly tools, separation technologies, and monitoring equipment, energy storage system recycling enables resource circulation, reduces environmental impact, complies with environmental regulations, and supports sustainable development.

When a parallel energy storage system is decommissioned, digital humans leverage LLMs to analyze global recycling regulations and material price trends, planning optimal disassembly routes (e.g., prioritizing the extraction of rare metals like lithium and cobalt). They simulate the economic and safety aspects of the recycling process through artificial systems to formulate the most environmentally friendly recycling solutions. Robotic humans are responsible for high-risk operations (e.g., battery crushing and hazardous material separation), using technologies like Computer Vision (CV) to sort reusable components. Additionally, they feed data from the decommissioned systems (e.g., equipment lifespan logs) back into the artificial system to inform the design of next-generation products. For example, after robotic humans disassemble used battery modules, LLMs automatically match buyers for recycled materials, while digital humans optimize the closed-loop supply chain to maximize resource recycling efficiency.

10.4 Key technologies

10.4.1 Large language models

Traditional Artificial Intelligence (AI) technologies, particularly Deep Neural Networks (DNNs), have been widely applied across various domains of energy systems, including modeling and simulation, system operation and maintenance, and operational control, among others [29,30]. With the rise of LLMs, exemplified by the GPT series [6], intelligent systems are undergoing profound transformations. LLMs drive the enhancement of intelligence levels through deep integration with complex systems [31]. In parallel energy storage systems, LLMs serve as data-driven core hubs, supporting efficient collaboration among digital humans, robotic humans, and biological humans within the system, thereby constructing a tightly integrated intelligent ecosystem.

Specifically, each digital human can be configured with different types of scalable LLM agents according to the diverse needs of the energy storage system. For example, tasks such as load forecasting, energy optimization, and equipment fault diagnosis can be efficiently performed by LLM agents. For tasks in specific vertical domains, general-purpose LLMs can be custom-optimized through instruction fine-tuning or prompt fine-tuning [32,26]. For instance, multiturn dialogues can be used to construct instruction-following datasets, and combined with efficient fine-tuning techniques like Low-Rank Adaptation (LoRA) [33], instruction fine-tuning can enable the model to more accurately interpret and execute complex operational commands. In prompt fine-tuning, using domain-specific pretraining data to finely optimize prompts can significantly enhance the model's performance in specific application scenarios. At the same time, the system designs specialized prompts or CoT templates based on the actual tasks to simplify operational processes, and enhances the agent's task understanding and execution accuracy through few-shot learning [34].

In robotic systems, the application of LLMs is key to achieving embodied intelligence. Each robot, as an embodied agent, is capable of flexibly responding to diverse tasks such as equipment monitoring, charging and discharging control, and system status evaluation. Currently, extensive research is dedicated to utilizing LLMs to enable the intelligence of robots, including robotic arms, drones, and others [35,36]. To ensure that robots operate efficiently in complex environments, the integrated LLMs must possess powerful multimodal perception and 3D spatial reasoning capabilities in certain task scenarios. In response to the demand for 3D spatial exploration, relevant research has laid a solid foundation [37]. For instance, the 3D-LLM model proposed by Hong et al. [38] has demonstrated excellent performance in handling complex 3D spatial tasks such as 3D scene Visual Question Answering (VQA), task decomposition, and navigation. Based on Visual Language Models (VLMs), robots can perform Visual Language Navigation (VLN) and Visual Language Action (VLA) execution, enabling them to effectively adapt to task changes in dynamic environments and ensure efficient execution. To improve task execution accuracy, the robotic system customizes pretraining or fine-tuning based on task characteristics, and adjusts the decision-making process by incorporating action context prompts to ensure the accuracy and rationality of the execution [39].

For biological humans (system operators), LLMs play a key role as decision-support tools in parallel energy storage systems. For example, by leveraging real-time data fusion and multiparameter forecasting, LLMs can provide accurate decision support for complex tasks such as load regulation, equipment maintenance, and risk warning, thereby significantly improving system operational efficiency. At the same time, biological humans are integrated into the system in a "Human-in-the-Loop" manner, responsible for ethical oversight and decision support. This includes, but is not limited to, reviewing the decisions made by digital human agents to ensure that the judgments of the LLMs align with ethical, safety, and compliance requirements.

10.4.2 Human-oriented operating systems

In modern energy systems, scheduling and control strategies are the core to ensuring efficient system operation. While traditional control strategies are effective in simple applications, their rigidity limits adaptability in dynamic, multidimensional systems. With technological advancements, intelligent control strategies such as Artificial Neural Networks (ANNs) and Model Predictive Controllers (MPCs) have been gradually introduced to address changes in complex system environments [40]. However, while these methods improve the system's adaptability, they also face challenges such as high data demands, difficulty in tuning parameters, and heavy real-time computational burdens. In this context, HOOS, as a key component of parallel energy storage systems, demonstrates great advantages and potential [6].

HOOS integrates foundational AI models and LLMs to build a human-machine interaction platform that supports intelligent decision-making and automated operations in energy storage systems [41–43]. Through the deep integration of LLMs, HOOS can alleviate the challenges faced by traditional intelligent control technologies. For instance, under high data demand conditions, LLMs significantly reduce the reliance on large training datasets through zero-shot learning; during the parameter tuning process, LLMs can automatically adjust control strategies through optimized prompts and the CoT mechanism, improving system debugging efficiency; in terms of real-time computation, the modular architecture of HOOS allows dynamic adjustment of functions and resource allocation, thus enhancing computational efficiency and system response speed when handling complex decision-making problems. Based on these advantages, HOOS can achieve more efficient scheduling and optimization, ensuring optimal energy utilization and system sustainability.

In addition, as a configurable and scalable modular system, HOOS can dynamically adjust its functions and interfaces according to the needs of the energy storage system. These modular designs enable HOOS to flexibly adapt to different user demands, providing personalized control and optimization solutions. By introducing standardized data interaction protocols such as the Model Context Protocol (MCP) [44], LLMs can interact efficiently with external data sources and tools in a standardized manner. This allows HOOS to obtain real-time multisource data, such as climate change, energy consumption, and load forecasting, and optimize the scheduling and management of energy storage systems based on this data.

HOOS bridges the gap between actual energy storage systems and virtual energy storage systems [45]. Modular encapsulation simplifies the operations of virtual scene construction, computational experiments, and parallel execution, enabling the parallel energy storage system to achieve sustainable development across various complex scenarios.

10.4.3 Scenarios engineering

Scenarios engineering is crucial in the construction of virtual energy storage systems. Its core goal is to flexibly design virtual scenarios based on the needs of the energy storage system, and to conduct computational experiments within these scenarios to formulate strategies and plans for optimizing the actual system. The construction of virtual energy storage systems does not require a one-to-one correspondence with the actual system, but rather involves constructing virtual models based on the characteristics of the specific application scenarios [46,47]. For example, in an energy dispatch scenario, it is not necessary to rebuild every detail of each energy storage unit; instead, by modeling its key features and behaviors, the system can simulate the energy storage system's response under different loads, thereby optimizing the overall dispatch strategy. The construction of virtual scenarios leverages technologies such as generative AI, digital twins, and software-defined approaches. Through the configurable and scalable capabilities provided by the HOOS architecture, the system is able to adapt to dynamic changes and the demands of complex tasks.

The first step in constructing virtual energy storage scenarios is to model based on the physical characteristics and data features of the energy storage system. This process relies on real data from energy storage devices (such as batteries, supercapacitors, etc.) and the system's operational environment, combined with multidimensional simulation models to simulate the dynamic behavior and performance of the energy storage system. To ensure the accuracy of the model and the reliability of the simulation results, commonly used simulation tools include MATLAB®/Simulink®, Modelica [48], and PSS/E, which provide powerful modeling and simulation support to help analyze complex physical processes and control strategies within the energy storage system [49–51].

In addition, robot simulation within the virtual energy storage system is also a key component. By utilizing advanced robot simulation platforms (such as NVIDIA Isaac) or generative AI physics engines (such as Genesis), robotic operations and task executions within the energy storage system can be simulated, including path planning, task allocation, and equipment handling. The robot simulation platform not only supports the efficient execution of robots in virtual environments but also optimizes the handling processes and operational strategies of devices within the energy storage system, enhancing the overall automation efficiency of the system [52]. At the same time, sensors and IoT technologies play an important role by collecting real-time operational data from the energy storage system and feeding it back into the virtual environment, enabling the synchronized evolution of both virtual and actual systems.

10.5 Future trends

10.5.1 Shared energy storage

With diversification in energy consumption patterns and the continued growth of distributed energy, traditional centralized energy storage models are increasingly revealing issues such as high costs and uneven resource allocation. Shared energy storage systems, as an important direction for the future development of energy systems, demonstrate immense potential in improving energy utilization and reducing costs [53,54]. By adopting a shared economy model, shared energy storage allows multiple users to share the same energy storage device or system, effectively distributing the cost burden among users and optimizing the usage efficiency of energy storage devices.

Cloud energy storage, as the core technology of this model, leverages cloud computing to virtualize and centrally manage distributed energy storage resources, enabling users to dynamically adjust energy storage resources through the cloud platform based on demand, achieving optimal configuration and efficient utilization of energy [55]. In this process, the application of IoT technology allows each energy storage unit to monitor its operational status in real-time, including the charge/discharge state of the batteries, equipment health, and more, providing accurate data support for the cloud platform. By combining with edge storage devices, cloud energy storage enables cloud-edge collaboration, further enhancing the flexibility of the system [56].

With the development of Decentralized Autonomous Organizations (DAOs), shared energy storage systems are expected to realize a more democratic and efficient governance model [57]. With the help of DAO's smart contracts and voting mechanisms, energy community members can participate directly in the allocation and decision-making of energy resources without relying on traditional power companies. This decentralized management model ensures the transparency and security of transactions through blockchain technology, while reducing costs through automated energy trading and scheduling, enabling more flexible resource allocation. Furthermore, by integrating LLM technology, energy storage strategies based on smart contracts will achieve a higher degree of automation, driving the development of shared energy storage systems towards a smarter, fairer, and more efficient future [58].

10.5.2 Multienergy Internet and energy router

With the large-scale application of renewable energy sources such as solar and wind power, the volatility and instability of energy are increasingly becoming the main bottleneck restricting the development of energy systems. The MultiEnergy Internet (MEI), by interconnecting various energy forms such as electricity, heat, and natural gas, can effectively coordinate the conversion between different energy sources [59]. For example, in power systems, surplus electricity can be converted into thermal energy or stored and then converted back into electricity to supply the market during peak demand periods, thereby reducing energy waste and improving energy utilization.

As a core component of MEI, energy routers can intelligently schedule and allocate energy according to real-time demand. With advanced intelligent algorithms, energy routers can coordinate the conversion between different energy forms, balance supply and demand, support the large-scale integration of renewable energy, and effectively address issues caused by energy fluctuations. This technology will drive the energy system toward higher levels of intelligence and adaptability [60,61].

The efficient utilization of parallel energy storage systems is expected to provide significant support for MEI and energy routers. Parallel energy storage, through dynamic interactions between virtual and actual energy storage systems, can optimize the conversion and configuration of energy across different forms in real-time. Through efficient data collection and analysis, parallel energy storage systems can predict changes in the supply and demand of various energy types, further optimizing the scheduling strategies of energy routers to ensure the stable supply of all types of energy.

10.5.3 Blockchain as a service

The introduction of blockchain technology brings a revolutionary transformation to future energy storage systems [62]. First, blockchain supports Peer-to-Peer (P2P) energy trading, enabling users of shared energy storage systems to directly exchange energy without relying on traditional power companies, thus improving efficiency and reducing costs while enhancing the fairness and transparency of the energy market. Second, the immutability and distributed ledger features of blockchain ensure the security and credibility of energy storage system operational data and transaction records, preventing data tampering and fraudulent activities. The introduction of smart contracts further automates tasks such as energy trading, equipment maintenance, and fault handling, reducing the need for human intervention [63].

Blockchain as a Service (BaaS) [64] provides an innovative architecture for energy storage systems, aimed at optimizing the management and operation of Battery Energy Storage Systems (BESSs). Through blockchain technology, BaaS platforms can enable data sharing and validation, supporting online battery health monitoring, supply chain management, and energy trading services for energy storage systems. BaaS enhances the security and operational efficiency of BESS, effectively addressing potential vulnerabilities in the networked physical environment.

Moreover, blockchain intelligence [65,66], as an integrated technology system, combines blockchain, smart contracts, AI, and DAO, and is expected to provide intelligent governance and decision-making support for energy storage systems. This technology system, which can enhance the intelligence level of various aspects of energy systems, on the basis of BaaS, such as energy scheduling, equipment management, and supply chain optimization, thus promoting the development of energy systems toward higher efficiency, transparency, and sustainability [67].

10.6 Conclusion

As the global energy landscape accelerates its transition toward renewable sources, energy storage systems have become increasingly pivotal in ensuring grid stability, balancing supply and demand, and advancing sustainable development. This chapter has comprehensively explored the LLM-based parallel energy storage framework, delineating its foundational architecture, operational processes, key enabling technologies, and future trajectories. By orchestrating the synergy among the storage layer, computing layer, and execution layer, this framework integrates the intelligent capabilities of biological humans, digital humans, and robotic humans, establishing an efficient, reliable, and adaptive energy ecosystem. The incorporation of parallel intelligence theory and CPSS provides both theoretical guidance and technical infrastructure to address the intricate interplay of technological, societal, and human dynamics within energy systems.

Technologically, LLMs significantly elevate the intelligence of parallel energy storage systems. Their prowess in language comprehension, logical reasoning, and multimodal data processing optimizes energy distribution and storage strategies while democratizing energy management by lowering the knowledge barriers for public participation. The HOOS, with its modular design and real-time data interaction, enhances system flexibility and responsiveness. Meanwhile, scenarios engineering, through virtual modeling and computational experimentation, offers a scientific basis for optimizing real-world systems. Looking ahead, emerging trends such as shared energy storage, the multienergy internet, and blockchain-as-a-service promise to further drive the decentralization, intelligence, and sustainability of energy systems, paving the way for equitable resource allocation and efficient utilization.

In conclusion, the LLM-based parallel energy storage system exemplifies a cutting-edge convergence of energy technology and AI. It not only provides innovative solutions to the intermittency challenges of renewable energy but also lays a robust foundation for future energy management. Through sustained technological innovation and interdisciplinary collaboration, this framework is poised to realize the 6S vision, thereby propelling the global energy paradigm toward a greener, smarter, and more inclusive future.

References

[1] Bogachan Gungor, Ibrahim Dincer, Development of a sustainable community with an integrated renewable and waste to energy system for multiple useful outputs, Journal of Cleaner Production 312 (2021) 127704.

[2] Ahmad Tanveer, Dongdong Zhang, A critical review of comparative global historical energy consumption and future demand: the story told so far, Energy Reports 6 (2020) 1973–1991.

[3] Xingxia Wang, Jing Yang, Yuhang Liu, Yutong Wang, Fei-Yue Wang, Mengzhen Kang, Yonglin Tian, Imre Rudas, Lingxi Li, Maria Pia Fanti, et al., Parallel intelligence in three decades: a historical review and future perspective on acp and cyber-physical-social systems, Artificial Intelligence Review 57 (9) (2024) 255.

[4] Jing Yang, Xiao Wang, Yonglin Tian, Fei-Yue Wang, Parallel intelligence in cpsss: being, becoming, and believing, IEEE Intelligent Systems 38 (6) (2023) 75–80.

[5] Xiao Wang, Yutong Wang, Jing Yang, Xiaofeng Jia, Lijun Li, Weiping Ding, Fei-Yue Wang, The survey on multi-source data fusion in cyber-physical-social systems: foundational infrastructure for industrial metaverses and industries 5.0, Information Fusion (2024) 102321.

[6] Yutong Wang, Xiao Wang, Xingxia Wang, Jing Yang, Oliver Kwan, Lingxi Li, Fei-Yue Wang, The chatgpt after: building knowledge factories for knowledge workers with knowledge automation, IEEE/CAA Journal of Automatica Sinica 10 (11) (2023) 2041–2044.

[7] Qihao Zhu, Daya Guo, Zhihong Shao, Dejian Yang, Peiyi Wang, Runxin Xu, Y. Wu, Yukun Li, Huazuo Gao, Shirong Ma, et al., Deepseek-coder-v2: breaking the barrier of closed-source models in code intelligence, arXiv preprint, arXiv:2406.11931, 2024.

[8] Sheng Jie, Zhang Kun, Wang Xihui, Hu Jinhao, Yao Xinrui, Huang Hualin, Application and prospect of deep learning in new energy power system, in: 2021 IEEE 5th Conference on Energy Internet and Energy System Integration (EI2), IEEE, 2021, pp. 461–465.

[9] Christopher W. Anderson, Joost R. Santos, Yacov Y. Haimes, A risk-based input–output methodology for measuring the effects of the August 2003 northeast blackout, Economic Systems Research 19 (2) (2007) 183–204.

[10] Moez Krichen, Yasir Basheer, Saeed Mian Qaisar, Asad Waqar, A survey on energy storage: techniques and challenges, Energies 16 (5) (2023) 2271.

[11] Joseph Anto, Thanga Raj Chelliah, A review of power electronic converters for variable speed pumped storage plants: configurations, operational challenges, and future scopes, IEEE Journal of Emerging and Selected Topics in Power Electronics 6 (1) (2017) 103–119.

[12] Zheming Tong, Zhewu Cheng, Shuiguang Tong, A review on the development of compressed air energy storage in China: technical and economic challenges to commercialization, Renewable and Sustainable Energy Reviews 135 (2021) 110178.

[13] Jisung Lee, Sangkwon Jeong, Young Hee Han, Byung Jun Park, Concept of cold energy storage for superconducting flywheel energy storage system, IEEE Transactions on Applied Superconductivity 21 (3) (2010) 2221–2224.

[14] Poulomi Mukherjee, V.V. Rao, Design and development of high temperature superconducting magnetic energy storage for power applications-a review, Physica C: Superconductivity and its applications 563 (2019) 67–73.

[15] Hongyu Zhu, Hui Hwang Goh, Dongdong Zhang, Tanveer Ahmad, Hui Liu, Shuyao Wang, Shenwang Li, Tianhao Liu, Hang Dai, Thomas Wu, Key technologies for smart energy systems: recent developments, challenges, and research opportunities in the context of carbon neutrality, Journal of Cleaner Production 331 (2022) 129809.

[16] Giorgio De Santillana, Alessandro volta, Scientific American 212 (1) (1965) 82–91.

[17] Khaled Itani, Alexandre De Bernardinis, Review on new-generation batteries technologies: trends and future directions, Energies 16 (22) (2023) 7530.

[18] Fei-Yue Wang, Huaiguang Jiang, Parallel battery: the framework and process for an intelligent and ecological battery system and related services, Chinese Journal of Intelligent Science and Technology 3 (2021), EL/JA-5D00-83475.

[19] Xiao Wang, Jing Yang, Xiaoshuang Li, Fei-Yue Wang, Parallel intelligence for cphs: an acp approach, in: Cyber–Physical–Human Systems: Fundamentals and Applications, 2023, pp. 145–169.

[20] Fangyuan Wang, Bo Xu, Shifted chunk encoder for transformer based streaming end-to-end asr, in: International Conference on Neural Information Processing, Springer, 2022, pp. 39–50.

[21] Alec Radford, Jong Wook Kim, Chris Hallacy, Aditya Ramesh, Gabriel Goh, Sandhini Agarwal, Girish Sastry, Amanda Askell, Pamela Mishkin, Jack Clark, et al., Learning transferable visual models from natural language supervision, in: International Conference on Machine Learning, PMLR, 2021, pp. 8748–8763.

[22] Rui Sun, Yumin Zhang, Tejal Shah, Jiahao Sun, Shuoying Zhang, Wenqi Li, Haoran Duan, Bo Wei, Rajiv Ranjan, From sora what we can see: a survey of text-to-video generation, arXiv preprint, arXiv:2405.10674, 2024.

[23] Alec Radford, Jong Wook Kim, Tao Xu, Greg Brockman, Christine McLeavey, Ilya Sutskever, Robust speech recognition via large-scale weak supervision, in: International Conference on Machine Learning, PMLR, 2023, pp. 28492–28518.

[24] Jean-Baptiste Alayrac, Jeff Donahue, Pauline Luc, Antoine Miech, Iain Barr, Yana Hasson, Karel Lenc, Arthur Mensch, Katherine Millican, Malcolm Reynolds, et al., Flamingo: a visual language model for few-shot learning, Advances in Neural Information Processing Systems 35 (2022) 23716–23736.

[25] Noam Wies, Yoav Levine, Amnon Shashua, The learnability of in-context learning, Advances in Neural Information Processing Systems 36 (2023) 36637–36651.

[26] Jason Wei, Xuezhi Wang, Dale Schuurmans, Maarten Bosma, Fei Xia, Ed Chi, Quoc V. Le, Denny Zhou, et al., Chain-of-thought prompting elicits reasoning in large language models, Advances in Neural Information Processing Systems 35 (2022) 24824–24837.

[27] Shunyu Yao, Jeffrey Zhao, Dian Yu, Nan Du, Izhak Shafran, Karthik Narasimhan, Yuan Cao, React: synergizing reasoning and acting in language models, in: International Conference on Learning Representations (ICLR), 2023.

[28] Patrick Lewis, Ethan Perez, Aleksandra Piktus, Fabio Petroni, Vladimir Karpukhin, Naman Goyal, Heinrich Küttler, Mike Lewis, Wen-tau Yih, Tim Rocktäschel, et al., Retrieval-augmented generation for knowledge-intensive nlp tasks, Advances in Neural Information Processing Systems 33 (2020) 9459–9474.

[29] Mohammad Mahdi Forootan, Iman Larki, Rahim Zahedi, Abolfazl Ahmadi, Machine learning and deep learning in energy systems: a review, Sustainability 14 (8) (2022) 4832.

[30] Janice Klaiber, Clemens Van Dinther, Deep learning for variable renewable energy: a systematic review, ACM Computing Surveys 56 (1) (2023) 1–37.

[31] Shanshan Han, Qifan Zhang, Yuhang Yao, Weizhao Jin, Zhaozhuo Xu, Chaoyang He, Llm multi-agent systems: challenges and open problems, arXiv preprint, arXiv:2402.03578, 2024.

[32] Haotian Liu, Chunyuan Li, Qingyang Wu, Yong Jae Lee, Visual instruction tuning, Advances in Neural Information Processing Systems 36 (2024).

[33] Edward J. Hu, Yelong Shen, Phillip Wallis, Zeyuan Allen-Zhu, Yuanzhi Li, Shean Wang, Lu Wang, Weizhu Chen, Lora: Low-rank adaptation of large language models, arXiv preprint, arXiv:2106.09685, 2021.

[34] Yisheng Song, Ting Wang, Puyu Cai, Subrota K. Mondal, Jyoti Prakash Sahoo, A comprehensive survey of few-shot learning: evolution, applications, challenges, and opportunities, ACM Computing Surveys 55 (13s) (2023) 1–40.

[35] Wenyuan Chen, Guangyong Li, Mingwei Li, Wenxue Wang, Peng Li, Xiujuan Xue, Xingang Zhao, Lianqing Liu, Llm-enabled incremental learning framework for hand exoskeleton control, IEEE Transactions on Automation Science and Engineering (2024).

[36] Yonglin Tian, Fei Lin, Yiduo Li, Tengchao Zhang, Qiyao Zhang, Xuan Fu, Jun Huang, Xingyuan Dai, Yutong Wang, Chunwei Tian, et al., Uavs meet llms: overviews and perspectives toward agentic low-altitude mobility, arXiv preprint, arXiv:2501.02341, 2025.

[37] Xianzheng Ma, Yash Bhalgat, Brandon Smart, Shuai Chen, Xinghui Li, Jian Ding, Jindong Gu, Dave Zhenyu Chen, Songyou Peng, Jia-Wang Bian, et al., When llms step into the 3d world: a survey and meta-analysis of 3d tasks via multi-modal large language models, arXiv preprint, arXiv:2405.10255, 2024.

[38] Yining Hong, Haoyu Zhen, Peihao Chen, Shuhong Zheng, Yilun Du, Zhenfang Chen, Chuang Gan, 3d-llm: injecting the 3d world into large language models, Advances in Neural Information Processing Systems 36 (2023) 20482–20494.

[39] Haoyu Zhen, Xiaowen Qiu, Peihao Chen, Jincheng Yang, Xin Yan, Yilun Du, Yining Hong, Chuang Gan, 3d-vla: a 3d vision-language-action generative world model, arXiv preprint, arXiv:2403.09631, 2024.

[40] Lee Wai Chong, Yee Wan Wong, Rajprasad Kumar Rajkumar, Rajpartiban Kumar Rajkumar, Dino Isa, Hybrid energy storage systems and control strategies for stand-alone renewable energy power systems, Renewable & Sustainable Energy Reviews 66 (2016) 174–189.

[41] Yuhang Liu, Yutong Wang, Yuhang Li, Chaoyue Dai, Fei-Yue Wang, Sensingagent: advancing vehicular sensing systems for spatiotemporal cognitive intelligence, IEEE Transactions on Intelligent Vehicles (2024).

[42] Xingxia Wang, Jing Yang, Yutong Wang, Qinghai Miao, Fei-Yue Wang, Aijun Zhao, Jian-Ling Deng, Lingxi Li, Xiaoxiang Na, Ljubo Vlacic, Steps toward industry 5.0: building "6s" parallel industries with cyber-physical-social intelligence, IEEE/CAA Journal of Automatica Sinica 10 (8) (2023) 1692–1703.

[43] Xingxia Wang, Yutong Wang, Jing Yang, Xiao Wang, Zonglin Meng, Zhaohai Liu, Fei-Yue Wang, A paradigm shift for modeling and operation of oil and gas: from industry 4.0 in cps to industry 5.0 in cpss, IEEE Transactions on Industrial Informatics (2024).

[44] Anthropic, Introducing the model context protocol [Online]. Available: https://www.anthropic.com/news/model-context-protocol, November 2024.

[45] Fei-Yue Wang, Jing Yang, Xingxia Wang, Juanjuan Li, Qing-Long Han, Chat with chatgpt on industry 5.0: learning and decision-making for intelligent industries, IEEE/CAA Journal of Automatica Sinica 10 (4) (2023) 831–834.

[46] Xiao Xue, Xiangning Yu, Deyu Zhou, Xiao Wang, Chongke Bi, Shufang Wang, Fei-Yue Wang, Computational experiments for complex social systems: integrated design of experiment system, IEEE/CAA Journal of Automatica Sinica 11 (5) (2024) 1175–1189.

[47] Xuan Li, Peijun Ye, Juanjuan Li, Zhongmin Liu, Longbing Cao, Fei-Yue Wang, From features engineering to scenarios engineering for trustworthy ai: I&i, c&c, and v&v, IEEE Intelligent Systems 37 (4) (2022) 18–26.

[48] Peter Fritzson, Vadim Engelson, Modelica—a unified object-oriented language for system modeling and simulation, in: ECOOP'98—Object-Oriented Programming: 12th European Conference, Brussels, Belgium, July 20–24, 1998, in: Proceedings, vol. 12, Springer, 1998, pp. 67–90.

[49] Tao Ma, Hongxing Yang, Lin Lu, Development of hybrid battery–supercapacitor energy storage for remote area renewable energy systems, Applied Energy 153 (2015) 56–62.

[50] Kaiying Qiu, Junlu Yang, Zhi Gao, Fusuo Xu, A review of modelica language in building and energy: development, applications, and future prospect, Energy and Buildings (2024) 113998.
[51] Chen-Cheng Lee, Yu-Min Hsin, Shang-Chun Dai, Cheng-Chien Kuo, Optimal modeling for dynamic response of energy storage systems, Applied Sciences 13 (8) (2023) 4943.
[52] Mayank Mittal, Calvin Yu, Qinxi Yu, Jingzhou Liu, Nikita Rudin, David Hoeller, Jia Lin Yuan, Ritvik Singh, Yunrong Guo, Hammad Mazhar, et al., Orbit: a unified simulation framework for interactive robot learning environments, IEEE Robotics and Automation Letters 8 (6) (2023) 3740–3747.
[53] Rui Dai, Rasul Esmaeilbeigi, Hadi Charkhgard, The utilization of shared energy storage in energy systems: a comprehensive review, IEEE Transactions on Smart Grid 12 (4) (2021) 3163–3174.
[54] Na Pei, Xiaoling Song, Zhe Zhang, Feniosky Peña-Mora, Optimizing the operation and allocating the cost of shared energy storage for multiple renewable energy stations in power generation side, Energy Conversion and Management 302 (2024) 118148.
[55] Shixu Zhang, Yaowang Li, Ershun Du, Chuan Fan, Zhenlong Wu, Yong Yao, Lurao Liu, Ning Zhang, A review and outlook on cloud energy storage: an aggregated and shared utilizing method of energy storage system, Renewable and Sustainable Energy Reviews 185 (2023) 113606.
[56] Ning Zhang, Haiyang Jiang, Yaowang Li, Pei Yong, Mingxuan Li, Huan Zhu, Song Ci, Chongqing Kang, Aggregating distributed energy storage: cloud-based flexibility services from China, IEEE Power & Energy Magazine 19 (4) (2021) 63–73.
[57] Juanjuan Li, Rui Qin, Fei-Yue Wang, The future of management: Dao to smart organizations and intelligent operations, IEEE Transactions on Systems, Man and Cybernetics: Systems 53 (6) (2022) 3389–3399.
[58] Wei Ma, Daoyuan Wu, Yuqiang Sun, Tianwen Wang, Shangqing Liu, Jian Zhang, Yue Xue, Yang Liu, Combining fine-tuning and llm-based agents for intuitive smart contract auditing with justifications, arXiv preprint, arXiv:2403.16073, 2024.
[59] Fangyuan Si, Jinkuan Wang, Yinghua Han, Qiang Zhao, Peng Han, Yan Li, Cost-efficient multi-energy management with flexible complementarity strategy for energy Internet, Applied Energy 231 (2018) 803–815.
[60] Yi Xu, Jianhua Zhang, Wenye Wang, Avik Juneja, Subhashish Bhattacharya, Energy router: architectures and functionalities toward energy Internet, in: 2011 IEEE International Conference on Smart Grid Communications (SmartGridComm), IEEE, 2011, pp. 31–36.
[61] Linfeng Sun, Wei Jiang, Seiji Hashimoto, Zhengyu Lin, Takahiro Kawaguchi, Multiport energy router for dc grid clusters, IEEE Journal of Emerging and Selected Topics in Power Electronics 12 (2) (2024) 1666–1682.
[62] Alysson A.P. Machado, Rodrigo Fiorotti, Rodolfo da, S. Villaça, Helder R.O. Rocha, Profit distribution through blockchain solution from battery energy storage system in a virtual power plant using intelligence techniques, Journal of Energy Storage 98 (2024) 113150.
[63] Dong Han, Chengzhenghao Zhang, Jian Ping, Zheng Yan, Smart contract architecture for decentralized energy trading and management based on blockchains, Energy 199 (2020) 117417.
[64] Justin J. Ochoa, Gomanth Bere, Indrasena R. Aenugu, Taesic Kim, Kim-Kwang Raymond Choo, Blockchain-as-a-service (baas) for battery energy storage systems, in: 2020 IEEE Texas Power and Energy Conference (TPEC), IEEE, 2020, pp. 1–6.
[65] Juanjuan Li, Rui Qin, Sangtian Guan, Jiachen Hou, Fei-Yue Wang, Blockchain intelligence: intelligent blockchains for web 3.0 and beyond, IEEE Transactions on Systems, Man and Cybernetics: Systems 54 (11) (2024) 6633–6642.
[66] Juanjuan Li, Fei-Yue Wang, The tao of blockchain intelligence for intelligent web 3.0, IEEE/CAA Journal of Automatica Sinica 10 (12) (2023) 2183–2186.
[67] Fei Lin, Qinghua Ni, Jing Yang, Juanjuan Li, Nan Zheng, Levente Kovács, Radu Prodan, Mariagrazia Dotoli, Qing-Long Han, Fei-Yue Wang, Securing the future after pagerbombs: lifecycle protection of smart devices via blockchain intelligence, IEEE/CAA Journal of Automatica Sinica 11 (12) (2024) 2355–2358.

Index

C

D

E

H

I

J

K

L

M

N

O

P

R

S

T

Z

9780323954198